AF389088

Optical Gas Sensing: Media, Mechanisms and Applications

Optical Gas Sensing: Media, Mechanisms and Applications

Editors

Krzysztof M. Abramski
Piotr Jaworski

MDPI • Basel • Beijing • Wuhan • Barcelona • Belgrade • Manchester • Tokyo • Cluj • Tianjin

Editors
Krzysztof M. Abramski
Wrocław University of Science
and Technology
Poland

Piotr Jaworski
Wrocław University of Science
and Technology
Poland

Editorial Office
MDPI
St. Alban-Anlage 66
4052 Basel, Switzerland

This is a reprint of articles from the Special Issue published online in the open access journal *Sensors* (ISSN 1424-8220) (available at: https://www.mdpi.com/journal/sensors/special_issues/Optical_Gas_Sensing_Applications).

For citation purposes, cite each article independently as indicated on the article page online and as indicated below:

LastName, A.A.; LastName, B.B.; LastName, C.C. Article Title. *Journal Name* **Year**, *Volume Number*, Page Range.

ISBN 978-3-0365-3479-4 (Hbk)
ISBN 978-3-0365-3480-0 (PDF)

Contents

About the Editors

Krzysztof M. Abramski (Prof.) received his MSc and PhD in electronics from Wrocław University of Technology, Poland, in 1971 and 1978, respectively. At that time, he worked on the frequency stabilization of gas laser radiation (He-Ne/CH 4) and many aspects of spectral properties of laser radiation. In 1983-84, he worked as the Research Fellow in Quantum Electronics Group, Twente University of Technology, the Netherlands (optogalvanic effects in CO2 , CO lasers, high pressure XeCl excimer lasers). In 1987, he spent half a year in the Department of Applied Physics, Hull University, England; after that, for over four years, till 1992, he worked as a visiting scientist at Optoelectrionics and Lasers Engineering Group, Heriott-Watt University, Edinburgh, working on different aspects of RF excited CO2 lasers (wavegudes, slab-waveguides, waveguide arrays, phase locking structures). He returned to Wroclaw in 1993, where he started creating the Laser and Fiber Electronics Group. In 2006, he initiated the popular teaching specialization "Optocommunications", and in 2011, he initiated the English MSc program "Advanced Applied Electronics" (about 20–30 student per course). In 2016, he initiated the English engineering program "Electronic and Computer Engineering" (about 45–50 students per course) at the Electronics Faculty. His scientific interests include optical fiber lasers (cw tunable, femtosecond, combs), microchip solid state diode pumped, optical fiber sensors, free-space and optical fiber communications, and laser spectroscopy. He has supervised 21 doctoral candidates. He currently holds the position of a full professor (and from 2018, of professor emeritus).

Piotr Jaworski (Dr.) received his PhD in Physics from Heriot-Watt University, Edinburgh, UK in 2015. During his PhD studies, he was working on the development of novel anti-resonant hollow-core fibers for the delivery of high-peak power-pulsed laser light in the near-IR and visible spectral range, and their application in precision micro-machining. After that, in 2015, he joined the Lukasiewicz Research Network - PORT Institute (before Wroclaw Research Center EIT+) as a research engineer in the Laser Sensing Laboratory. Since that time, his research work has been focused on laser-based gas sensing using both bulk optics-based and hollow-core fiber-based gas spectrometers. In 2018, he became the head of the Laser Sensing Laboratory in PORT. In 2019, he joined the Laser and Fiber Electronics Group at Wroclaw University of Science and Technology focusing his research on gas sensing using hollow-core micro-structured fibers. Piotr Jaworski is an author and co-author of 56 research articles, conference proceedings and conference presentations. He has been a main investigator in eight research projects and principal investigator in one research grant. He has also developed two laser-based gas sensing systems for industrial enterprises from Poland.

Review

A Review of Antiresonant Hollow-Core Fiber-Assisted Spectroscopy of Gases

Piotr Jaworski

Laser and Fiber Electronics Group, Faculty of Electronics, Wroclaw University of Science and Technology, Wybrzeze Wyspianskiego 27, 50-370 Wroclaw, Poland; piotr.jaworski@pwr.edu.pl

Abstract: Antiresonant Hollow-Core Fibers (ARHCFs), thanks to the excellent capability of guiding light in an air core with low loss over a very broad spectral range, have attracted significant attention of researchers worldwide who especially focus their work on laser-based spectroscopy of gaseous substances. It was shown that the ARHCFs can be used as low-volume, non-complex, and versatile gas absorption cells forming the sensing path length in the sensor, thus serving as a promising alternative to commonly used bulk optics-based configurations. The ARHCF-aided sensors proved to deliver high sensitivity and long-term stability, which justifies their suitability for this particular application. In this review, the recent progress in laser-based gas sensors aided with ARHCFs combined with various laser-based spectroscopy techniques is discussed and summarized.

Keywords: antiresonant hollow core fibers; laser spectroscopy; wavelength modulation spectroscopy; tunable diode laser absorption spectroscopy; photothermal spectroscopy; photoacoustic spectroscopy; fiber gas sensors

Citation: Jaworski, P. A Review of Antiresonant Hollow-Core Fiber-Assisted Spectroscopy of Gases. *Sensors* **2021**, *21*, 5640. https://doi.org/10.3390/s21165640

Academic Editor: Bernhard Wilhelm Roth

Received: 29 July 2021
Accepted: 19 August 2021
Published: 21 August 2021

Publisher's Note: MDPI stays neutral with regard to jurisdictional claims in published maps and institutional affiliations.

1. Introduction

The end of the previous century has brought a new type of optical fiber, the so-called hollow-core fiber (HCF), which due to its unique structure and ability to guide light in the air via the photonic bandgap effect, rather than via the conventional total internal reflection phenomenon, revolutionized the development and application areas of optical fiber technology [1]. Further development of the HCF structure and exploration of different guidance mechanisms of light in the air have enabled access to the HCFs, which deliver a superb ability to efficiently guide laser light, especially in the mid-infrared (mid-IR) spectral band, where conventional solid-core fibers suffer from high attenuation of the glass material [2,3]. Currently, three major types of HCFs have been proposed, fabricated, and successfully used in various applications [4–7], amongst which the laser-based spectroscopy of gases has attracted significant attention of researchers around the world [8–10]: the hollow-core photonic bandgap fiber (HC-PBGF) [1], the Kagome HCF [11] and the Antiresonant Hollow-Core Fiber (ARHCF) [8]. Benefiting from an empty core, which can be filled with the target gas, HCFs can be utilized as low-volume absorption gas cells, forming versatile light-gas molecules interaction paths with the desired length within a sensor setup [8]. Since the sensitivity of the majority of laser-based gas sensors can be relatively simply and significantly enhanced by increasing the interaction path length, access to non-complex and long optical paths is highly desired. Hence, the incorporation of HCF-based absorption cells into laser-based gas detectors could lead to high sensor detection capability and less complex design in comparison with commonly used bulk optics-based solutions, e.g., utilizing multipass cells [12,13]. Multipass cells delivering optical paths with several tens of meters length require advanced optical arrangements for coupling into them the laser beam in a way allowing for obtaining the proper number of light passes, hence the desired path length. Since optical and optomechanical components are sensitive to vibrations and temperature changes, which negatively affect their long-term stability, even a slight misalignment of the coupling optics disturbs the light propagation inside the multipass cell. Unfortunately, optics-free coupling into a multipass cell is not possible. This leads to the increase in the noise level, the reduced amplitude of the measured signal, hence a significant drop in the detection capability of the sensor. On the other hand, light guidance in HCFs can be efficiently excited via an optics-free butt-coupling approach of the laser beam. Furthermore, the multipass cells based on the use of optical mirrors (e.g., Herriot- or White-type) mounted in e.g., metal optomechanical housings are sensitive to temperature changes due to thermal expansion of the material, which additionally affects the stability of the gas sensor. This can be minimized by using materials with lower thermal expansion coefficient, e.g., invar, however at cost of a significant increase in the sensor's price, especially when multipass cells delivering several tens of meters long paths lengths are used in the setup.

HC-PBGFs have been successfully used in various gas sensor configurations, however, they target transitions of different gases in the wavelength range not exceeding 3.4 μm [9,14,15]. It was established that the main issues connected with the use of this particular type of fiber that significantly limits the sensitivity and versatility of fiber-based gas sensors arise from the multimode nature of these fibers and their maximum operational wavelength range [9,16]. Multimode guidance leads to the intermodal interference between the fiber-supported fundamental mode and the higher-order modes, which negatively impacts the noise level in the sensor [16]. This can be minimized by combining HC-PBGFs with spectroscopic techniques that have a built-in capability of reducing the impact of the fringe noise on the measured signal, e.g., Chirped Laser Dispersion Spectroscopy (CLaDS) or Photothermal Spectroscopy (PTS) [15,16]. Moreover, despite guidance in air, this fiber is still characterized by a relatively high overlap between its glass structure and the guiding light, which limits the transmission bandwidth to approximately 3 μm spectral band [17]. Furthermore, due to the small core size (typically up to 20 μm), the gas filling time of HC-PBGFs can reach even several hours, which severely limits the response time of the sensing systems utilizing these fibers [16].

A partial solution to the issues that are present in HC-PBGFs comes with the aid of the Kagome type HCFs, which guide light via the inhibited coupling mechanism [11]. As a result of a modified fiber structure and different light guidance principles, these fibers can efficiently transmit light in the near-infrared (near-IR) and mid-IR [11,18]. Furthermore, the core size of the Kagome fibers is a few times greater (116 μm) in comparison with the HC-PBGFs guiding light within the same spectral band, which results in the reduction of the gas filling time down to several seconds [18]. Nevertheless, it was indicated in [10,18] that the problem connected with the multimode guidance is also present in the Kagome HCFs, which was identified as the main limiting factor in the performance of the gas sensors utilizing these fibers.

ARHCFs, in which light transmission is realized by the Antiresonant Reflecting Optical Waveguiding (ARROW) principle [19], can deliver low loss in both near- and mid-IR spectral bands, fast gas exchange time, and single-transversal mode operation if a proper fiber structure is designed [8,20]. Currently, ARHCFs have been successfully used in gas sensors utilizing a variety of laser-based sensing techniques, i.e., Tunable Diode Laser Absorption Spectroscopy (TDLAS), Wavelength Modulation Spectroscopy (WMS), PTS, and Photoacoustic Spectroscopy (PAS) [8,21–23]. Examples of the ARHCFs used in gas sensing applications are depicted in Figure 1. Researchers have shown that the ARHCF-aided gas sensors can target molecules with transitions in the wavelength range up to 5.26 μm, which is unreachable with the use of other types of HCFs [20,24]. Furthermore, benefiting from their ability to simultaneously guide laser radiation within two dissimilar spectral bands, the ARHCF-based detectors can be used to analyze gas mixtures that contain molecules having transitions in both near- and mid-IR [8]. Similar to the Kagome HCFs, ARROW-guiding fibers are characterized by the core size in the range of several tens of micrometers, which in combination with a proper gas delivery system allows obtaining gas exchange times in the range of several seconds [20]. The combination of ARHCF-based gas absorption cells with, e.g., the PTS technique enables obtaining superb long-term stability of the sensor, giving a promising perspective for their future application in out-of-lab conditions [25]. The sensors utilizing such fibers have been demonstrated to provide detection capability even at a level comparable to the bulk optics-based setups, indicating that the fiber-based configuration of the sensors can form a new branch of sensitive, selective, and non-complex gas sensing platforms.

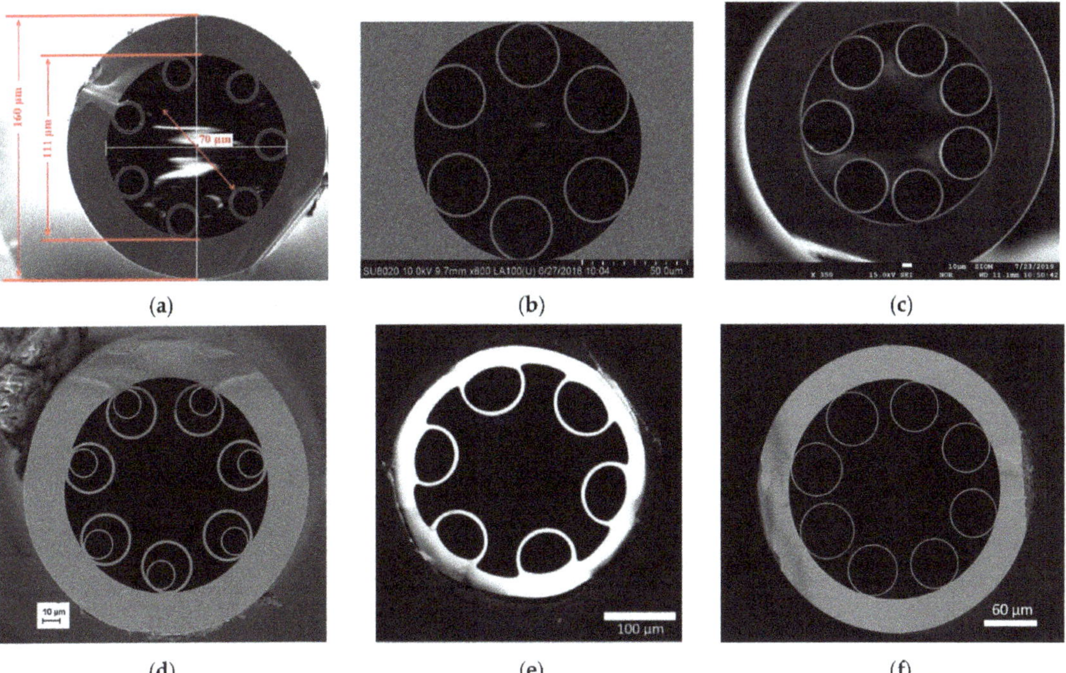

Figure 1. Examples of ARHCFs used in gas sensing systems. (**a**) Silica-based ARHCF is designed to operate in the 2 μm wavelength range with a core size of 70 μm. Reprinted with permission from [21] © The Optical Society. (**b**) Six-capillary cladding silica-based ARHCF with a core diameter of 65 μm for guidance at ~1.55 μm and 3.34 μm. Reprinted with permission from [26] © The Optical Society. (**c**) Seven-capillary cladding silica-based ARHCF with a core size of 84 μm providing low-loss transmission in the near- and mid-IR. Reprinted with permission by MDPI from [8]. (**d**) Silica-based ARHCF with nested capillary cladding and a core diameter of 65 μm for guidance at ~4.54 μm. Reprinted with permission from [27] © The Optical Society. (**e**) Tellurite ARHCF enabling light guidance ~5 μm inside a hollow core with a 139 μm diameter. Reprinted with permission from [28] © The Optical Society. (**f**) 5.26 μm-guiding borosilicate glass-based ARHCF with a core size of 122 μm.

In this review, the recent progress in ARHCF-based gas sensors utilizing the aforementioned gas sensing techniques will be discussed. Several different sensor configurations are presented and their advantages along with main limiting factors are reviewed. Section 2 of this review aims at explaining the light guidance properties of the ARHCFs. Section 3 is devoted to the implementation of the ARHCFs into TDLAS-based gas sensors. Section 4 is focused on the WMS gas sensors aided with different types of ARHCFs. Section 5 presents the PTS technique supported by the ARHCFs and explains how the few-moded guidance of the fiber can be transferred to the high sensor stability and sensitivity together with an introduction to the new gas sensing method in ARHCFs, the so-called Photoacoustic Brillouin Spectroscopy (PABS) [23]. Section 6 summarizes the performance of the reported up-to-date ARHCF-aided gas sensor configurations.

2. Light Guidance in Antiresonant Hollow-Core Fibers

ARHCFs are a new type of HCFs, which light guidance mechanism can be explained by the means of the ARROW model as shown in Figure 2a. According to this, the core boundary area of the ARHCF can be treated as a Fabry–Perot resonant cavity, as it is formed by low and high refractive index layers (e.g., air and glass) as presented in Figure 2b [8]. This Fabry–Perot cavity enables only the transmission of the optical frequencies, which are not in resonance with the core wall (capillary walls). These optical frequencies are reflected back to the fiber core where they propagate with low loss. On the other hand, the resonant optical frequencies cannot be confined within the fiber core and leak away to the cladding area where they experience high leakage and material loss [8]. The antiresonant wavelength range supported by an ARHCF can be calculated according to the following formula [19]:

$$\lambda_{antires} = \frac{4y}{(2m+1)} \sqrt{n_2^2 - n_1^2}, \ m = 0, 1, 2, \ \dots \tag{1}$$

where y is the core wall thickness (capillary wall thickness), $n1$ and $n2$ are the refractive indices of the core and cladding, respectively. The resonant wavelength range can be defined as [19]:

$$\lambda_{res} \sim \frac{2y}{m} \sqrt{n_2^2 - n_1^2}, \ m = 0, 1, 2, \ \dots \tag{2}$$

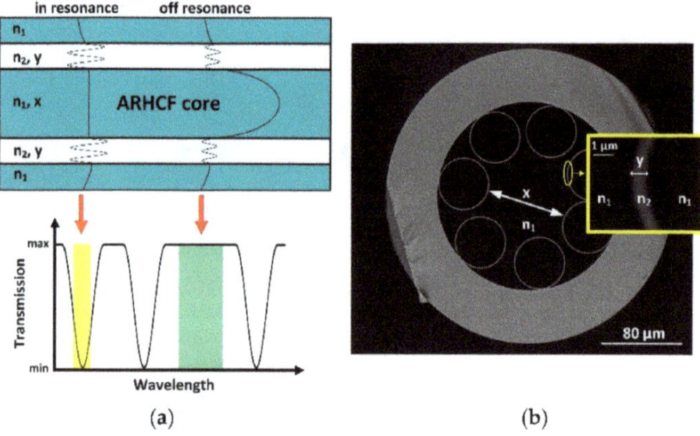

Figure 2. Light guidance mechanism in ARHCF. (**a**) 2D representation of the ARHCF (top) showing light transmission (bottom) in the core while the coupled light wavelength is in resonance and off resonance with the core wall. When the optical frequency (wavelength) does not match the resonant frequency of the Fabry–Perot cavity, the transmission of light in the core reaches its maximum. (**b**) SEM image of the ARHCF designed to operate at ~3.4 µm wavelength with a core wall thickness of ~1 µm. The inset shows the core boundary layer forming the resonant cavity. n_1 and n_2—refractive indices of the air region (core, gaps between capillaries and inner parts of the capillaries) and capillary walls, respectively, x—core diameter, y—capillary wall thickness.

Based on the above equations, it can be concluded that the transmitted wavelength, and thus the position of the transmission window, depends mainly on the thickness of the capillary walls and not on the core size. However, it was reported in [29] that the hollow-core diameter of the ARHCF and diameter of the cladding capillaries have a strong impact on the bending properties and single-mode guidance of this particular fiber type. It was shown that dimensions of both have to be carefully selected to match the optimum ratio of core/capillary diameter of ~0.65 that enables single-mode transmission within the fiber low-loss window as a result of increased loss ratio between the fundamental mode and the higher-order modes supported by the fiber [8,29].

ARHCFs are fabricated with the aid of the commonly used stack-and-draw technique [30]. In majority, these fibers are drawn down from high purity fused silica glass (e.g., Suprasil F300) [8], however, due to the high material absorption of this material at the wavelengths above 5 μm, several successful attempts have been reported on fabricating ARHCFs from borosilicate and telluride glass allowing these fibers to efficiently guide light beyond the aforementioned wavelength range [28,31]. Thanks to the unique structure and light guidance properties, ARHCFs deliver better performance and versatility in comparison to other types of HCFs, especially in the area of fiber-aided gas sensing. A comparison of the parameters of the most commonly used HCFs in gas sensing is presented in Table 1.

Table 1. Comparison of the performance of the HCFs used in gas sensing applications.

Fiber Type	Wavelength	Light Guidance	Core Size Loss @ 3.4 μm	Loss @ ~3 μm	Min. Gas Loss Filling Time for ~1 m Fiber
ARHCF	up to 5.26 μm [20]	Single-mode with proper fiber structure [20]	84 μm [8]	0.03 dB/m [8]	5 s [21]
Kagome HCF	up to 3.4 μm [18]	few-moded [18]	116 μm [18]	~0.1 dB/m [18]	<10 s [18]
HC-PBGF	up to 3.4 μm [9]	few-moded [9]	40 μm [9]	2.6 dB/m [9]	1200 s [16]

3. Tunable Diode Laser Absorption Spectroscopy

One of the simplest and easiest methods used for laser-based gas sensing is the TDLAS [32,33]. In TDLAS, the information about the molecular concentration within a defined measurement path is retrieved based on the analysis of an interaction between the laser radiation and the gas molecules. The interaction leads to the absorption of light by the gas molecules, which are excited at the wavelength corresponding to the selected molecular transition. This phenomenon is governed by the Beer–Lambert law and expressed by the following formula [34,35]:

$$\frac{I_p}{I_0} = \exp(-\varepsilon(\lambda)L), \tag{3}$$

where I_p corresponds to the light intensity after passing through the gas sample, I_0 is the incident light intensity, ε represents the absorption coefficient of the gas molecules, λ is the wavelength of the light expressed in wavenumbers and L is the light-gas molecules interaction path length. In TDLAS-based gas sensors, the molecules of the target gas are typically illuminated by light delivered from a narrow linewidth laser, e.g., a distributed feedback diode laser (DFB) or a quantum cascade laser (QCL). The level of absorption of the gas molecules excitation light is observed as a drop in the signal intensity registered by a photodiode, while the laser beam is passing through the gas sample and its wavelength is tuned across the gas transition or kept at its peak. According to Equation 1, the sensitivity of the sensors relying on this method can be easily and efficiently enhanced by increasing the interaction path length within the sensor's setup. This is commonly realized by implementing bulk optics-based absorption cells or multipass cells, e.g., Herriot-, White- or toroidal-type, which are filled with the measured gas sample [12,36,37]. This approach indeed results in the improved sensor's performance, however, at the cost of the significantly increased complexity of its configuration and reduced immunity to, e.g., vibrations, temperature drifts, etc. Therefore, the application of the HCFs, especially the ARHCFs, seems to be a promising way to deliver low-volume, robust, and long optical paths. A successful demonstration of TDLAS-based gas sensors aided with ARHCFs has been already demonstrated by several research groups justifying the viability of this approach [21,24,35,38].

Nikodem et al. reported in [21] a very simple carbon dioxide (CO_2) sensor configuration utilizing a silica-based 7 capillary ARHCF as depicted in Figure 3. CO_2 molecules were excited using a fiber-coupled discrete mode diode laser targeting their strong absorption line at 2.004 µm. The absorption cell within the sensor setup was formed by a 1.35 m long ARHCF with a hollow core diameter of 70 µm as depicted in Figure 1a. The fiber output end facet was placed in an air-tight housing used as a gas-filling cell, which was closed with a photodetector. The light from the laser was directly coupled into the ARHCF using a simple butt-coupling method. Subsequently, the fiber-delivered beam was directed onto the photodetector using a similar approach. Hence, the sensing part of the sensor was constructed in an all-fiber configuration. The ARHCF was filled with the target gas using a slight overpressure of 100 Torr. Despite a significantly reduced sensor complexity, the proposed system was characterized by a very poor detection capability. It allowed registering clear spectroscopic signals arising from only high (at the level of 1.5%) concentrations of CO_2 inside the ARHCF. This was a direct result of the high background noise level induced by the optical fringes arising from the light coupling method used, intermodal interference in the ARHCF, and its non-uniform guidance characteristic. On the other hand, the responsivity of the sensor was at the level of several seconds, which is more than two orders of magnitude faster in comparison to the detectors based on the use of conventional HC-PBGFs [39].

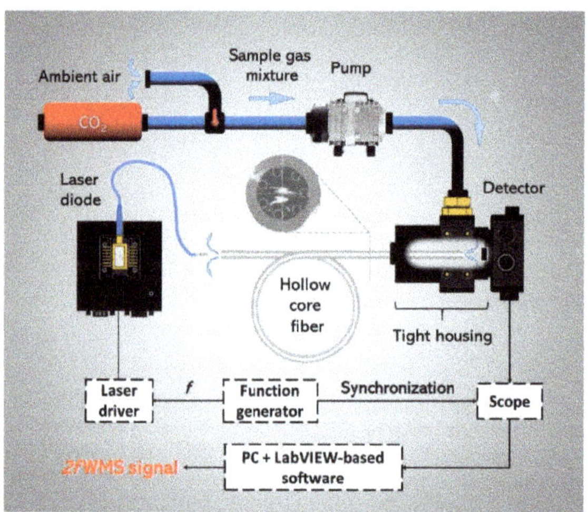

Figure 3. A schematic representation of the TDLAS/WMS gas sensor utilizing a 1.35 m long absorption cell based on an ARHCF. Reprinted with permission from [21] © The Optical Society.

Another interesting TDLAS-based gas sensor configuration shown in Figure 4 was reported by Yao et al. in [38]. The gas absorption cell in the setup was formed by a 0.85 m long ARHCF with an air-core diameter of 40 µm, which both ends were placed in air-tight gas filling cells closed with calcium fluoride (CaF_2) windows. The ARHCF was filled with carbon monoxide (CO) using an overpressure of 0.8 bar, which resulted in the gas exchange time of the sensor at the level of 5 s. CO molecules inside the ARHCF core were excited with the aid of a 2.3 µm DFB laser, which was coupled into the fiber core using a set of properly selected lenses with a coupling efficiency of 90%. The minimum detection limit (MDL) in this particular sensor configuration reached 13 parts-per-million by volume (ppmv) of CO, yielding the noise equivalent absorption (NEA) of 5.2×10^{-6} cm^{-1}. Similar to the earlier described work, the main limiting factor of the sensor was related to the presence of modal noise in the fiber, indicating not sufficient suppression of the higher-order modes along the relatively short fiber length and not entirely optimized light coupling conditions into the ARHCF core.

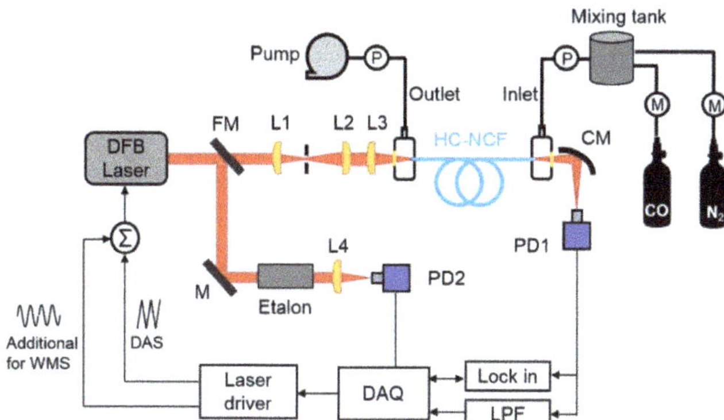

Figure 4. A setup of the CO sensor relying of the use of the TDLAS technique and an ARHCF-based gas absorption cell. L—lenses, M—mirror, CM—concave mirror, FM—flip mirror, P—pressure gauge, LPF—low-pass filter, PD1/PD2—photodetectors, DAQ—data acquisition card. Reprinted from [38] with permission from Elsevier.

Further work in this area reported by Yao et al. in [35] was focused on developing a TDLAS-based sensor targeting a strong transition of nitrous oxide (N_2O) in the mid-IR spectral band. The sensor configuration was similar to the one presented in Figure 4, however, with the main difference in the type of the ARHCF and the used laser. In this case, the N_2O molecules were excited at 2778.37 cm^{-1} (~3.6 µm) using an interband cascade laser (ICL), which output was coupled into the gas-filled ARHCF with a coupling efficiency of 66%. The gas molecules-light interaction path was formed by a six-capillary cladding ARHCF with a core size of 65 µm and a length of 120 cm, which was filled with N_2O via a pair of gas cells placed at its ends aided with an overpressure. Despite low transmission loss at the level of 0.6 dB/m at the considered wavelength, the fiber was characterized by a few-moded behavior that influenced the overall performance of the sensor. It was noticed that both the fundamental mode and the higher-order modes were simultaneously excited in the fiber, leading to the intermodal interference, which directly impacted the noise level in the registered spectroscopic signal, hence reducing the sensors' detection capability. Nevertheless, the authors have shown that this parasitic effect can be minimized by properly selecting the light coupling conditions into the ARHCF and reducing the pressure of the gas inside the fiber core. As a consequence, the sensor reached an NEA of 2.5×10^{-7} cm^{-1}.

The most recent work published by Yao et al. in [24] was focused on developing a TDLAS-based gas sensor utilizing a tellurite glass-based ARHCF enabling efficient light guidance above 5 µm wavelength. The 21 cm long ARHCF consisted of six non-adjacent capillaries forming its cladding and defining the hollow core region with a diameter of 75 µm, similar to the fiber shown in Figure 1e. The input end facet of the ARHCF forming the absorption cell was placed inside a pressure-tight gas cell closed with a CaF$_2$ wedged window, which was implemented to reduce parasitic interference. The gas cell was mechanically modified in a way that allowed it to be easily connected to the cage system rods, hence properly and efficiently align the fiber end placed in the cell with respect to the focusing lens and a QCL. This approach provided a robust and stable coupling between the laser and the fiber. Similar to the previously described work, the ARHCF was filled with the target gas using an overpressure, which allowed obtaining the response time of the sensor below 1 s. The nitric oxide (NO) molecules inside the hollow core of the fiber were pumped at 5.26 µm using a continuous wave (CW) light from a QCL. The registered TDLAS signal from 100 ppmv NO in the ARHCF was characterized by the presence of strong background noise, which resulted from the multimode nature of the fiber, and could not be eliminated by simple signal averaging or the usage of different light coupling conditions. However, the authors minimized its influence on the sensor's performance by introducing proper electrical filtering of the signal at the frequencies where the fringe noise was dominating. Thanks to this, the sensor reached an MDL of 1.2 ppmv, which corresponds to an NEA of 2.1×10^{-5} cm^{-1}.

4. Wavelength Modulation Spectroscopy

WMS technique is a modification of the conventional TDLAS method, which enables reducing the influence of noise on the measured spectroscopic signal. In WMS, the wavelength of the laser that is used to excite gas molecules is (in comparison to TDLAS) additionally modulated with a sinusoidal signal with a strictly defined frequency and modulation depth, both dependent on the target gas transition characteristic [40,41]. The modulated laser frequency and the spectral absorbance in WMS are described by the following equations [41]:

$$\upsilon(t) = \bar{\upsilon} + A\cos(2\pi ft), \tag{4}$$

$$-\alpha[\bar{\upsilon} + A\cos(2\pi ft)] = \sum_{k=0}^{\infty} H_k(\bar{\upsilon}, A)\cos(k2\pi ft), \tag{5}$$

$$H_k(\bar{\upsilon}, A) = \frac{P X_i L}{\pi} \int_{-\pi}^{\pi} \sum_j S_j \varphi_j (\bar{\upsilon} + A\cos\theta)\cos k\theta d\theta, \tag{6}$$

where $\upsilon(t)$ is the modulated laser frequency in function of time, $\bar{\upsilon}$ is the center laser frequency, A corresponds to the modulation depth, f is the modulation frequency, α is the spectral absorbance, P is the total gas pressure, X_i is the mole fraction of the absorbing gas sample, S_j is the j-th absorption line strength function and φ_j is the j-th absorption line shape function and L defines the interaction path length. Typically, the modulation frequency is in the range of a few to a few tens of kHz with a modulation depth equal to ~2.2 × full width at half maximum (FWHM) of the selected gas absorption line. In WMS-based sensors, the sinusoidally modulated laser beam experiences a nonlinear interaction with the gas molecules. This leads to the rise of additional components in the signals registered by the photodetector at frequencies corresponding to the harmonics of the fundamental modulation frequency. The amplitude of the even harmonics is proportional to the concentration of gas molecules within the measurement path length, hence the sensitivity of such sensors can be effectively increased by elongating the gas-laser interaction path. The harmonic components can be efficiently retrieved using a phase-sensitive lock-in amplifier-based approach [20,40]. As the lock-in amplifier allows demodulation of the measured signal at the desired frequency with a limited demodulation bandwidth, the noise level, which manifests itself especially in the lower frequency range, can be reduced. Therefore, the signal-to-noise ratio (SNR) of the sensor can be significantly increased in comparison to the TDLAS-based technique, which directly enhances the detection capability of the gas spectrometers [24].

It has already been demonstrated by various research groups that a combination of the WMS technique with ARHCFs leads to a significant improvement in the sensor's detection capability, which results from the reduction of the fringe noise [21,24,38]. When the configurations of the sensors described in Section 3 of this manuscript were modified to allow WMS-based signal acquisition, the obtained NEA values were decreased by even two orders of magnitude compared to sensors operating in the pure TDLAS regime [24]. This enabled the ARHCF-based gas sensors to reach detection limits at a level comparable to the state-of-the-art bulk-optics-based setups.

Especially interesting work focused on WMS-based gas sensing aided with ARHCFs concerns the recent development of these fibers, which enabled them to guide light above 4.5 μm wavelength range. Nikodem et al. reported in [27] the first experimental demonstration of an ARHCF-based system capable of targeting a very strong N_2O absorption line located at 2203.7 cm^{-1}. As presented in Figure 5, the sensor utilized a QCL as a gas excitation source, which wavelength was tuned to the center of the selected N_2O transition and subsequently coupled via an off-axis parabolic mirror into an absorption cell formed by a 3.2 m long nested ARHCF (shown in Figure 1d). The gas delivery system and method were similar to the ones described earlier and allowed filling the fiber core within the 23 s period. The sensor in this configuration reached an MDL of 5.4 ppbv at 1 s integration time, which corresponds to a minimum fractional absorption (MFA) of 1.2×10^{-4} (NEA $\sim 3.7 \times 10^{-7}$ cm^{-1}). The obtained detector's sensitivity was not at the record level, mainly due to the transmission characteristic of the fiber at the considered wavelength range. In this configuration, the QCL wavelength was placed at the edge of the low-loss transmission band of the fiber, where ARHCFs are typically characterized by the few-moded behavior [4]. Nevertheless, due to its unique structure, the fiber was characterized by an exceptional immunity to bending, which indicates the excellent robustness, compactness, and versatility of the ARHCF-based absorption cells delivering a few meters long interaction path.

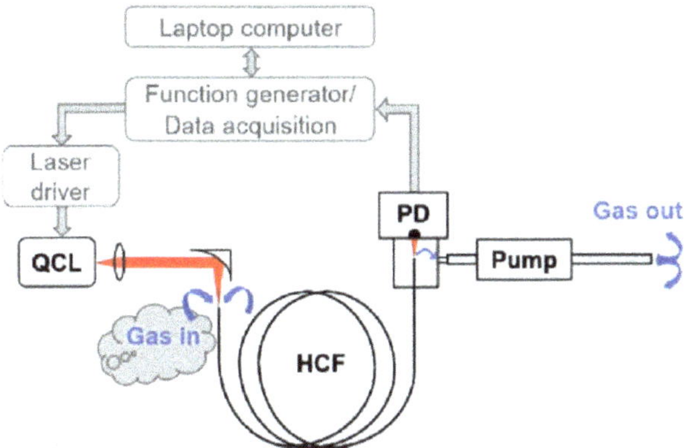

Figure 5. Experimental setup of the ARHCF-based N$_2$O sensor utilizing WMS technique and a 4.54 µm QCL. Reprinted with permission from [27]. © The Optical Society.

The operational wavelength range of the ARHCF-based gas sensors was significantly increased with the development of borosilicate-glass- and telluride-glass-based fibers [28,31], which broke the barrier of 5 µm wavelength range, where the attenuation of silica glass increases rapidly [3,42]. Thanks to this unique feature, the ARHCF technology could be implemented in NO detectors. Jaworski et al. reported in [20] the first experimental demonstration of a WMS-based NO sensor utilizing a 1.15 m long borosilicate glass ARHCF, as depicted in Figure 6a. The sensor targeted a strong NO doublet located in the vicinity of 1900.08 cm^{-1}, which was registered with the aid of a QCL. The gas filling method was similar to the one described earlier. Thanks to the large core size of the ARHCF (122 µm diameter), the sensor was characterized by the filling time of less than 10 s as shown in Figure 6b. As a result of the low-loss and single-transversal-mode guidance of the fiber, the sensor reached an MDL of 20 parts-per-billion by volume (ppbv) for 70 s integration time, which yields an NEA of 2.0×10^{-5} cm^{-1} and allowed registering clear spectra of 2f WMS signals from 100 ppmv NO inside the fiber as plotted in Figure 6c. Figure 6d shows a photograph of the sensor, which length does not exceed 75 cm. It is expected that the size of the sensor could be further reduced by decreasing the size of the electronic and optomechanical components used together with tightened bending of the fiber-forming the absorption cell. The authors indicated that the obtained MDL was less than an order of magnitude worse in comparison to a sensor utilizing the more advanced and complex quartz enhanced PAS technique [43]. This result was further improved by Yao et al. as reported in [24], where the WMS-operating sensor utilizing a tellurite ARHCF reached an MDL of 6 ppbv for 30 s integration time.

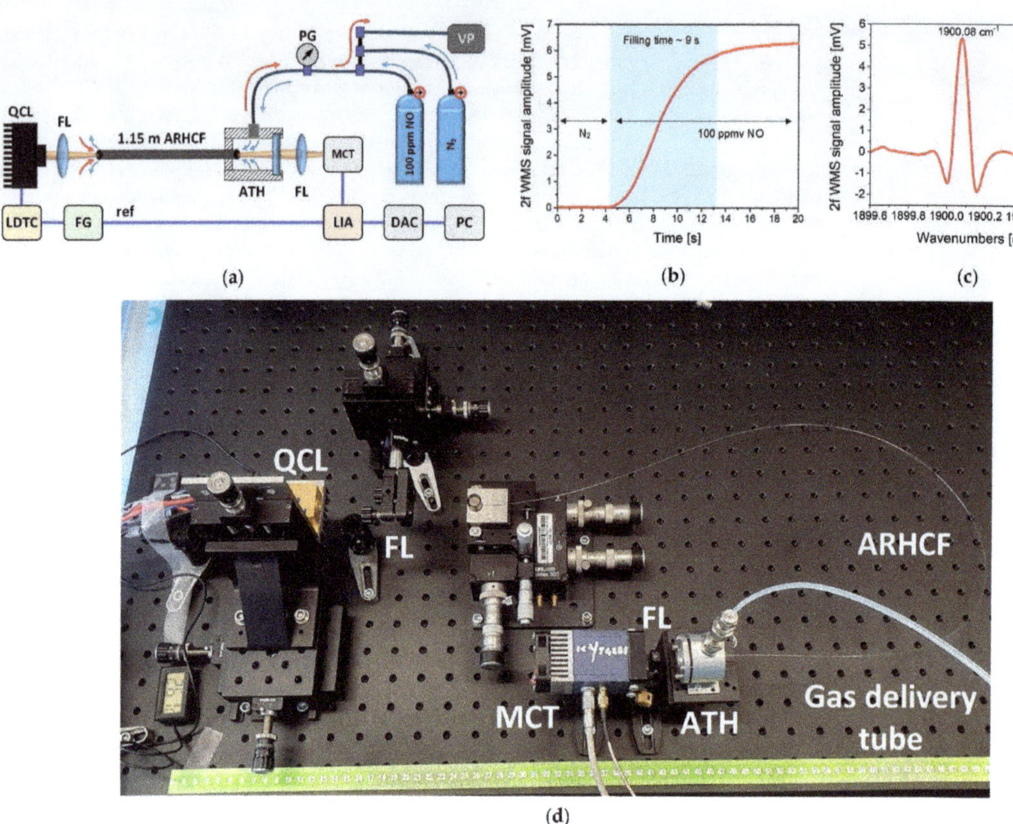

Figure 6. NO sensor operating at 5.26 μm based on the use of the WMS technique and a borosilicate ARHCF. (**a**) Experimental setup. QCL—quantum cascade laser, LDTC—laser driver, FL—focusing lens, FG—function generator, ATH— air tight housing, MCT—mercury-cadmium-telluride photodetector, PG—pressure gauge, VP—vacuum pump, LIA—lock-in amplifier, DAC—data acquisition card, PC—computer. (**b**) Gas filling profile of the ARHCF using an overpressure-assisted gas delivery method. (**c**) 2f WMS signal spectrum of the NO doublet for 100 ppmv NO inside 1.15 m ARHCF. Reprinted with permission from [20] © The Optical Society. (**d**) Photograph of the sensor setup.

Another interesting and highly advantageous feature of the ARHCFs concerns their unique ability to transmit with low-loss light in several dissimilar wavelength bands. Jaworski et al. utilized this phenomenon for gas sensing and for the first time demonstrated the simultaneous detection of CO_2 and methane (CH_4) inside the ARHCF, targeting the transitions of these gases in the near- and mid-IR spectral bands [8]. The experimental setup shown in Figure 7 consisted of a difference frequency generation (DFG) and DFB sources, which operated at 3.334 μm and 1.574 μm, respectively. The DFG source was used to excite molecules of CH_4, while the DFB laser targeted the CO_2 transition. Both lasers were simultaneously coupled into a 1 m long ARHCF (shown in Figure 1c) filled with a mixture of the aforementioned gases through a gas filling cell. Thanks to the low loss and near single-mode guidance of the fiber at both wavelengths, the sensor reached an MDL of 24 ppbv for 40 s integration time and 144 ppmv for 1.5 s integration time for CH_4 and CO_2, respectively. The obtained MDLs yielded NEA coefficients of 1.6×10^{-7} cm^{-1} (CH_4) and 1.17×10^{-7} cm^{-1} (CO_2). The sensitivity of the sensor beats the performance of the WMS-based setups utilizing Kagome HCFs and HC-PBGFs, confirming the versatility of this approach [14,18]. The authors indicated that the further improvement of the developed sensor's sensitivity could be obtained by introducing a longer fiber (i.e., with several tens of meters in length) with a properly modified structure, allowing obtaining pure single-mode transmission and uniform loss within the guidance windows.

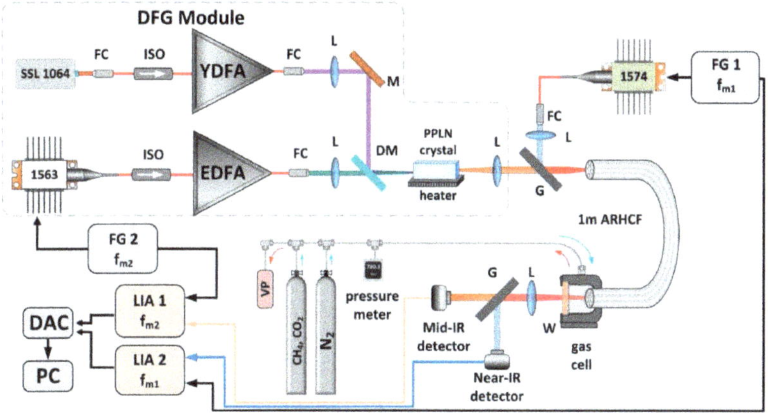

Figure 7. Schematic of the dual-gas sensor based on the near- and mid-IR guiding ARHCF and WMS technique. SSL—diode pumped solid state laser, FC—fiber collimator, ISO—isolator, YDFA/EDFA—Ytterbium- and Erbium-doped fiber amplifiers, L—lenses, M—mirror, DM—dichroic mirror, G—germanium window, W—CaF$_2$ wedge, FG—function generator, LIA—lock-in amplifier, DAC—data acquisition card, PC—computer, VP—vacuum pump. Reprinted from [8] with permission from MDPI.

5. Photothermal and Photoacoustic Spectroscopy

PTS is a technique where the spectroscopic signal retrieval is directly related to the heating of gas molecules with the aid of laser light [44,45]. In PTS, the gas molecules are excited with a laser source (*pump*) that emits radiation at a wavelength matching the selected gas transition (similarly to TDLAS and WMS). However, the retrieved signal is not connected with the intensity drop of the laser due to its gas-induced absorption, but with the local refractive index (RI) change that results from the increased temperature of the gas sample due to non-radiative relaxation of the molecules illuminated by the *pump* light [44]. The observed change of the RI can be determined based on the following equation [46]:

$$\Delta n = \frac{(n-1)\varepsilon P_{exc}}{T_0 4\pi a^2 \rho C_p f},$$ (7)

where n and ε are the refractive index and absorption coefficient of the gas sample, respectively, P_{exc} is the intensity of the *pump* light, T_0 is the absolute temperature, a is the *pump* beam diameter, ρ is the gas sample density, C_p corresponds to the specific heat of the gas sample, f is the modulation frequency of the *pump* light. The photothermal-induced RI change is typically retrieved using the second laser—*probe* (with a wavelength different from the *pump* light), usually using an interferometric approach. The *probe* light due to RI change experiences a phase shift according to the following formula [44]:

$$\Delta \varphi = \frac{(2\pi L \Delta n)}{\lambda},$$ (8)

where L is the laser-gas molecules' interaction path length and λ is the wavelength of the *probe* light. The unique feature of the PTS is the fact that the *probe* beam wavelength can significantly differ from the *pump* wavelength, hence the PTS signal readout can be achieved using inexpensive and widely available electronic, fiber, and optical components developed for the so-called telecom spectral band (i.e., ~1.55 μm). In addition, when the *pump* light is modulated with a sinewave signal, the spectroscopic information can be conveniently retrieved using the WMS-based technique [47–49]. Furthermore, the RI modulation can be also encoded into the frequency deviation of the beat note of the *probe* beam by using the heterodyne detection [49]. In such a configuration, the signal analysis in the frequency, not amplitude domain gives the PTS sensors immunity to the negative influence of, e.g., optical fringes, which results in the baseline-free signal retrieval, hence very high sensor sensitivity. It can be seen from the aforementioned equations that the PTS signal can be linearly enhanced with the increase in the *pump* power density over the interaction path length, however, a perfect overlap between the *pump* and the *probe* beams is mandatory, but not simple to achieve using bulk optics-based components. The perfect solution to this problem comes with the aid of ARHCFs. These fibers are characterized with mode field diameters (MFD) typically in the range of a few tens of micrometers, which means that a small beam size, hence high power density can be efficiently confined and maintained throughout the entire fiber length. Furthermore, the ability to guide light in dissimilar wavelength regions allows ARHCFs to transmit simultaneously both the *pump* and *probe* light in the gas-filled core, providing the perfect overlap between them. So far, several configurations of the PTS sensors utilizing ARHCFs as absorption cells have been developed based on the Mach-Zehnder (MZI) and Fabry–Perot (FPI) interferometer setups and shown to provide an exceptional detection capability [22,26,50–54].

5.1. MZI PTS in ARHCF

In an MZI-based PTS sensor configuration, the modulation of the RI induced by the gas molecules excitation is observed as the difference in the optical path length (hence the phase difference), that is experienced by the *probe* beam propagating in two arms of the interferometer [55]. In a typical MZI PTS sensor setup, the sensing arm of the interferometer consists of an absorption cell while the second is used as a reference and is free of gas molecules.

An ARHCF-aided configuration of such a sensor was demonstrated by Yao et al. in [50] and was aimed at detecting CO. The experimental setup of the sensor is presented in Figure 8. In this configuration, the *pump* laser operated at 2327 nm, which corresponds to the R(10) transition of CO in the $2v_1$ band, while the *probe* laser wavelength was set to 1533 nm. The *probe* beam was divided into two arms of the MZI and coupled together with the *pump* light into a 0.85 m long gas-filled hollow-core negative curvature fiber (HC-NCF) placed in the sensing arm. The dichroic mirror in the sensing arm was used to separate the remaining *pump* light from the *probe* beam that contained the information of the induced RI modulation. The MZI was set to operate in the quadrature point by implementing a piezo-electric transducer with a piece of a conventional single-mode fiber coiled on it. The *probe* beams leaving both arms of the MZI are combined using a fiber coupler, and subsequently, the beat note signal was detected by a photodetector. The interferometric signal contained information about the phase change of the *probe* light after passing through the heated gas sample in the fiber core. With the additional sinewave modulation applied to the *pump* laser injection current, the spectroscopic signal was retrieved using the well-known WMS method. The registered signal was free from the intermodal interference noise in the fiber, which allowed the sensor to obtain a normalized noise equivalent absorption coefficient (NNEA) at the level of 4.4×10^{-8} cm^{-1} WHz$^{-1/2}$ (90 ppmv), which gives an order of magnitude improvement in comparison to the similar sensor configuration utilizing a hollow-core capillary tube [55].

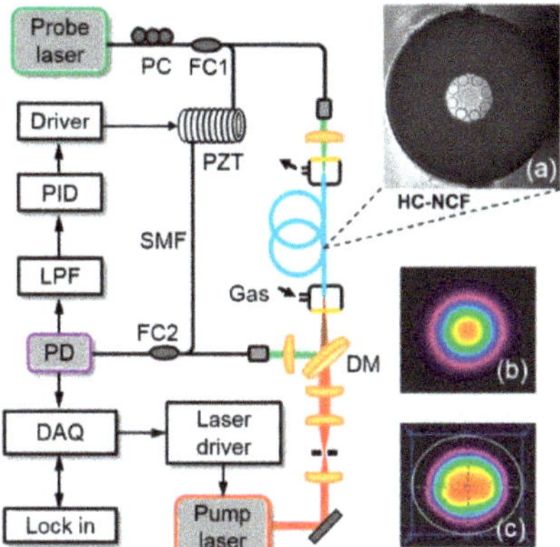

Figure 8. Schematic diagram of the MZI PTS sensor utilizing an HC-NCF as an absorption cell used to detect CO at 2327 nm. The fiber was filled with the gas analyte using gas filling cells placed at both ends. HC-NCF—hollow core negative curvature fiber (ARHCF), PID—proportional-integral-derivative controller, PD—photodetector, LP—Flow-pass filter, FCfiber coupler, PCpolarization controller, PZT—piezo-electric transducer, DM—dichroic mirror. (**a**) Cross-section of the HC-NCF used in the experiment. (**b**) HC-NCF-delivered beam profile. (**c**) Profile of the beam emitted by the *pump* laser. Reprinted with permission from [50] © The Optical Society.

Further development of ARHCF-aided MZI PTS sensors was reported in [51]. The authors benefited from the unique property of the ARHCFs, which enables them to guide light with low loss both in the near- and mid-IR spectral band. The developed sensor configuration was similar to the one presented in Figure 8, however, the *pump* light was emitted from an ICL operating at 2778.48 cm^{-1}, which corresponded to the strong transition of formaldehyde (CH$_2$O), while the *probe* beam was delivered from a telecom DFB laser emitting light at 1.56 µm. The gas-laser interaction path in the sensing arm of the MZI was formed by a 1.2 m long ARHCF having an air core with a diameter of 65 µm, which was characterized by the attenuation of 0.6 dB/m and 0.43 dB/m at the *pump* and *probe* wavelengths, respectively. Hence, both wavelengths could be simultaneously coupled into the fiber and transmitted through it with low loss. The ICL was modulated with a sinusoidal signal with a frequency of 8 kHz, which enabled performing WMS-based spectroscopic signal retrieval at the harmonics of the modulation frequency. The authors proved that the demodulation of the photodetector signal at the 1st harmonic (1f detection) provided greater signal amplitude while maintaining a low noise level in comparison to the 2f signal when the sinewave modulation frequency was greater than 6 kHz. Furthermore, the background free behavior of the sensor was maintained utilizing the 1f detection scheme. The obtained SNR for the 1f signal reached 163, which was 2.4 times better compared to the value obtained for the 2f signal. The obtained SNR yielded an MDL of 0.18 ppmv, which corresponds to an NNEA of 4×10^{-9} cm^{-1} WHz$^{-1/2}$.

5.2. FPI PTS in ARHCF

Another approach to measuring the photothermal effect refers to the application of an FPI, which enables efficient detection of the phase change of the propagating *probe* beam after passing through the heated gas sample, hence experiencing the locally modulated RI. The intensity of the beam exiting the Fabry–Perot cavity can be determined based on the equation [44,56]:

$$I_t = I_0 \frac{1}{1 + \left(\frac{2F}{\pi}\right)^2 \sin^2\left(\frac{\Delta\varphi}{2}\right)},\qquad(9)$$

where I_t is the transmitted beam intensity, I_0 is the intensity of the beam before entering the Fabry–Perot cavity, F is the cavity finesse, $\Delta\varphi$ corresponds to the phase change, which can be further defined by [44]:

$$\Delta\varphi = \frac{2\pi}{\lambda_p}2nL\cos\theta, \tag{10}$$

where λ_p is the probe laser wavelength, n is the RI, L is the cavity length, and $\cos\theta$ defines the angle of incidence. It can be seen that due to the photothermal effect inside the cavity, the modulation of the RI has a direct impact on the change in the intensity of the transmitted radiation, which combined with the PTS effect can be used to determine the molecular concertation of the measured sample. Furthermore, in comparison to the homodyne MZI PTS detection scheme, the FPI PTS sensor can achieve long-term repeatability via a non-complex stabilization of the *probe* laser wavelength to the quadrature of the FPI using a proportional-derivative-integral (PID) based approach [52]. Several configurations of the FPI PTS sensors utilizing the ARHCFs have already been demonstrated and will be discussed in this subsection.

Chen et al. reported in [26] an FPI PTS gas sensor targeting ethane (C_2H_6) at 3.348 µm, which setup is presented in Figure 9a. The absorption cell was formed by a 14 cm long HC-NCF with a core diameter of 65 µm. The molecules of the target gas were excited using an ICL operating at the aforementioned wavelength, while the *probe* beam was delivered from a 1.55 µm fiber laser. The FPI cavity layout is shown in Figure 9b. The *probe* light was coupled into the HC-NCF directly from a conventional single-mode fiber (SMF) using a butt-coupling approach. The same technique was used to couple the *pump* light, however, the ICL beam delivery fiber was an InF_3 (indium fluoride) mid-IR guiding SMF. The FPI cavity was realized based on the ~4% probe light reflections at the HC-NCF/SMF and HC-NCF/InF_3 SMF interfaces. The end facets of each fiber were glued into the ceramic ferrules, which mechanically stabilized the coupling between them and were used to deliver the gas sample into the HC-NCF core. The *probe* laser wavelength was locked and stabilized at the quadrature point of the interference fringe using a servo loop, which allowed converting the induced phase change into the intensity change at the output of the FPI. The *pump* wavelength was additionally modulated with a sinewave signal to perform WMS-based signal readout at the 2nd harmonic of the modulation frequency. The remaining *pump* light leaving the HC-NCF was filtered out by the 1.55 µm SMF. The stabilization was mandatory to obtain efficient operation of the sensor and its long-term stability, which was experimentally verified by registering the 2f signal amplitude over an 8 hour period. The system allowed obtaining an MDL of 2.6 ppbv for 410 s integration time, which gives an NEA of 2.0×10^{-6} cm^{-1}.

Krzempek et al. presented in [52] an FPI PTS sensor configuration aided with a borosilicate ARHCF (shown in Figure 1f), which pushed for the first time the operational wavelength of such sensors beyond the 5 µm range. The experimental setup of the sensor is shown in Figure 10a. In this configuration, the *pump* light was delivered from a QCL operating at 1900.09 cm^{-1}, which provided access to a strong transition of NO. The *probe* beam came from a standard telecom DFB laser delivering 1.55 µm output. The *probe* laser wavelength was stabilized using the PID-based approach to keep it operating at the quadrature point of the FPI. The 1550 nm light after passing the circulator was coupled into the ARHCF using the butt coupling technique. After leaving the ARHCF, the *probe* beam was reflected back to the fiber from a germanium window and directed via the circulator to the near-IR photodetector, which combined with a lock-in amplifier allowed 2f WMS-based signal readout. The FPI cavity in the sensor was formed by the *probe* beam reflections from the SMF28 end facet (R1) and the germanium window (R2) as depicted in Figure 10b. The absorption cell was constructed based on a 25 cm long ARHCF with a core size of 122 µm. The ARHCF was filled with the target gas mixture through a set of femtosecond laser processed microchannels, which provided direct access to the core region and eliminated the need of using gas filling cells. The fiber was glued into a steel tube equipped with a gas delivery port, which enabled the authors to efficiently fill the fiber with NO. The registered 2f signal spectrum shown in Figure 10c perfectly matches the simulated signal and confirms the baseline-free characteristic of the sensor. The unique sensor configuration combined with the ARHCF ability to simultaneously guide near- and mid-IR light resulted in an MDL of 11 ppbv for 144 s (NEA ~1.68×10^{-7} cm^{-1}) integration time with an NNEA of 4.29×10^{-7} cm^{-1} WHz$^{-1/2}$.

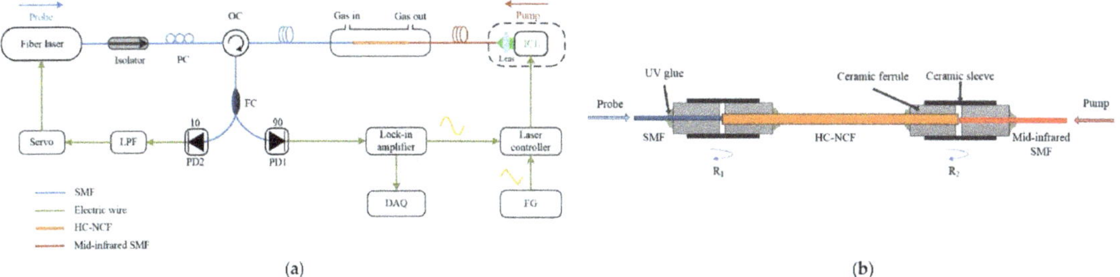

(a)   (b)

Figure 9. FPI PTS C$_2$H$_6$ gas sensor based on the use of a mid-IR ARHCF. (**a**) Experimental setup. PC—polarization controller, OC—circulator, ICL—interband cascade laser, FG—function generator, DAQ—data acquisition card, PD—photodetector, LPF—low-pass filter, SMF—single-mode fiber, HC-NCF—hollow core negative curvature fiber. (**b**) The absorption cell is formed by the gas-filled HC-NCF (ARHCF), which is butt-coupled with a conventional SMF and mid-IR SMF. R$_1$ and R$_2$ indicate the reflections in the FPI cavity formed by the interfaces between the coupling points of these fibers. Reprinted with permission from [26] © The Optical Society.

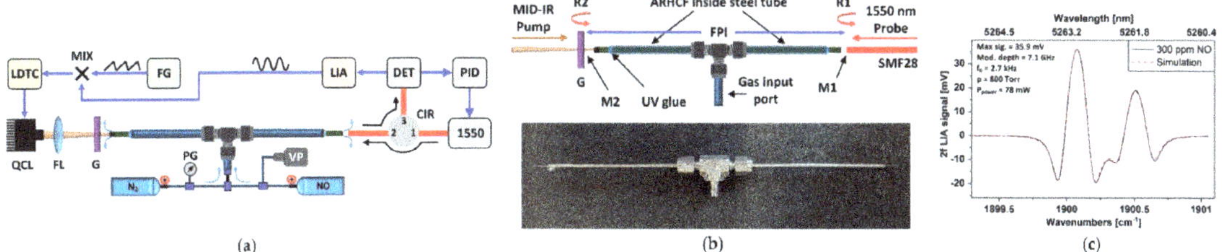

(a)   (b)   (c)

Figure 10. FPI PTS NO sensor utilizing a borosilicate ARHCF. (**a**) Schematic of the sensor setup. LDTC—laser driver, MIX—frequency mixer, FG—function generator, LIA—lock-in amplifier, DET—photodetector, PID—proportional-derivative-integral controller, CIR—circulator, G—germanium window, FL—focusing lens, QCL—quantum cascade laser, VP—vacuum pump, PG—pressure gauge. (**b**) FPI cavity (top). R1 and R2—reflections in the FPI, M1, and M2—FPI mirrors. 25 cm long ARHCF-based absorption cell (bottom) glued into a steel tube inserted into a gas delivery t-junction port. (**c**) Registered 2f signal spectrum from 300 ppmv NO in the ARHCF (black trace) and simulated (red trace). Reprinted from [52] with permission from Elsevier.

FPI PTS sensor aided with ARHCFs was also designed and developed to operate only in the near-IR spectral band, where both the *pump* and the *probe* beam were transmitted within the same low-loss window of the ARHCF. Bao et al. reported in [54] an acetylene (C$_2$H$_2$) sensor that utilized a 5.5 cm long ARHCF that was *pumped* with 1532.5 nm using a DFB laser. The FPI cavity was similar to the one shown in Figure 9b. The *probe* light at 1551.3 nm was provided by the external cavity diode laser (ECDL) and was separated from the unabsorbed *pump* light in the gas-filled ARHCF using a wavelength division multiplexer (WDM) before being directed onto a photodetector and further analyzed. The spectroscopic signal retrieval was realized using the same approach as described above. The sensor reached an MDL of 2.3 ppbv for 670 s integration time and an NEA of 2.3×10^{-9} cm^{-1}.

5.3. MPD PTS in ARHCF

In the majority, the multimode characteristic of the ARHCFs is not desired in the sensor configurations where the intermodal interference induces a significant increase in the baseline noise in the sensor. However, with a specific sensor design and signal retrieval method, the few-moded behavior of the ARHCF can be very beneficial, especially in combination with the PTS.

Zhao et al. proposed in [25] a novel approach to the PTS technique, the so-called mode-phase-difference photothermal spectroscopy (MPD-PTS), which measures the PT-induced differential phase change of the LP_{01} and LP_{11} modes of the *probe* beam propagating through the gas-filled ARHCF core, which can be expressed by [25]:

$$\delta\varphi = \Delta\varphi_{01} - \Delta\varphi_{11} = k^*(\eta, f)\left(1 - e^{-\alpha(\lambda_{pump})CL}\right)P_{pump} \approx k^*(\eta, f)\alpha(\lambda_{pump})CLP_{pump}, \tag{11}$$

where $\Delta\varphi_{01}$ and $\Delta\varphi_{11}$ are the phase modulation for LP_{01} and LP_{11} modes of the *probe* light, respectively, k^* is the differential phase modulation coefficient, which is a function of the fractional *pump* power in the LP_{01} mode—η and sinewave modulation frequency—f of the *pump* light, $\alpha(\lambda_{pump})$ is the absorption coefficient for the relative concentration of 100%, C corresponds to the target gas concentration, L is the ARHCF length and P_{pump} defines the average *pump* beam power over the fiber length. In principle, the *pump* light transmitted in the gas-filled ARHCF core in the LP_{01} and LP_{11} mode heats the gas sample in a way corresponding to the intensity distribution of each mode as shown in Figure 11a,b. This results in the periodically varying RI modulation following the temperature change trend over the entire core length as presented in Figure 11c. This leads to the coherent mixing of the LP_{01} and LP_{11} *pump* modes with the spatial period corresponding to their beat length [25]. If the *probe* beam is simultaneously coupled to the fundamental mode and the higher-order mode, it experiences the RI modulation induced by the corresponding *pump* light modes. This results in the different phase modulation of each *probe* mode, which can be detected in a sensor configuration based on an in-line dual-mode interferometer as presented in Figure 11d. Here, the hollow-core fiber can act as an absorption cell, which can be filled with the gas analyte through the gaps between its input/output end facets and the SMFs placed at both ends or via a laser-drilled microchannel along its length. The *pump* and *probe* beams were coupled into the ARHCF using the butt coupling approach from SMF, which provided excitation of the LP_{01} and LP_{11} modes. The interferometer cavity was implemented in the setup shown in Figure 10e, which enabled C_2H_2 sensing based on the 2f signal readout. The gas molecules were *pumped* at 1532.83 nm using a DFB laser and the induced RI modulation was *probed* with an ECDL operating at 1550 nm. The ARHCF forming absorption cell was designed to operate in the ~1.5 μm wavelength band. The length of the fiber was 4.67 m with an air core diameter of 28 μm. The WDM couplers provided perfect separation of the *probe* beam from the *pump* light before directing it to the photodetector and subsequently to a lock-in amplifier for 2f signal demodulation. The authors have demonstrated that the MPD-PTS sensor can reach an MDL at the level of 15 parts-per-trillion by volume (pptv) for 100 s integration time with an NEA of 1.6×10^{-11} cm^{-1}. Furthermore, the long-term stability tests of the developed sensor confirmed its excellent robustness, greater in comparison to the ARHCF-aided sensors utilizing the MZI setup.

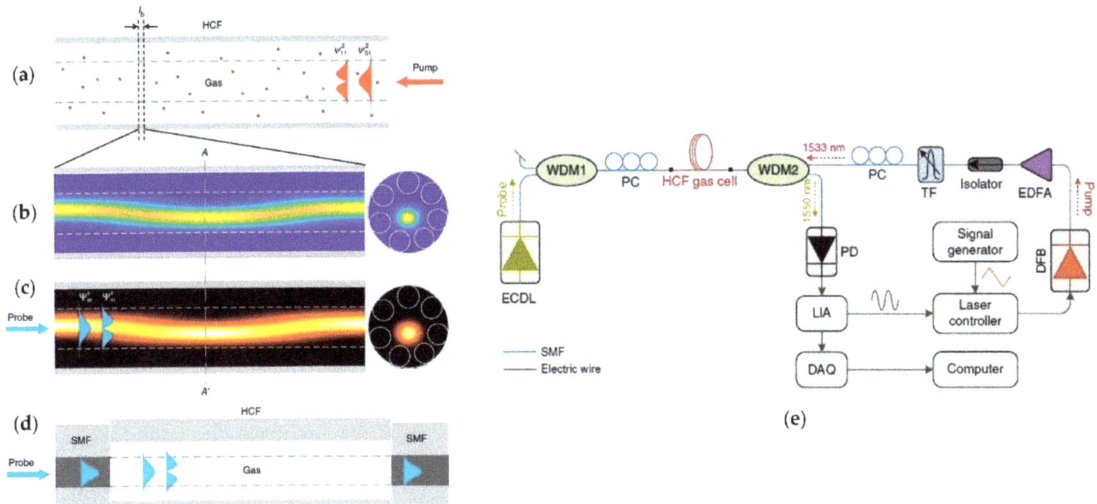

Figure 11. Principle of the MPD-PTS in ARHCF for gas sensing. (**a**) Intensity profiles of the *pump* $LP_{01}\left(\psi_{01}^2\right)$ and $LP_{11}\left(\psi_{11}^2\right)$ modes in the gas-filled ARHCF. (**b**) *Pump* intensity change within the modal beat length. (**c**) The temperature change, which induces the RI modulation in the gas-filled ARHCF core while the gas molecules are excited by the *pump* light. It can be seen that the *probe* modes $LP_{01}\left(\Psi_{01}^2\right)$ and $LP_{11}\left(\Psi_{11}^2\right)$ are experiencing different RI changes. (**d**) A schematic of the in-line dual-mode interferometer. (**e**) Experimental setup of the MPD-PTS sensor utilizing an ARHCF as a C_2H_2 absorption cell. HCF—hollow-core fiber, SMF—single-mode fiber, I_b—modal beat length, WDM—wavelength division multiplexer, PC—polarization controller, ECDL—external cavity diode laser, TF—tunable optical filter, EDFA—erbium-doped fiber amplifier, LIA—lock-in amplifier, DAQ—data acquisition card Reprinted from [25] with permission from Springer Nature.

The calculations performed by Zhao et al. in [22] have indicated that the differential phase modulation in an MPD-PTS gas sensor reaches its maximum if the *pump* light power is coupled into the LP_{01} mode. However, at the same time, the *probe* light must be transmitted within the LP_{01} and LP_{11} modes simultaneously so the induced RI modulation by the heated inside ARHCF gas molecules can introduce the phase difference between these modes, which can be subsequently analyzed. Excitation of the LP_{01} *pump* mode only and two *probe* modes at the same time is not trivial and almost impossible to realize by offsetting the SMFs with respect to the ARHCF core as it was reported by the Authors in their previous work [25]. According to this, the MPD-PTS gas sensor performance did not reach its maximum. To address this problem the Authors inscribed a long period grating (LPG) in a negative curvature – hollow core fiber (NC-HCF) forming an absorption cell as shown in Figure 12. The LPG enabled the excitation of the LP_{01} and LP_{11} modes at the *probe* wavelength maintaining *pump* light transmission in the LP_{01} mode only. This modification of the sensor resulted in the maximization of the differential phase modulation induced by the photothermal effect and reduced the complexity of the sensor setup. The approach was tested in an experimental configuration of an MPD-PTS detector targeting C_2H_2 at 1532.83 nm in an 85 cm long NC-HCF with a core size of ~30 μm. The induced RI modulation was *probed* at 1620 nm. The setup of the sensor and measurement procedure was similar to the described above. The sensor reached an MDL of 600 pptv at 100 s integration time with an NEA of 6.3×10^{-10} cm^{-1}, maintaining a less complex configuration than reported in [25].

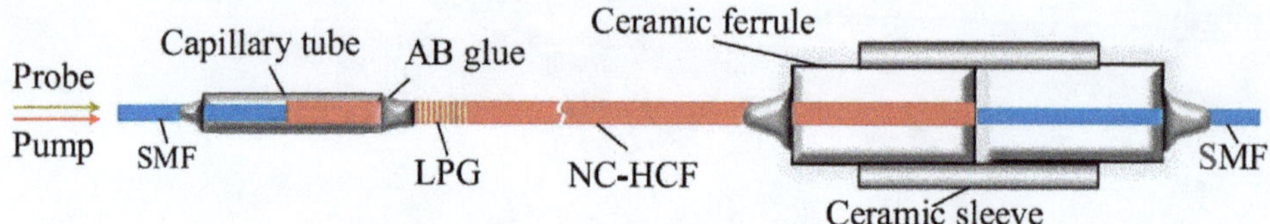

Figure 12. Schematic of the absorption cell formed by an NC-HCF with LPG inscribed in it. The absorption cell forms an in-line dual-mode interferometer. The *probe* and *pump* beams were delivered to and outcoupled from the NC-HCF using the butt-coupling approach with SMFs. Reprinted with permission from [22] © The Optical Society.

Zhao et al. reported in [53] an MPD-PTS CH_4 sensor utilizing an ARHCF with an inscribed LPG. The CH_4 molecules were *pumped* at ~1653.7 nm with the aid of a DFB laser, which was additionally amplified using a fiber-based Raman amplifier to maximize the photothermal effect. The *probe* light at 1550 nm was delivered from a fiber laser equipped with a wavelength stabilization unit. The absorption cell was realized based on a 2.4 m long ARHCF with a core size of ~30 μm, which was characterized by an attenuation of 0.16 dB/m and 0.25 dB/m at the *probe* and *pump* wavelengths, respectively. The sensor setup was built in a configuration similar to the one shown in Figure 11e and the spectroscopic signal retrieval was based on the 2f signal demodulation. In the proposed configuration, a part of the ARHCF forming the absorption cell was placed inside a column oven to investigate the influence of the temperature change on the induced differential phase modulation of the *probe* beam. The performed experiments indicated that the operating point of the dual-mode interferometer (the quadrature point of the interference fringe) is highly sensitive to the temperature deviations, hence a proper compensation of the temperature drift is necessary to improve the sensor stability. It was possible to minimize the photothermal signal variations from ~9.4% to ~2.1% using a linear temperature compensation scheme and obtain an MDL of approximately 4.3 ppbv for 100 s integration time. The NEA coefficient reached the level of 1.6×10^{-9} cm^{-1}.

5.4. PABS in ARHCF

PAS is a spectroscopic technique similar to the PTS, however, in this case, the modulation of the phase of the *probe* light along the gas-molecules interaction path length results from an increase of the local pressure gradient due to gas molecules excitation by the *pump* laser light [57]. When the *pump* laser beam intensity is additionally modulated at a certain frequency, the pressure will change periodically, producing an acoustic wave having a frequency equal to the modulation frequency of the *pump* light [57]. The photoacoustic effect has been widely used in laser-based spectroscopy, delivering an exceptional sensitivity of the PAS-based gas sensors [58–60].

The PAS technique was recently combined with gas sensing inside an ARHCF by Zhao et al. in [23]. The authors have shown that it is possible to efficiently excite acoustic modes inside the gas-filled fiber core as a result of the photoacoustic effect and the structure of the ARHCF. This modulates the RI of the gas sample and subsequently the phase of the *probe* beam in the fiber. The interaction between the optical and acoustic modes is explained by means of the Forward Brillouin Scattering (FBS) phenomenon [61]. The silica cladding of the ARHCF forms an acoustic resonator and the gas-filled part of the fiber acts as a region for acoustic wave generation. As a result, the modulated light-excited and heated gas molecules induce an acoustic wave, which is resonantly enhanced inside the ARHCF. When the *probe* light propagates inside the fiber, it experiences a phase modulation, which can be measured using the earlier described dual-mode interferometer method. Since an ARHCF supports capillary and core acoustics modes as shown in Figure 13a,b, the optical LP_{01} and LP_{11} modes differently overlap with them, which introduces different phase modulation of each of these modes. The observed differential phase modulation can be described by the following formula [23]:

$$\Delta\varnothing(\Omega\lambda_p) = \xi(\Omega)\alpha(\lambda_p)CL_{eff}P_p, \tag{12}$$

where Ω is the modulation frequency, λ_p is the *pump* wavelength, ξ defines the normalized phase modulation coefficient, α is the absorption coefficient, C is the gas concentration in the ARHCF, L_{eff} is the effective length of the ARHCF (directly depending on the molecular concentration, absorption coefficient and the actual fiber length), and P_p is the *pump* power. The gas sensing using PABS in the ARHCF was experimentally verified in the setup depicted in Figure 13c. A DFB laser operating at 1532.83 nm was used as the *pump* source for C_2H_2 molecules excitation in a 30 cm long ARHCF-based absorption cell. The *pump* light was modulated with a sinewave signal at the frequency Ω, which corresponds to the frequency of the selected acoustic mode with the aid of an acousto-optic modulator (AOM). The *probe* beam was delivered from an ECDL and simultaneously coupled with the *pump* beam into the gas-filled ARHCF. Via the lateral offset coupling to the ARHCF, it was possible to excite LP_{01} and LP_{11} modes, hence forming a dual-mode interferometer. The ARHCF was filled with the measured gas via a gas filling cell. The optical signal was directed to a balanced photodetector to minimize the intensity noise and subsequently demodulated at $f = \Omega/2\pi$ using a lock-in amplifier. The system was characterized to provide an MDL of 8 ppbv for 100 s integration time. Operation of the system at frequencies in the range of MHz allows it to minimize the negative influence of the $1/f$ noise, which is not obtainable in PTS-based gas sensors.

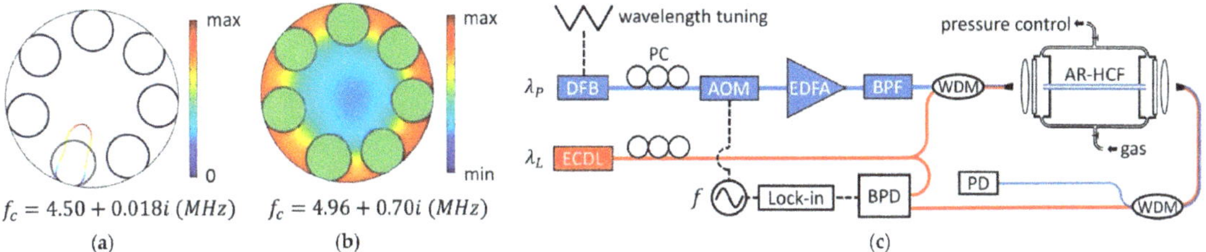

Figure 13. (a) A capillary acoustic mode in the ARHCF. (b) A core acoustic mode in the ARHCF. f_c is the eigenfrequency of each mode. (c) Experimental setup of the PABS sensor. DFB—distributed feedback diode laser, ECDL—external cavity diode laser, PC—polarization controller, AOM—acousto-optic modulator, f is the sinewave modulation frequency generator, EDFA—erbium-doped fiber amplifier, BPF—optical bandpass filter, WDM—wavelength division multiplexer, PD—photodetector, BPD—balanced photodetector. Reprinted with permission from [23] © The Optical Society.

6. Summary

In this review, recent advancements in the ARHCF-based gas sensors were presented and discussed. It was shown that the properly designed ARHCFs could be successfully used as low volume, compact, and long absorption cells in sensing systems utilizing various laser spectroscopic techniques, i.e., TDLAS, WMS, PTS (in several dissimilar configurations), and PABS. The ARHCF absorption cells forming the gas molecules' interaction path length within the sensor can be efficiently and at a reasonably fast time filled with the target gas mixture using an overpressure induced gas flow, which results in an acceptable response time of such a sensor. A summary of the performance of the selected ARHCF-aided gas sensors is presented in Table 2.

The ARHCF-aided gas sensors utilizing the TDLAS technique indeed proved their capability to detect molecules of various gases in a simple and non-complex way, however, did not provide sufficient suppression of the background noise resulting from, i.e., intermodal interference. This drawback severely limits the sensitivity and long-term stability of such sensors, hence different spectroscopic techniques have to be combined with ARHCFs to fully benefit from the connection of both.

Table 2. Comparison of the performance of the ARHCF-based gas sensors.

Configuration	Gas	Wavelength	ARHCF Length	Filling Time	MDL	NEA [cm^{-1}]	NNEA [cm^{-1} WHz$^{-1/2}$]	Integration Time
TDLAS [21]	CO_2	2004 nm	1.35 m	5 s	1.5%	-	-	-
TDLAS [38]	CO	2326.8 nm	0.85 m	5 s	13 ppmv	5.2×10^{-6}	-	-
TDLAS [35]	N_2O	3599.2 nm	1.20 m	-	-	2.5×10^{-7}	-	40 s
TDLAS [24]	NO	5262.9 nm	0.21 m	0.3 s	1.2 ppmv	2.1×10^{-5}	-	-
WMS [21]	CO_2	2004 nm	1.35 m	5 s	5 ppmv	7.4×10^{-7}	-	3 s
WMS [38]	CO	2326.8 nm	0.85 m	5 s	0.4 ppmv	1.6×10^{-7}	-	30 s
WMS [24]	NO	5262.9 nm	0.21 m	0.3 s	6 ppbv	1.0×10^{-7}	-	30 s
WMS [27]	N_2O	4537.8 nm	3.20 m	23 s	5.4 ppbv	3.7×10^{-7}	-	1 s
WMS [20]	NO	5262.9 nm	1.15 m	9 s	20 ppbv	2.0×10^{-5}	-	70 s
WMS [8]	CO_2	1574 nm	1.0 m	-	144 ppmv	1.17×10^{-7}	-	1.5 s
WMS [8]	CH_4	3334 nm	1.0 m	19 s	24 ppbv	1.6×10^{-7}	-	40 s
MZI PTS [50]	CO	2327 nm	0.85 m	-	90 ppmv	-	4.4×10^{-8}	-
MZI PTS [51]	CH_2O	3599.1 nm	1.20 m	-	0.18 ppmv	-	4.0×10^{-9}	-
FPI PTS [26]	C_2H_6	3348.15 nm	0.14 m	300 s	2.6 ppbv	2.0×10^{-6}	-	410 s
FPI PTS [52]	NO	5262.9 nm	0.25 m	63 s	11 ppbv	1.68×10^{-7}	4.29×10^{-7}	144 s
FPI PTS [54]	C_2H_2	1532.5 nm	0.055 m	52 s	2.3 ppbv	2.3×10^{-9}	-	670 s
MPD PTS [25]	C_2H_2	1532.83 nm	4.67 m	-	15 pptv	1.6×10^{-11}	-	100 s
MPD PTS [22]	C_2H_2	1532.83 nm	0.85 m	-	600 pptv	6.3×10^{-10}	-	100 s
MPD PTS [53]	CH_4	1653.7 nm	2.40 m	-	4.3 ppbv	1.6×10^{-9}	-	100 s
PABS [23]	C_2H_2	1532.83 nm	0.30 m	20 s	8 ppbv	-	-	100 s

The combination of ARHCFs with the WMS technique unquestionably provides a significant improvement in the noise level reduction in the ARHCF-assisted gas sensor configurations, which directly leads to a better sensor sensitivity. However, such sensors are still not immune to the uncontrolled fluctuations of the registered signal. Since especially short lengths of the ARHCFs are characterized by the presence of higher-order modes or longer pieces from the non-uniformity of the structure, the amplitude-based signal retrieval is sensitive to the parasitic changes of the analyzed light intensity, which significantly impacts the overall performance of this type of gas sensors. Hence, not only proper optimization of the fiber structure but also a change of the gas sensing method can provide the desired improvement in the operation of the ARHCF-based sensors.

The other approach for gas sensing aided from ARHCFs comes with the PTS technique, where the spectroscopic signal retrieval is based on the analysis of the *probe* light phase modulation induced by a local change in the RI due to heating of gas molecules. The phase modulation can be precisely investigated based on the interferometric signal measurement. The ARHCFs were implemented in the PTS sensors utilizing MZI, FPI, and MPD sensor setups. The main drawback of the MZI PTS gas sensors is the necessity of actively and precisely stabilizing the sensor's setup to maintain its operation at the quadrature point of the interferometer. This is not simple to realize, and the long-term stability of such sensors is very difficult to be obtained, which severely limits their ability to operate in real application conditions. The less complex sensor design is delivered by the application of an FPI. In an FPI PTS gas sensor utilizing ARHCFs, stabilizing the cavity at quadrature can be realized in a much more accessible and reliable way, by stabilizing the wavelength of the *probe* beam, e.g., with an aid of PID-based technique. With this technique, the sensitivity of the ARHCF-assisted PTS sensors is improved in comparison to the sensors based on the MZI. Until now, the most sensitive configuration of the ARHCF-based gas sensors was based on the MPD PTS technique. The MPD PTS sensors were shown to provide significantly better long-term stability and immunity to the negative influence of external factors than, e.g., MZI PTS. However, it can be assumed that even simple WMS-based sensors aided with several tens of meters long ARHCFs could provide comparable or even better sensitivity benefiting from long interaction path lengths. The main requirements which have to be fulfilled are the pure single-mode guidance of the ARHCF and the uniformity of its structure, which both are mandatory to eliminate the fringe noise and the parasitic fluctuations of the signal amplitude. Furthermore, the application of heterodyne signal retrieval in the MZI PTS configurations instead of the homodyne technique will result in the possibility of analyzing the signal in the frequency, not amplitude domain which could provide the ultimate immunity of the sensor to the fringe noise (resulting from the intermodal interference in the ARHCFs) and the amplitude noise (e.g., residual amplitude modulation) [49].

A very interesting approach to gas sensing inside the ARHCFs was presented in [23], where the sensing system was based on the PABS technique. The interaction between the fiber-supported acoustics and optical modes enables signal retrieval in the MHz frequency range, which minimizes the $1/f$ noise. It was stated that with the further optimization of the fiber structure, hence the structure of the formed in it an acoustic resonator, ARHCF-based PABS sensor could deliver significantly greater sensitivity than reported by the authors.

The ARHCF-based gas sensors have already proved their excellent suitability for this application and opened a way to the new branch of sensitive, low-volume, and versatile detectors. It is expected that further development of the hollow-core fiber technology will result in the possibility of fabricating several tens of meters long fibers, maintaining uniform structure along their entire length, which will enable low-loss and single-mode transmission not only in the near- but also in the mid-IR spectral band. The utilization of different than fused silica materials, such as telluride or chalcogenide glasses should extend the operational wavelength range of these fibers to the spectral bands above 5.26 µm, hence allowing the detecting of various gases over a significantly broader than currently available range. The improvement in the transmission properties of the ARHCFs should also allow them to be successfully used in the broadband spectroscopy applications, i.e., in the frequency comb spectroscopy [62], which will enable an in-depth analysis of the complex mixture of gaseous substances, maintaining a low-volume of the sensing unit with high detection capability. Currently, the ARHCF-based gas sensors are in majority realized in laboratory conditions, which results in their still large size (despite the low-volume formed by the absorption cells) in comparison to field-deployable bulk-optics based sensors [63] and most commonly used in real-life applications non-optical gas detectors [64,65]. However, further minimization of the electronic components, accompanied with optics-free light coupling into the fiber and the possibility of bending the fiber with bend radius in the range of a few centimeters [29,66] should result in a significant reduction in the size of ARHCF-aided gas sensors, even beating the dimensions of the sensors utilizing multipass cells delivering several tens of meters long optical path lengths. Moreover, proper laser-based modification of the fiber structure, enabling loss-free access to the fiber core for gas filling purposes [67], should result in pure diffusion-based gas exchange in such sensors, significantly enhancing their versatility and usefulness together with further reduction in the overall size of the entire sensing unit. The combination of a low-volume, non-complex design, selectivity, excellent detection capability, and calibration-free operation should lead the ARHCF-aided gas sensing approach to the development of the sensors that could in future form a new branch of gas spectrometers with the parameters comparable or even beating the currently used devices.

Funding: This research was funded by Narodowe Centrum Nauki (NCN), grant number UMO−2018/30/Q/ST3/00809. The APC was funded by MDPI Sensors journal.

Institutional Review Board Statement: Not applicable.

Informed Consent Statement: Not applicable.

Data Availability Statement: Not applicable.

Acknowledgments: The author would like to thank Viktoria Hoppe for taking the SEM image of the borosilicate ARHCF. The author would like to acknowledge the financial support of MDPI Sensors journal for generously covering the publication fee.

Conflicts of Interest: There is no conflict of interest to be declared.

Nomeclature

List of Acronyms	Definition
AOM	acousto-optic modulator
ARHCF	Antiresonant Hollow-Core Fiber
ARROW	Antiresonant Reflecting Optical Waveguiding
CLaDS	Chirped Laser Dispersion Spectroscopy
CW	continuous wave
DFB	distributed feedback diode laser
DFG	difference frequency generation
ECDL	external cavity diode laser
FBS	Forward Brillouin Scattering
FPI	Fabry-Perot interferometer
FWHM	full width at half maximum
HCF	hollow-core fiber
HC-NCF	hollow-core negative curvature fiber
HC-PBGF	hollow-core photonic bandgap fiber
ICL	interband cascade laser
LPG	long period grating
MDL	minimum detection limit
MFA	minimum fractional absorption
MFD	mode field diameter
mid-IR	mid-infrared
MPD-PTS	Mode-phase-difference photothermal spectroscopy
MZI	Mach-Zehnder interferometer
NC-HCF	negative curvature–hollow core fiber
NEA	noise equivalent absorption
near-IR	near-infrared
NNEA	normalized noise equivalent absorption
PABS	Photoacoustic Brillouin Spectroscopy
PAS	Photoacoustic Spectroscopy
PID	proportional-derivative-integral
ppbv	parts-per-billion by volume
ppmv	parts-per-million by volume
PTS	Photothermal Spectroscopy
pptv	parts-per-trillion by volume
QCL	quantum cascade laser
RI	refractive index
SMF	single-mode fiber
SNR	signal-to-noise ratio
TDLAS	Tunable Diode Laser Absorption Spectroscopy
WDM	wavelength division multiplexer
WMS	Wavelength Modulation Spectroscopy

References

1. Cregan, R.F.; Mangan, B.J.; Knight, J.C.; Birks, T.A.; Russell, P.S.J.; Roberts, P.J.; Allan, D.C. Single-Mode Photonic Band Gap Guidance of Light in Air. *Science* **1999**, *285*, 1537–1539. [CrossRef]
2. Urich, A.; Maier, R.R.J.; Yu, F.; Knight, J.C.; Hand, D.P.; Shephard, J.D. Flexible Delivery of Er:YAG Radiation at 2.94 Mm with Negative Curvature Silica Glass Fibers: A New Solution for Minimally Invasive Surgical Procedures. *Biomed. Opt. Express* **2013**, *4*, 193–205. [CrossRef]
3. Humbach, O.; Fabian, H.; Grzesik, U.; Haken, U.; Heitmann, W. Analysis of OH Absorption Bands in Synthetic Silica. *J. Non Cryst. Solids* **1996**, *203*, 19–26. [CrossRef]
4. Jaworski, P.; Yu, F.; Maier, R.R.J.; Wadsworth, W.J.; Knight, J.C.; Shephard, J.D.; Hand, D.P. Picosecond and Nanosecond Pulse Delivery through a Hollow-Core Negative Curvature Fiber for Micro-Machining Applications. *Opt. Express* **2013**, *21*, 22742–22753. [CrossRef] [PubMed]
5. Emaury, F.; Dutin, C.F.; Saraceno, C.J.; Trant, M.; Heckl, O.H.; Wang, Y.Y.; Schriber, C.; Gerome, F.; Südmeyer, T.; Benabid, F.; et al. Beam Delivery and Pulse Compression to Sub-50 Fs of a Modelocked Thin-Disk Laser in a Gas-Filled Kagome-Type HC-PCF Fiber. *Opt. Express* **2013**, *21*, 4986–4994. [CrossRef] [PubMed]
6. Hong, Y.; Sakr, H.; Taengnoi, N.; Bottrill, K.R.H.; Bradley, T.D.; Hayes, J.R.; Jasion, G.T.; Kim, H.; Thipparapu, N.K.; Wang, Y.; et al. Multi-Band Direct-Detection Transmission Over an Ultrawide Bandwidth Hollow-Core NANF. *J. Lightwave Technol.* **2020**, *38*, 2849–2857. [CrossRef]
7. Cui, Y.; Huang, W.; Wang, Z.; Wang, M.; Zhou, Z.; Li, Z.; Gao, S.; Wang, Y.; Wang, P. 4.3 µm Fiber Laser in CO_2-Filled Hollow-Core Silica Fibers. *Optica* **2019**, *6*, 951–954. [CrossRef]
8. Jaworski, P.; Kozioł, P.; Krzempek, K.; Wu, D.; Yu, F.; Bojęś, P.; Dudzik, G.; Liao, M.; Abramski, K.; Knight, J. Antiresonant Hollow-Core Fiber-Based Dual Gas Sensor for Detection of Methane and Carbon Dioxide in the Near- and Mid-Infrared Regions. *Sensors* **2020**, *20*, 3813. [CrossRef]
9. Kornaszewski, Ł.; Gayraud, N.; Stone, J.M.; MacPherson, W.N.; George, A.K.; Knight, J.C.; Hand, D.P.; Reid, D.T. Mid-Infrared Methane Detection in a Photonic Bandgap Fiber Using a Broadband Optical Parametric Oscillator. *Opt. Express* **2007**, *15*, 11219–11224. [CrossRef]
10. Krzempek, K.; Abramski, K.; Nikodem, M. Kagome Hollow Core Fiber-Based Mid-Infrared Dispersion Spectroscopy of Methane at Sub-Ppm Levels. *Sensors* **2019**, *19*, 3352. [CrossRef]
11. Wang, Y.Y.; Wheeler, N.V.; Couny, F.; Roberts, P.J.; Benabid, F. Low Loss Broadband Transmission in Hypocycloid-Core Kagome Hollow-Core Photonic Crystal Fiber. *Opt. Lett.* **2011**, *36*, 669–671. [CrossRef]
12. Herriott, D.R.; Schulte, H.J. Folded Optical Delay Lines. *Appl. Opt.* **1965**, *4*, 883–889. [CrossRef]
13. Wysocki, G.; Bakhirkin, Y.; So, S.; Tittel, F.K.; Hill, C.J.; Yang, R.Q.; Fraser, M.P. Dual Interband Cascade Laser Based Trace-Gas Sensor for Environmental Monitoring. *Appl. Opt.* **2007**, *46*, 8202–8210. [CrossRef] [PubMed]
14. Mejia Quintero, S.M.; Guedes Valente, L.C.; De Paula Gomes, M.S.; Gomes da Silva, H.; Caroli de Souza, B.; Morikawa, S.R.K. All-Fiber CO_2 Sensor Using Hollow Core PCF Operating in the 2 Mm Region. *Sensors* **2018**, *18*, 4393. [CrossRef] [PubMed]
15. Jin, W.; Cao, Y.; Yang, F.; Ho, H.L. Ultra-Sensitive All-Fibre Photothermal Spectroscopy with Large Dynamic Range. *Nat. Commun.* **2015**, *6*, 6767. [CrossRef] [PubMed]
16. Jaworski, P. Molecular Dispersion Spectroscopy in a CO_2-Filled All-Fiber Gas Cells Based on a Hollow-Core Photonic Crystal Fiber. *Opt. Eng.* **2019**, *58*, 026112. [CrossRef]
17. Shephard, J.D.; MacPherson, W.N.; Maier, R.R.J.; Jones, J.D.C.; Hand, D.P.; Mohebbi, M.; George, A.K.; Roberts, P.J.; Knight, J.C. Single-Mode Mid-IR Guidance in a Hollow-Core Photonic Crystal Fiber. *Opt. Express* **2005**, *13*, 7139–7144. [CrossRef]
18. Nikodem, M.; Krzempek, K.; Dudzik, G.; Abramski, K. Hollow Core Fiber-Assisted Absorption Spectroscopy of Methane at 3.4 µm. *Opt. Express* **2018**, *26*, 21843–21848. [CrossRef]
19. Litchinitser, N.M.; Abeeluck, A.K.; Headley, C.; Eggleton, B.J. Antiresonant Reflecting Photonic Crystal Optical Waveguides. *Opt. Lett.* **2002**, *27*, 1592–1594. [CrossRef]
20. Jaworski, P.; Krzempek, K.; Dudzik, G.; Sazio, P.J.; Belardi, W. Nitrous Oxide Detection at 5.26 Mm with a Compound Glass Antiresonant Hollow-Core Optical Fiber. *Opt. Lett.* **2020**, *45*, 1326–1329. [CrossRef]
21. Nikodem, M.; Gomółka, G.; Klimczak, M.; Pysz, D.; Buczyński, R. Laser Absorption Spectroscopy at 2 µm inside Revolver-Type Anti-Resonant Hollow Core Fiber. *Opt. Express* **2019**, *27*, 14998. [CrossRef]
22. Zhao, P.; Zhao, P.; Zhao, P.; Ho, H.L.; Ho, H.L.; Jin, W.; Jin, W.; Jin, W.; Fan, S.; Fan, S.; et al. Gas Sensing with Mode-Phase-Difference Photothermal Spectroscopy Assisted by a Long Period Grating in a Dual-Mode Negative-Curvature Hollow-Core Optical Fiber. *Opt. Lett.* **2020**, *45*, 5660–5663. [CrossRef]
23. Zhao, Y.; Zhao, Y.; Qi, Y.; Qi, Y.; Ho, H.L.; Ho, H.L.; Gao, S.; Wang, Y.; Jin, W.; Jin, W. Photoacoustic Brillouin Spectroscopy of Gas-Filled Anti-Resonant Hollow-Core Optical Fibers. *Optica* **2021**, *8*, 532–538. [CrossRef]
24. Yao, C.; Hu, M.; Ventura, A.; Hayashi, J.; Poletti, F.; Ren, W. Tellurite Hollow-Core Antiresonant Fiber-Coupled Quantum Cascade Laser Absorption Spectroscopy. *J. Light. Technol.* **2021**, 1. [CrossRef]
25. Zhao, P.; Zhao, Y.; Bao, H.; Ho, H.L.; Jin, W.; Fan, S.; Gao, S.; Wang, Y.; Wang, P. Mode-Phase-Difference Photothermal Spectroscopy for Gas Detection with an Anti-Resonant Hollow-Core Optical Fiber. *Nat. Commun.* **2020**, *11*, 847. [CrossRef]
26. Chen, F.; Chen, F.; Jiang, S.; Jin, W.; Jin, W.; Bao, H.; Bao, H.; Ho, H.L.; Ho, H.L.; Wang, C.; et al. Ethane Detection with Mid-Infrared Hollow-Core Fiber Photothermal Spectroscopy. *Opt. Express* **2020**, *28*, 38115–38126. [CrossRef] [PubMed]

27. Nikodem, M.; Gomółka, G.; Klimczak, M.; Pysz, D.; Buczyński, R.; Buczyński, R. Demonstration of Mid-Infrared Gas Sensing Using an Anti-Resonant Hollow Core Fiber and a Quantum Cascade Laser. *Opt. Express* **2019**, *27*, 36350–36357. [CrossRef]

28. Ventura, A.; Hayashi, J.G.; Cimek, J.; Jasion, G.; Janicek, P.; Janicek, P.; Slimen, F.B.; White, N.; Fu, Q.; Xu, L.; et al. Extruded Tellurite Antiresonant Hollow Core Fiber for Mid-IR Operation. *Opt. Express* **2020**, *28*, 16542–16553. [CrossRef] [PubMed]

29. Carter, R.M.; Yu, F.; Wadsworth, W.J.; Shephard, J.D.; Birks, T.; Knight, J.C.; Hand, D.P. Measurement of Resonant Bend Loss in Anti-Resonant Hollow Core Optical Fiber. *Opt. Express* **2017**, *25*, 20612–20621. [CrossRef] [PubMed]

30. Benabid, F. Hollow-Core Photonic Bandgap Fibre: New Light Guidance for New Science and Technology. *Philos. Trans. R. Soc. A Math. Phys. Eng. Sci.* **2006**, *364*, 3439–3462. [CrossRef] [PubMed]

31. Belardi, W.; Sazio, P.J. Borosilicate Based Hollow-Core Optical Fibers. *Fibers* **2019**, *7*, 73. [CrossRef]

32. Schiff, H.I.; Mackay, G.I.; Bechara, J. The Use of Tunable Diode Laser Absorption Spectroscopy for Atmospheric Measurements. *Res. Chem. Intermed.* **1994**, *20*, 525–556. [CrossRef]

33. Curl, R.F.; Capasso, F.; Gmachl, C.; Kosterev, A.A.; McManus, B.; Lewicki, R.; Pusharsky, M.; Wysocki, G.; Tittel, F.K. Quantum Cascade Lasers in Chemical Physics. *Chem. Phys. Lett.* **2010**, *487*, 1–18. [CrossRef]

34. Swinehart, D.F. The Beer-Lambert Law. *J. Chem. Educ.* **1962**, *39*, 333. [CrossRef]

35. Yao, C.; Gao, S.; Wang, Y.; Wang, P.; Jin, W.; Ren, W. Silica Hollow-Core Negative Curvature Fibers Enable Ultrasensitive Mid-Infrared Absorption Spectroscopy. *J. Light. Technol.* **2020**, *38*, 2067–2072. [CrossRef]

36. Chang, H.; Feng, S.; Qiu, X.; Meng, H.; Guo, G.; He, X.; He, Q.; Yang, X.; Ma, W.; Kan, R.; et al. Implementation of the Toroidal Absorption Cell with Multi-Layer Patterns by a Single Ring Surface. *Opt. Lett.* **2020**, *45*, 5897–5900. [CrossRef] [PubMed]

37. Stachowiak, D.; Jaworski, P.; Krzaczek, P.; Maj, G.; Nikodem, M. Laser-Based Monitoring of CH_4, CO_2, NH_3, and H_2S in Animal Farming—System Characterization and Initial Demonstration. *Sensors* **2018**, *18*, 529. [CrossRef] [PubMed]

38. Yao, C.; Xiao, L.; Gao, S.; Wang, Y.; Wang, P.; Kan, R.; Jin, W.; Ren, W. Sub-Ppm CO Detection in a Sub-Meter-Long Hollow-Core Negative Curvature Fiber Using Absorption Spectroscopy at 2.3 µm. *Sens. Actuators B Chem.* **2020**, *303*, 127238. [CrossRef]

39. Parry, J.P.; Griffiths, B.C.; Gayraud, N.; McNaghten, E.D.; Parkes, A.M.; MacPherson, W.N.; Hand, D.P. Towards Practical Gas Sensing with Micro-Structured Fibres. *Meas. Sci. Technol.* **2009**, *20*, 075301. [CrossRef]

40. Rieker, G.B.; Jeffries, J.B.; Hanson, R.K. Calibration-Free Wavelength-Modulation Spectroscopy for Measurements of Gas Temperature and Concentration in Harsh Environments. *Appl. Opt.* **2009**, *48*, 5546–5560. [CrossRef] [PubMed]

41. Li, H.; Rieker, G.B.; Liu, X.; Jeffries, J.B.; Hanson, R.K. Extension of Wavelength-Modulation Spectroscopy to Large Modulation Depth for Diode Laser Absorption Measurements in High-Pressure Gases. *Appl. Opt.* **2006**, *45*, 1052. [CrossRef]

42. Kolyadin, A.N.; Kosolapov, A.F.; Pryamikov, A.D.; Biriukov, A.S.; Plotnichenko, V.G.; Dianov, E.M. Light Transmission in Negative Curvature Hollow Core Fiber in Extremely High Material Loss Region. *Opt. Express* **2013**, *21*, 9514–9519. [CrossRef]

43. Dong, L.; Spagnolo, V.; Lewicki, R.; Tittel, F.K. Ppb-Level Detection of Nitric Oxide Using an External Cavity Quantum Cascade Laser Based QEPAS Sensor. *Opt. Express* **2011**, *19*, 24037–24045. [CrossRef]

44. Krzempek, K. A Review of Photothermal Detection Techniques for Gas Sensing Applications. *Appl. Sci.* **2019**, *9*, 2826. [CrossRef]

45. Bialkowski, S.; Astrath, N.; Proskurnin, M. *Photothermal Spectroscopy Methods*, 2nd ed.; John Wiley & Sons: Hoboken, NJ, USA, 2019; ISBN 978-1-119-27910-5.

46. Krzempek, K.; Dudzik, G.; Abramski, K. Photothermal Spectroscopy of CO_2 in an Intracavity Mode-Locked Fiber Laser Configuration. *Opt. Express* **2018**, *26*, 28861. [CrossRef]

47. Waclawek, J.P.; Bauer, V.C.; Moser, H.; Lendl, B. 2f-Wavelength Modulation Fabry-Perot Photothermal Interferometry. *Opt. Express* **2016**, *24*, 28958–28967. [CrossRef]

48. Krzempek, K.; Hudzikowski, A.; Głuszek, A.; Dudzik, G.; Abramski, K.; Wysocki, G.; Nikodem, M. Multi-Pass Cell-Assisted Photoacoustic/Photothermal Spectroscopy of Gases Using Quantum Cascade Laser Excitation and Heterodyne Interferometric Signal Detection. *Appl. Phys. B* **2018**, *124*, 74. [CrossRef]

49. Krzempek, K.; Dudzik, G.; Abramski, K.; Wysocki, G.; Jaworski, P.; Nikodem, M. Heterodyne Interferometric Signal Retrieval in Photoacoustic Spectroscopy. *Opt. Express* **2018**, *26*, 1125–1132. [CrossRef] [PubMed]

50. Yao, C.; Wang, Q.; Lin, Y.; Jin, W.; Xiao, L.; Gao, S.; Wang, Y.; Wang, P.; Ren, W. Photothermal CO Detection in a Hollow-Core Negative Curvature Fiber. *Opt. Lett.* **2019**, *44*, 4048–4051. [CrossRef] [PubMed]

51. Yao, C.; Gao, S.; Wang, Y.; Wang, P.; Jin, W.; Ren, W. MIR-Pump NIR-Probe Fiber-Optic Photothermal Spectroscopy With Background-Free First Harmonic Detection. *IEEE Sens. J.* **2020**, *20*, 12709–12715. [CrossRef]

52. Krzempek, K.; Jaworski, P.; Kozioł, P.; Belardi, W. Antiresonant Hollow Core Fiber-Assisted Photothermal Spectroscopy of Nitric Oxide at 5.26 µm with Parts-per-Billion Sensitivity. *Sens. Actuators B Chem.* **2021**, *345*, 130374. [CrossRef]

53. Zhao, P.; Ho, H.L.; Jin, W.; Fan, S.; Gao, S.; Wang, Y. Hollow-Core Fiber Photothermal Methane Sensor with Temperature Compensation. *Opt. Lett.* **2021**, *46*, 2762. [CrossRef] [PubMed]

54. Bao, H.; Bao, H.; Hong, Y.; Hong, Y.; Jin, W.; Jin, W.; Ho, H.L.; Ho, H.L.; Wang, C.; Wang, C.; et al. Modeling and Performance Evaluation of In-Line Fabry-Perot Photothermal Gas Sensors with Hollow-Core Optical Fibers. *Opt. Express* **2020**, *28*, 5423–5435. [CrossRef]

55. Li, Z.; Wang, Z.; Yang, F.; Jin, W.; Ren, W. Mid-Infrared Fiber-Optic Photothermal Interferometry. *Opt. Lett.* **2017**, *42*, 3718. [CrossRef]

56. Reider, G.A. *Photonics: An Introduction*; Springer International Publishing: Berlin/Heidelberg, Germany, 2016; ISBN 978-3-319-26074-7.

57. Patimisco, P.; Scamarcio, G.; Tittel, F.K.; Spagnolo, V. Quartz-Enhanced Photoacoustic Spectroscopy: A Review. *Sensors* **2014**, *14*, 6165–6206. [CrossRef]
58. Ma, Y.; Yu, G.; Zhang, J.; Yu, X.; Sun, R.; Tittel, F.K. Quartz Enhanced Photoacoustic Spectroscopy Based Trace Gas Sensors Using Different Quartz Tuning Forks. *Sensors* **2015**, *15*, 7596–7604. [CrossRef]
59. Kosterev, A.A.; Bakhirkin, Y.A.; Curl, R.F.; Tittel, F.K. Quartz-Enhanced Photoacoustic Spectroscopy. *Opt. Lett.* **2002**, *27*, 1902–1904. [CrossRef] [PubMed]
60. Dello Russo, S.; Sampaolo, A.; Patimisco, P.; Menduni, G.; Giglio, M.; Hoelzl, C.; Passaro, V.M.N.; Wu, H.; Dong, L.; Spagnolo, V. Quartz-Enhanced Photoacoustic Spectroscopy Exploiting Low-Frequency Tuning Forks as a Tool to Measure the Vibrational Relaxation Rate in Gas Species. *Photoacoustics* **2021**, *21*, 100227. [CrossRef] [PubMed]
61. Shelby, R.M.; Levenson, M.D.; Bayer, P.W. Guided Acoustic-Wave Brillouin Scattering. *Phys. Rev. B* **1985**, *31*, 5244–5252. [CrossRef]
62. Picqué, N.; Hänsch, T.W. Frequency Comb Spectroscopy. *Nat. Photon.* **2019**, *13*, 146–157. [CrossRef]
63. Yacovitch, T.I.; Herndon, S.C.; Roscioli, J.R.; Floerchinger, C.; McGovern, R.M.; Agnese, M.; Pétron, G.; Kofler, J.; Sweeney, C.; Karion, A.; et al. Demonstration of an Ethane Spectrometer for Methane Source Identification. *Environ. Sci. Technol.* **2014**, *48*, 8028–8034. [CrossRef] [PubMed]
64. Pineda-Reyes, A.M.; Herrera-Rivera, M.R.; Rojas-Chávez, H.; Cruz-Martínez, H.; Medina, D.I. Recent Advances in ZnO-Based Carbon Monoxide Sensors: Role of Doping. *Sensors* **2021**, *21*, 4425. [CrossRef] [PubMed]
65. Wang, S.; Liu, B.; Duan, Z.; Zhao, Q.; Zhang, Y.; Xie, G.; Jiang, Y.; Li, S.; Tai, H. PANI Nanofibers-Supported Nb2CTx Nanosheets-Enabled Selective NH3 Detection Driven by TENG at Room Temperature. *Sens. Actuators B Chem.* **2021**, *327*, 128923. [CrossRef]
66. Gao, S.-F.; Wang, Y.-Y.; Liu, X.-L.; Ding, W.; Wang, P. Bending Loss Characterization in Nodeless Hollow-Core Anti-Resonant Fiber. *Opt. Express* **2016**, *24*, 14801–14811. [CrossRef] [PubMed]
67. Kozioł, P.; Jaworski, P.; Yu, F.; Krzempek, K.; Wu, D.; Dudzik, G.; Liao, M.; Abramski, K. Microdrilling of channels in antiresonant hollow-core fiber using femtosecond laser pulses. In Proceedings of the Laser Congress 2020 (ASSL, LAC) (2020), Virtual Conference, 13 October 2020; Optical Society of America: Washington, DC, USA, 2020; p. JTh2A.3.

Review

Integrated Nanophotonic Waveguide-Based Devices for IR and Raman Gas Spectroscopy

Sebastián Alberti *, Anurup Datta and Jana Jágerská

Department of Physics and Technology, UiT the Arctic University of Norway, 9019 Tromsø, Norway;
anurup.datta@uit.no (A.D.); jana.jagerska@uit.no (J.J.)
* Correspondence: sebastian.alberti@uit.no

Abstract: On-chip devices for absorption spectroscopy and Raman spectroscopy have been developing rapidly in the last few years, triggered by the growing availability of compact and affordable tunable lasers, detectors, and on-chip spectrometers. Material processing that is compatible with mass production has been proven to be capable of long low-loss waveguides of sophisticated designs, which are indispensable for high-light–analyte interactions. Sensitivity and selectivity have been further improved by the development of sorbent cladding. In this review, we discuss the latest advances and challenges in the field of waveguide-enhanced Raman spectroscopy (WERS) and waveguide infrared absorption spectroscopy (WIRAS). The development of integrated light sources and detectors toward miniaturization will be presented, together with the recent advances on waveguides and cladding to improve sensitivity. The latest reports on gas-sensing applications and main configurations for WERS and WIRAS will be described, and the most relevant figures of merit and limitations of different sensor realizations summarized.

Keywords: integrated sensors; waveguides; absorption spectroscopy; Raman spectroscopy; gas sensing

Citation: Alberti, S.; Datta, A.; Jágerská, J. Integrated Nanophotonic Waveguide-Based Devices for IR and Raman Gas Spectroscopy. *Sensors* **2021**, *21*, 7224. https://doi.org/10.3390/s21217224

Academic Editors: Krzysztof M. Abramski and Piotr Jaworski

Received: 17 September 2021
Accepted: 26 October 2021
Published: 30 October 2021

Publisher's Note: MDPI stays neutral with regard to jurisdictional claims in published maps and institutional affiliations.

1. Introduction

Both IR absorption spectroscopy and Raman scattering spectroscopy are nowadays routine characterization techniques that are available in most organic and material processing labs, as well as in the industry. These techniques provide specific information about molecules or chemical functional groups in a fast, non-invasive, and reliable manner and have been used to identify compounds, follow reactions, and track absorption processes. Their wide variety of applications include environmental monitoring, (i.e., not only the monitoring of pollutants and greenhouse gases but also real-time monitoring of anesthetics and respiratory gases during surgery), explosives detection, medical diagnostics, and even the authentication of paintings, aside from their widespread use in research and industry [1–4].

Absorption and Raman spectroscopy, although providing similar information, are complementary techniques as the rotational-vibrational signal, silent in Raman scattering, can be highly noticeable in absorption experiments, and vice versa. It is therefore of no surprise that both techniques have been developed simultaneously for similar purposes. They exhibit different configurations according to the nature of the sample, i.e., for liquids, thin films, powders, or gases. Gas detection based on Raman and absorption spectroscopy relies heavily on the boost in sensitivity through the use of resonant cavities and multipass cells that increase the path length and, hence, the interactions of the beam with the analyte. Such cells and cavities have, almost exclusively, been realized by free-space beams and bulk optics; as a consequence, the standard high-end spectroscopy instruments still remain bulky, and samples often need to be collected and analyzed in the laboratory.

Recently, portable tunable laser absorption spectroscopy (TLAS) instruments for trace gas analysis have been developed by Aerodyne Research Inc., LI-COR, Picarro, IRsweep, and others. These, being typically packaged as 19-inch rack modules or of the size of a large suitcase, have been used for in situ monitoring in mobile vehicles, including airborne field campaigns. Shoebox-sized and only 3 kg in weight, instruments offered by Aeris Technologies followed, as well as a compact, 2.1 kg and battery-powered sensor developed by Empa for measurements aboard unmanned aerial vehicles (UAVs) [5]. Alternative portable devices include quartz-enhanced photoacoustic spectroscopy (QEPAS) [6], a variant of photoacoustic spectroscopy (PAS) in which the microphone is replaced by a quartz tuning fork. The instrumentation of Raman spectroscopy includes handheld Raman spectrometers, produced by Bruker, Thermo Fisher Scientific, and other companies.

Integrated on-chip devices appear to be the next logical steps to decreasing the size even further while keeping the advantages of molecular selectivity and sensitivity as offered by IR and Raman spectroscopy techniques. This will ultimately require the monolithic integration of a laser and a detector, with a photonic chip replacing a classical gas cell. In most cases, this photonic chip is constituted by a high-finesse photonic cavity, or a long, single-mode waveguide often curled into a spiral and tightly patterned on a photonic chip. Both these configurations enable long optical pathlengths for high sensitivity while keeping a minimal footprint, e.g., a chip size of the order of one square centimeter. In such waveguide-based sensors, the guided light penetrates the evanescent field outside the waveguide core and probes a sample close to the waveguide surface. Molecules present within the evanescent field will absorb light or generate Raman scattered light that will couple back into the waveguide modes. The respective interaction pathways give rise to analytical techniques that are also known as waveguide infrared absorption spectroscopy (WIRAS) and waveguide-enhanced Raman spectroscopy (WERS).

In this review, we will discuss the major advances made in waveguide-based absorption and Raman spectroscopy for gas detection. The first section will address the main components developed so far to achieve the miniaturization of integrated sensing devices: light sources, waveguides, cladding, and detectors. In particular, the materials and designs that are proposed to decrease losses while increasing light interaction with the surrounding environment will be described. Additionally, a brief description of cladding as a strategy to improve sensitivity will be provided. The second section will focus on the integrated Raman and IR absorption sensors reported so far. The main sensor configurations will be introduced, and the latest applications will be discussed, discriminating between air-clad and functionalized/clad waveguides. Finally, a technology map will compare the performance of individual WIRAS and WERS sensors reported to date.

2. Efforts toward Miniaturization

2.1. Light Sources

2.1.1. IR Absorption Spectroscopy

One of the most widely used IR absorption spectroscopic techniques is based on tunable laser absorption spectroscopy (TLAS), which relies on a narrow-band light source such as a single-mode laser, where the wavelength can be carefully tuned to overlap with an absorption peak of the target analyte. This strategy has been applied most particularly in high-end trace gas sensors in the traditional gas cell configurations and their associated derivations involving optical cavities (such as cavity ringdown spectroscopy, cavity-enhanced absorption spectroscopy, and noise-immune cavity-enhanced optical-heterodyne molecular spectroscopy). Advances in MIR photonics over the last two decades have brought about high-quality laser diode sources based on interband cascade lasers (ICLs) [7,8], quantum cascade lasers (QCLs) [9,10], vertical-external-cavity surface-emitting lasers (VECSEL) [11–13], and frequency comb lasers [14,15]. The possibility of integrating light sources into chip devices has made them particularly suited for use in waveguide-based spectroscopy devices [11,16,17]. Excellent stability, tunability and the narrow line-width of these lasers have enabled IR laser absorption spectroscopy of unprecedented sensitivity and specificity, making TLAS gas sensors a powerful alternative to conventional FTIR and NDIR spectroscopy. While IR laser absorption spectroscopy has traditionally focused on single-species detection, extending the tunable wavelength range enables multispecies detection with a single laser source. Dual-wavelength distributed feedback (DFB) QCLs or lasers implementing Vernier-effect tuning are integrated light sources that have turned this into a reality, thus expanding the range and applicability of such lasers [18–20]. Further extension of the tunable range of QCLs has used a Fabry–Perot QCL chip in an external-cavity (EC) system, where the laser could be tuned across the whole gain curve. In addition, DFB QCL arrays have been used to extend the tuning range, making it possible to electrically switch between emission frequencies [21–23].

On the other hand, broadband-coherent sources, such as supercontinuum lasers [24,25], have also been used for MIR spectroscopy. The development of the waveguide-based generation of a supercontinuum [26–29] has provided the perfect impetus for the realization of compact on-chip light sources. The broadband nature of these supercontinuum sources allows the simultaneous probing of several analytes but this also translates to low selectivity; hence, complex post-processing algorithms are needed to demarcate between the overlapping absorption spectra across different analytes [30,31]. Often, configurations in conjunction with wavelength filters or spectrometers are needed in order to maintain selectivity. While these laser sources have good coherence and adequate power output, making them well-suited for detecting trace quantities of gas, they are generally very complex to fabricate and are hence quite expensive as well.

Some state-of-the-art light sources include nanolasers based on plasmonic structures [32] and metamaterials [33], which hold a lot of promise regarding the realization of compact light sources. Many of these metamaterial-based light sources offer an easy way to control the wavelength of the emission through scaling the unit cell design to longer or shorter wavelengths, as compared to other light sources [34,35]. In contrast to this, incoherent light sources, such as MIR light-emitting diodes (LEDs), have also received attention due to their small size and low power consumption [36–38]. Among the MIR LEDs, super-luminescent light-emitting diodes (sLEDs) offer a unique combination of high brightness, good beam directionality and broadband capability [39]. However, they have been limited to wavelengths smaller than 5 μm due to the poor efficiency of light emission at longer wavelengths [40]. In particular, on-chip LEDs on SOI that is fabricated through the heterogeneous integration of InP membranes help to couple the light efficiently to a single-mode waveguide and help to avoid high coupling losses and high packaging costs [41]. Another approach is the direct material integration of active emitters, such as quantum dots, within the waveguide itself. As demonstrated in a silicon nitride platform with embedded quantum dots, this represents an elegant solution for generating waveguide source light with high-mode coupling [42].

Thermal emitters have emerged as one of the latest and most promising means of generating MIR radiation. In this case, plasmonics and metamaterials principles are used to design nanostructures with high spectrally selective absorptivity. Kirchoff's law then requires the radiation from these structures to emit in the same spectrally selective region. Certain MEMS-based structures and micro hot plates have been demonstrated as offering a good light, suitable for analyte sensing, due to their energy efficiency, fast modulation capability, and CMOS-compatible processing steps [43–45].

2.1.2. Raman Spectroscopy

Excitation light sources for Raman spectroscopic systems are typically high-power monochromatic laser sources, operating with more than 50 mW output power in the visible or in the near-infrared, typically at 785 nm or 1064 nm wavelengths. While external laser sources have primarily been used for Raman spectroscopy, recent advances in silicon photonics have made possible a wide variety of compact and chip-based lasers [46], which would ultimately pave the way for an integrated Raman system. However, to date, there are only a few examples of miniaturized laser sources combined with on-chip Raman systems. The difficulties in limiting the realizations of such systems arise from the requirement of a high-power monochromatic light source, and the difficulty in separating the pump and the scattered Raman signal with a high-enough extinction ratio on a chip. The strict requirement of high laser power is motivated by the inherent weak scattering efficiency of the Raman signal [47]. Several strategies have been proposed to increase the strength of the Raman signal, such as confining the light to a small volume, particularly through the use of nanophotonic platforms, such as enhanced hotspot formation through the employment of metallic nanostructures in surface-enhanced Raman configuration [48,49] or through the use of slot waveguide platforms [50,51]. Working in a lower wavelength regime also helps to increase the scattered signal intensity since the scattering varies inversely as the fourth power of wavelength. However, this is often accompanied by the presence of unwanted fluorescence background [52].

Large-throughput spectrometers have traditionally been a major and integral part of a Raman spectrometer setup (see Section 2.4). A recent work by Atabaki et al. has demonstrated the concept of using tunable lasers as a way of eliminating the spectrometer from a Raman setup, through a method called swept-source Raman spectroscopy (SRSS) [53]. In this case, an excitation laser with only a few mWs of power was tuned in combination with a narrow bandpass filter on the detector side, which resulted in significantly high optical throughput compared to benchtop and compact handheld dispersive Raman spectrometers. A MEMS tunable laser was used, based on the concept of vertical-cavity surface-emitting lasers (VCSELs), thus demonstrating the suitability of VCSEL lasers for use in miniature Raman sources, as was also proposed earlier [54,55]. This work represents a major step toward the realization of miniature light sources for Raman spectroscopy.

2.2. Waveguides

Waveguides for sensing, including those for WERS and WIRS, have been evolving to guarantee minimum losses and high light–analyte interaction. Losses can be divided into the absorption of the material, leakage to the substrate, bending losses, and scattering losses due to fabrication or material imperfections (inhomogeneity or crystal grains). To minimize them, a proper choice of both the material and the waveguide design is crucial.

Aside from the transparency of the waveguide material in the targeted wavelength range, refractive index, photo-stability up to high intensities, low level of fluorescence or Raman background, toxicity, availability/cost, and ease of production are the main selection criteria. From the great variety of materials proposed, materials compatible with CMOS/mass production, such as silicon, germanium, silicon oxide and silicon nitride, are the most common [56]. Nevertheless, a wide range of other materials has been reported, including polymers (i.e., photoresists and Teflon) [57], halides, chalcogenides (i.e., CaF2, NaBr and ZnSe) [28,58], oxides (i.e., alumina, titania and tantala) [59,60], diamond (due to its advantages for quantum photonics) [61,62], or InGaAs [63]. Recently, a review on waveguide materials has been published by Yadav and Agarwal [64].

Doped silica (UV-written) and silicon nitride over a silica bottom cladding have been the main materials used for on-chip waveguiding in the visible- to the NIR range; however, alternatives such as tantalum pentoxide have recently emerged. These materials have also been used, to a lesser extent, in the MIR, despite the presence of residual O-H and N-H groups that limit their transparency in certain frequency bands. For MIR applications, silicon on silica (i.e., silicon-on-insulator, SOI) and, less frequently due to increased costs, germanium on silica (germanium-on-insulator, GOI), have been the materials of choice due to their transparency at longer wavelengths. SiGe alloys possess the highest refractive indices of all the CMOS-compatible materials mentioned above and can, in addition, be doped in order to tailor their refractive index. The high refractive index has also been shown to be highly useful in avoiding mode leakage into the substrate, enhancing the electric field at the waveguide interface and, thus, the light–analyte interaction. Besides SOI and GOI, silicon on nitride, silicon on alumina, and germanium on silicon (or silicon-germanium alloy on silicon) have been proposed as novel waveguide alternatives in MIR, due to their capability of avoiding absorption by silica bottom cladding, especially above 3.5 μm [65,66]. Diamond also appeared recently on the scene as an ideal material with a transparency range from 0.22–20 µm; nevertheless, its applications are limited due to the difficulty of processing and high cost [67].

The processing of these materials into photonic waveguides has also developed greatly during the last decades, not only to account for more complex designs and profiles but additionally to decrease losses. While the homogeneity of the materials is highly dependent on the deposition technique and post-treatment, the surface and mainly the sidewall roughness are subject to the etching protocol that is followed. The former will decrease bulk scattering, the latter, the surface scattering. A great number of deposition and etching protocols are already available in the literature and depend greatly on the materials, the etch rate, selectivity and profiles, or the design. We will not go further into the topic, but we do encourage our readers to find further information in the work of William et al. [68].

Finally, the waveguides' design is crucially important when high sensitivity to the surrounding environment is targeted. Unlike waveguides developed for communication purposes, waveguides for gas sensing need to ensure high light–analyte interaction, assuming the strong presence of the optical field outside the solid waveguide core. The amount of this interaction can be described using the evanescent field confinement factor, Γ, which is defined as in [69]:

$$\Gamma = \left(\frac{n_g}{\mathrm{Re}\{n_{cl}\}}\right) \iint_{cl} \frac{\varepsilon |E|^2 dxdy}{\iint_{-\infty}^{\infty} \varepsilon |E|^2 dxdy} \tag{1}$$

The absorption along the waveguide length is then given by a modified Lambert-Beer law:

$$I = I_0 \exp[-\alpha \Gamma\, L] \tag{2}$$

Here, n_g is the group index, n_{cl} is the cladding's refractive index (equal to approx. 1 in the air), $\varepsilon(x, y)$ is the permittivity, $E(x, y)$ is the electric field, α is the bulk absorption, and L is the length of the waveguide. It is important to stress that the absorption not only depends on the evanescent field fraction but also on the waveguide dispersion through the group index n_g. Reporting only the evanescent field fraction and omitting the effect of dispersion is a common misconception in the literature, making it difficult to quantify and compare the light–analyte interaction across the different waveguide platforms reported in the literature.

Besides the confinement factor Γ, the sensitivity of the waveguide is also determined by the physical path-length of the waveguide L that is typically limited by the waveguide loss. Therefore, the ratio between the evanescent field confinement factor and the propagation loss was introduced by Kita et al. [70] as an additional figure of merit that fully determines the sensing performance of the waveguide. Both the confinement factor and the losses will be dependent on the material and the processing, as well as on the waveguide design [71].

The most common waveguides reported for sensing can be classified into five different designs: rib, strip, slot waveguides, sub-wavelength gratings and photonic waveguides [72]. Rib and strip waveguides can be realized swiftly in one step with UV lithography and easy-etching protocols. The former is characterized by a shallow step defining the waveguide, with a small side-wall area and, therefore, little surface scattering compared to other designs. The strip waveguide (Figure 1a) is etched all the way down to the bottom cladding and exhibits more scattering loss but, unlike rib waveguides, it allows scientists to confine light tightly in the horizontal axis, resulting in minimal bending loss. According to Kita et al., who compared the performance of strip, rib, and slot waveguides for sensing, the strip waveguide is the preferred geometry for bulk absorption sensing and refractometry and is comparable in performance to other, more complicated, geometries for surface-sensitive refractometry and absorption sensing [70].

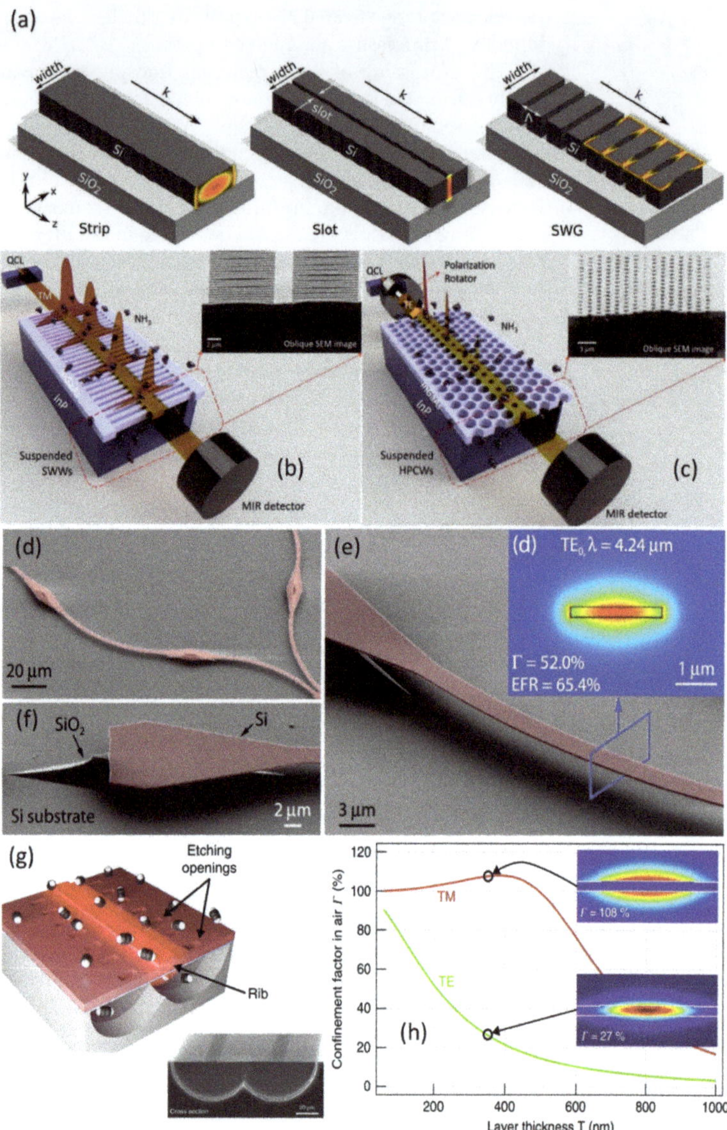

Figure 1. Different waveguide designs reported for WIRAS: (**a**) Strip, slot and subwavelength grating waveguides (SWG) (reproduced with permission from reference [70] © 2021 Optical Society of America). (**b**) Suspended subwavelength grating waveguide and (**c**) suspended photonic crystal waveguide. The insets show SEM images of the fabricated structures (reprinted with permission from reference [63], copyright 2020 American Chemical Society). (**d–f**) Suspended waveguides on pedestals; the inset in image (**e**) shows the cross-section and the electrical field distribution (reproduced with permission from reference [73], (© 2021 Optical Society of America). (**g**) Schematic and SEM image of a self-standing rib waveguide. (**h**) Simulated confinement factor of the waveguide in (**g**) at TM and TE polarizations, as a function of layer thickness, while the inset shows the electric field distribution at a thickness of 350 nm. Reproduced from reference [69], licensed under a Creative Commons Attribution 4.0 International License http://creativecommons.org/licenses/by/4.0/ assessed on 9 September 2021.

Slot, subwavelength grating (SWG), and photonic crystal waveguides have been reported as alternative designs that are capable of increasing the interaction with the surroundings by several times. Slot waveguides consist of two strips of high-refractive-index materials, separated by a subwavelength-scale low-refractive-index slot region that strongly confines light (Figure 1a). This design presents a light–analyte interaction more than 5 times larger than strip waveguides, which is highly desirable for gas sensing, while material losses are reduced due to the low intensity of the electric field in the material [70,74,75]. This design presents a good compromise between simplicity, air confinement factor, losses, and costs, and has been tested experimentally many times for both IR and Raman spectroscopy [76,77]. Despite the advantages, this design requires electron beam lithography in most cases to pattern the slot, which is typically of the order of 100 nanometers, and additional care needs to be taken during etching to guarantee a well-defined slot and little roughness [78].

A SWG waveguide (Figure 1a,b) is based on a periodic arrangement of two different materials having a period that is much smaller than the wavelength of light. It is characterized by field distribution with an air confinement factor 4–5 times higher compared to the strip waveguide [70]. Although, theoretically, no losses are expected from the design, the experimental propagation loss is normally above 2 dB/cm, due to imperfections arising during fabrication, particularly surface roughness and variability in the size of the waveguide segments [79].

Photonic crystal waveguides are distinguished by their ability to slow down light, i.e., to reduce the group velocity of the propagating waveguide mode as a result of coherent scattering on the photonic crystal lattice. Group velocity reduction by factors of between 1.5 and approx. 100 have been reported, and a corresponding increase in the interaction with the analyte has been observed in sensing experiments, both at NIR [80] and MIR wavelengths [81]. These waveguides are commonly formed by a linear defect in a photonic crystal lattice, patterned into a high-index dielectric membrane (see Figure 1c). The main drawback can be attributed to the difficulty of fabrication, a sensitivity to disorder that may lead to spectrally uneven enhancement, and an increased surface area that brings greater surface scattering and reflections [82]. Using slow light increases analyte-field interaction but, at the same time, it increases the interaction with the material, including absorption and scattering. In most demonstrations, the waveguide lengths are thus limited to hundreds of micrometers or, at most, millimeters.

To improve the light–analyte interaction even further, the bottom cladding can be removed either partially or completely. Partial bottom cladding removal was used to realize air-suspended waveguides supported by pedestals [49,83] or pillars [73,84] (Figure 1d–f). Complete cladding removal results in self-standing rib waveguides (Figure 1g), subwavelength grating waveguides (Figure 1b) [85], and photonic crystal waveguides (Figure 1c) [63]. Air-suspended structures appeared recently in the literature, exhibiting the largest reported confinement factors surpassing 100% (Figure 1h) [69] and reduced propagation losses [69,73]. By etching away the material beneath the waveguide, absorption due to the bottom cladding can be completely removed, leakage to the substrate avoided, and the volumetric interactions with the surrounding analyte increased [86]. Additionally, the lack of bottom cladding is well-suited for TM polarization which, in thin suspended waveguide designs, has minimal electric field overlap with the core material. This further increases the evanescent field confinement and decreases losses attributed to absorption in the constituent materials. Despite these advantages, the waveguide processing is complex and requires several lithography and etching steps; the structures are rather fragile, necessitating careful handling; and monolithic integration with laser sources and detectors appears more challenging than for waveguides supported by solid bottom cladding.

In order to achieve the lowest possible detection limits, the increase in sensitivity has to go hand-in-hand with the reduction of noise. An important noise source in integrated photonic circuits is a so-called interferometric noise, arising due to reflections from facets or defects, manifesting itself as spectral fringing in transmission. Such noise may interfere significantly with the recorded spectrum. For this purpose, antireflection coatings on the waveguide facet [87], the use of subwavelength gratings on the waveguide facet [88], or the use of appropriate signal processing algorithms [89] have been proposed to reduce the effect of the fringes. Substantial fringe reduction has also been achieved in air-suspended waveguides, characterized by strongly delocalized guided modes with an effective mode index close to unity. This automatically minimizes reflections at the facets or structural defects of the waveguide, leading to clear spectral transmission that is free from interferometric noise [69].

Another important point to consider in the context of high-index contrast waveguides for IR absorption spectroscopy is the possibility of saturation of the absorption signal, due to the intrinsically high intensity of the strongly confined guided modes. This typically occurs at high laser powers, in combination with intense absorption lines, where the excitation rate of the molecules can become faster than their relaxation rate [90]. To date, the literature mentions only sporadically the effects of saturation in waveguides, and detailed theoretical description is entirely absent. However, a waveguide design with a strongly delocalized field would mitigate this effect.

2.3. Cladding

Functional coatings, ranging from monolayers to films of several micrometers thick, have been used for decades as a route to increasing the sensitivity and/or specificity of integrated (bio)sensors. These layers work as molecular recognition coatings, serving as a solid-phase enrichment matrix for the targeted analyte, while simultaneously excluding undesirable molecules and avoiding unspecific binding. In other words, these layers enhance the signal to noise ratio by decreasing the background signal due to unspecific binding, they reduce the cross-interference with other molecules, and, at the same time, they increase the concentration of the targeted analyte relative to that of the surrounding media (i.e., solutions, atmosphere). The design of the coating layer can be adapted for a specific molecule through recognition sites, partitioning, and charge- or size exclusion [91]. This strategy has become of the utmost importance for highly sensitive transducers with low selectivity or specificity and has been adopted in a number of chemical-sensing devices such as opto-chemical, electro-chemical, plasmonic sensors, and refractive index waveguide-based sensors [92–96]. In the latter case, the analyte either increases the refractive index of the cladding or motivates a change in the thickness of the cladding itself [97–99], which is then detected by a phase-sensitive device such as a ring resonator, Mach-Zehnder interferometer, or a Bragg grating [95,97,100]. The advantages of functional cladding have also proven to be highly useful for already selective transducers, such as Raman and IR spectroscopy sensors, as a means to increase the sensitivity by analyte up-concentration and the reduction of the background signal in complex matrices.

The enrichment cladding layers can be oxides, polymers, silanes, specific biological molecules, or composite materials where more than one element is present. Polymers have been widely used due to simple processing, availability, tuneability (functional groups, molecular weight, ramification, backbone structure, or crosslinking degree), and their behaviour as extraction materials, with their enrichment properties mainly dependent on their polarity, free volume, pore size, and pore distribution [101]. A great variety of polymeric materials have already been investigated, including polyisobutylene [102], ethylene/propylene copolymer [103], low-density polyethylene [104], Teflon®AF [105], poly(dimethylsiloxane) [106,107], poly(acrylonitrile-co-butadiene) [107], poly(styrene-co-butadiene) [107], poly(vinyl chloride) [108], polystyrene, and poly(methyl methacrylate) [83,109,110]. Among these, fluorinated polymers have shown good transparency up to MIR wavelengths due to the substitution of C-H bonds, high free volume, and outstanding thermal and mechanical properties [57]. Although mostly used on waveguide-based refractive index sensors and ATR crystals, some specific polymers have been tested on integrated single-mode waveguides for Raman and IR absorption spectroscopy (see Sections 3.2.2 and 4.2.2).

Mesoporous inorganic and mesoporous hybrid inorganic-organic cladding, based on sol-gel chemistry, represent a robust alternative to polymers [111]. These materials are equally capable of providing partitioning for custom analytical tasks, while they exhibit advantageous optical, dielectric and thermal properties [112–115]. Mesoporous inorganic-based materials show robust mechanical and chemical stability capable of sustaining harsh environmental conditions such as high-energy radiation, acid, or alkaline media, as well as oxidative chemicals. Furthermore, these materials have tunable pore volume that can surpass 50% and a decreased response time (in the order of seconds) in comparison to many polymers (typically in the order of minutes). Optical losses in VIS-NIR in these materials are generally low, due to their amorphous structure and small pore size [111]. Nevertheless, the transparency at longer wavelengths suffers due to OH groups and the adsorption of water on the large surface area of the pore network. In addition, clad sensors normally need calibration due to variations in the material properties attributed to minimal changes in the process/environment.

Although cladding brings about improvements in specificity and sensitivity, calibration is mandatory for clad systems as the up-concentration factors are difficult to quantify analytically. Cladding properties may also change over time, due to phenomena such as thermal instability, dehydration, or reconfiguration. Increased response time and aging are among other limitations of clad sensor systems, as well as the potentially reduced reversibility of the system after exposure to the analyte, discriminating between disposable and reusable devices.

2.4. Detectors: Single Pixel, Arrays, Spectrometers

IR absorption spectroscopic systems primarily use single-pixel detectors in combination with monochromatic tunable laser sources. Here, the spectral selectivity comes from scanning over an absorption feature with a narrow linewidth of the laser rather than from a spectrally sensitive detection unit.

Among chip-integrated single-pixel detectors, group IV materials like silicon or germanium or a combination of III–V materials, such as InGaAs, InGaAsSb, InAsSb, PbTe, GaSb, and InP have been used as platforms for NIR and MIR sensing [64,65]. While Si can only absorb up to about 1.1 µm, ion implantation or introducing lattice defects or external agents can improve the detection range of Si-based photodetectors up to 2.4 µm [64]. The entire MIR spectral range can be covered by mercury cadmium telurite (MCT) detectors, which are also most widely used for MIR sensing in bulk spectrometers. However, due to a strong lattice mismatch, they are unsuitable for integration in chip-based systems. Besides this, 2D material-based detectors, such as black phosphorous, for up to about 4 µm [116–119] and semi-metals, like graphene, for a longer wavelength [120] have also been used as photodetectors. Lead chalcogenides, in particular PbTe, have also received significant attention for MIR photodetection due to their easy deposition process, excellent stability, and low cost. However, the above-listed photocarrier generation-based techniques for IR light detection inevitably suffer from high dark current, particularly when they are biased and, thus, they require significant cooling to achieve high sensitivity. This issue has been partially addressed by detectors with an engineered band structure such as superlattice detectors, quantum cascade detectors (QCDs), and interband cascade detectors (ICDs), which have received much attention due to their high detection sensitivity at room temperature [121–123].

The recent trend in the integrated detector domain has moved toward MEMS-based IR detectors, such as thermopiles, microbolometers, and pyroelectrics, which can work even at room temperatures and have detection capabilities across a longer wavelength range [124–126]. However, they still lag behind photodetectors in terms of sensitivity and have slower response times.

For broadband sources, which have been used for both IR spectroscopy and Raman spectroscopic systems, spectral discrimination is done at the detection side through one of the following systems: (i) miniaturized dispersive optics; (ii) narrowband filters; (iii) Fourier transform-based detection devices; and (iv) reconstructive spectrometers [127].

Dispersive optics has been the most widely used detection scheme, where dispersive optical elements, such as gratings, slits or arrays, spectrally separate the incoming broadband light and send it to either a single pixel detector or a detector array. While the usage of a single pixel detector is low cost, the setup requires the capability to scan across the spectrum, making the process slow and often prone to complex alignment requirements. The usage of an array of detectors makes the scan significantly faster but at a higher cost. The majority of dispersive spectrometers operate in the visible wavelength range, which makes them suitable for use in a Raman spectroscopic system. Infrared spectroscopy, on the other hand, most often relies on the principle of Fourier transform spectroscopy, its advantages being the ability to use only one detector, the simultaneous collection of spectral information (also known as Fellgett's advantage), and higher optical throughput due to the elimination of optical slits. Compact Fourier transform spectrometers have been realized through miniaturized Michelson or Fabry–Perot interferometers [128–130], micro-electromechanical systems (MEMS), or micro-opto-electromechanical systems (MOEMS) [131], MEMS-based digital micromirror devices (DMD) or linear variable filter arrays [132]. For photonic integrated sensors, spectral discrimination can be implemented using waveguide-based devices. Typical on-chip dispersion schemes include the use of photonic crystals, holographic elements, transmission gratings, self-focusing transmission gratings, arrayed waveguide gratings, and metasurfaces, and have been discussed in a recent review [127]. These typically suffer from low resolution due to small path lengths, compared to their traditional, bulky counterparts. A popular on-chip alternative to dispersive spectrometers is narrowband filters, where the filter wavelength is very often easily tunable. They can be ultra-compact, due to their planar nature and negligible path length. Both plasmonic and dielectric filter implementations in the near-infrared and mid-infrared have been shown in [133,134]. Of particular interest are ring resonator filters, as already demonstrated in [135]. A typical ring resonator spectrum contains several resonance peaks, separated by the free spectral range, which limits the spectral bandwidth of the filter. One recent demonstration of the integration of a ring resonator with a distributed Bragg reflector has shown the feasibility of isolating a single ring resonance line and making it more robust against thermal drift [136]. For the Raman spectroscopic system, arrayed waveguide gratings and low-loss microrings have been proposed, to function as UV spectrometers [137]. Hartmann et al. demonstrated a waveguide integrated broadband spectrometer, based on random scattering events in disordered medium, whose functionality extends through both the visible and NIR regions [138] and could be used in integration into a Raman spectroscopic system.

Integrated alternatives to Fourier transform spectrometers, based on waveguides, allow not only for system miniaturization but also for the complete elimination of moving parts. The first approach was to introduce multiple Mach-Zender interferometers (MZIs) of different pathlengths to generate a spatially varying interference pattern, also known as spatial heterodyne spectrometers (SHS) [139,140]. These have been demonstrated through the use of an array of tightly coiled spiral waveguides [141]. Other implementations include photonic circuits governed by spatial multiplexing among different interferometers with increasing varied path length, culminating in outputs coupled to a linear detector array [142,143]. A similar detection scheme has been shown using a single MZI and, hence, a single detector, where the electro-optic modulation in lithium niobate waveguides or thermo-optic modulation by using micro heaters was used to tune the optical pathlength of the interferometric arm [144]. Another approach toward miniaturized FTS was to use counter-propagating beams from two waveguides and allow them to interfere to generate a standing interference pattern, also known as stationary wave-integrated FTS, as first demonstrated by Coarer et al. [145]. In this scheme, the spatial interference pattern across the entire length of the waveguide is recorded with the help of many detection elements placed in close proximity to the waveguide. Since the resolution of such a spectrometer depends on the number of the detector elements and the length of the waveguide, difficulty in placing a sufficient number of detector elements results in low resolution. Nie et al. mitigated this issue by generating the interferogram by overlapping the evanescent fields of two co-propagating waveguide modes, thus effectively stretching the interferogram and allowing high resolution with large bandwidth and a low footprint [146].

3. Waveguide-Enhanced IR Absorption Spectroscopy

3.1. Configurations and Integration

To date, the most common configuration for optical sensing involves free-space coupling between the laser, waveguide, and detector by using bulk optics or microlenses. While conceptually simple, free-spacing coupling suffers from low coupling efficiency due to mode mismatch and is sensitive to vibrations and misalignment, thus affecting the robustness and compactness of the sensor. The usage of spot size converters or couplers has addressed some of the aforementioned problems [147]. At the same time, fiber-based coupling methods have also been used to improve the robustness and stability of the setup. This includes fiber pigtailed lasers or the direct coupling of a free-space beam into a fiber through a fiber collimator. Nonetheless, such techniques still fall short of an idealized compact and integrated version of the sensor that would fit completely on a single photonic chip and be suitable for mass production at a low cost.

First approaches for integration include hybrid integration, heterogeneous integration and monolithic integration. Even in integrated setups, inefficient coupling has been the major problem. This typically stems from the mode mismatch and misalignment between the active and passive sections of the chips, i.e., the laser cavity and the passive waveguide. For improving the coupling efficiency between the waveguide and the light sources, the usage of distributed Bragg reflectors (DBR) or appropriate mode profile engineering has been proposed [148]. Separate prototypes of integrations of the laser and the waveguide, the waveguide and the detector, the laser and detector are found in the literature; however, complete integration of all the three components in a single chip is still an active and ongoing pursuit, with very few actual demonstrations. The following sections describe recent efforts in this area.

3.1.1. On-Chip Light Sources and Passive Waveguide Integration

Silicon, as a platform for integrated sensors, has been the most frequent choice due to the existing mature process technology and the inherent compatibility with silicon waveguides. However, due to the absence of efficient light sources based on silicon, the heterogeneous integration of light sources, based on III–V materials, has been the preferred choice. This involves attaching the laser chip to a separate chip, with pre-patterned waveguides and other optical components for sensing and manipulation purposes. Heterogeneously integrated devices have been reported both in the NIR and MIR, and this has been discussed in detail in recent reviews [149,150]. Recent works have also focused on InP integration platforms, demonstrated through QCL integration with InGaAs passive waveguides [151] (Figure 2a) and photonic crystal-based laser source integration with silicon waveguides [152]. As a rare example of monolithic integration, QCLs integrated with plasmonic waveguides were demonstrated by Schwarz et al. [153]. In addition, Consani et al. [154] demonstrated the integration of a waveguide sensor with a MIR emitter for CO_2 sensing, showing that it is possible to abstain from using expensive laser sources and instead use a cheap thermal source (Figure 2b). In their case, the emitted light was broadband, requiring the use of filters on the detector side, but recent advances in narrowband filters and metamaterials emitters can eliminate the use of additional filters.

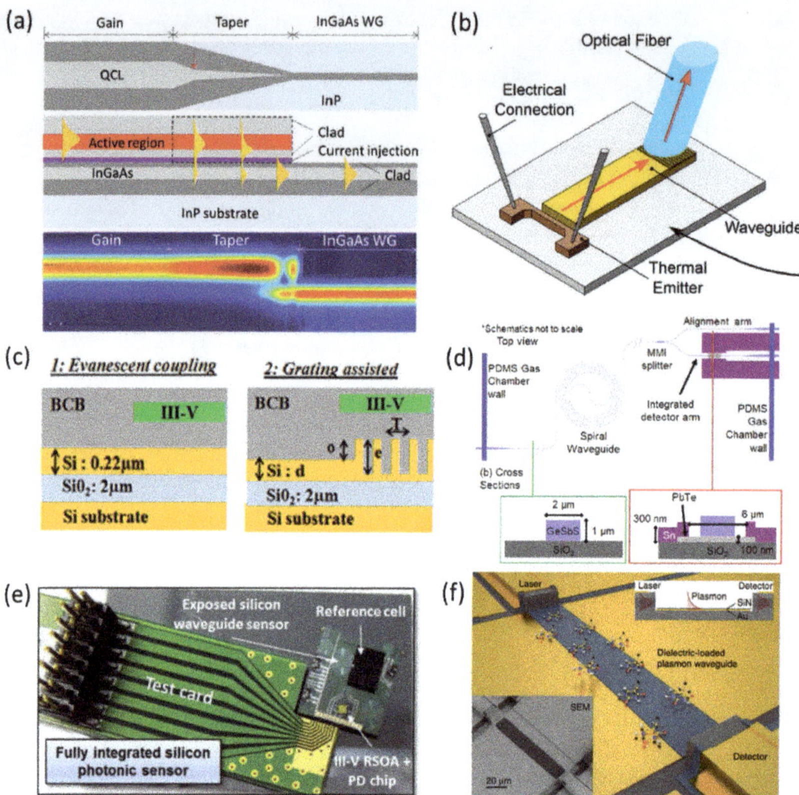

Figure 2. Integration schemes for waveguide-based IR absorption spectroscopy. (**a**) Integrated QCL source with a passive InGaAs waveguide (reproduced from [151] © 2021 Optical Society of America). (**b**) Integration of a waveguide sensor with a MIR thermal emitter (reproduced with permission from [154]). (**c**) Integration of a p-i-n-based GaInAsSb photodetector with a passive waveguide through evanescent and grating-assisted coupling (reproduced from [155] © 2021 Optical Society of America). (**d**) Monolithic integration of a PbTe detector, coupled with a spiral waveguide (reproduced with permission from [156]). (**e**) Prototype of a fully integrated silicon photonic sensor with laser, waveguide, and detector for sensing in the near-infrared (reproduced with permission from [157]). (**f**) Quantum cascade laser and detector, integrated with a dielectric-loaded plasmonic waveguide for sensing in the mid-infrared (copyright © 2021, the author(s) [153]).

3.1.2. Passive Waveguide and Detector Integration

Many demonstrations of integrated systems combining a detector and a passive waveguide involve integration of the active detector element, either monolithically or by the hybrid bonding or transfer of the active detector material. In particular, p-i-n photodiodes and 2D materials have been used for the active detector element to achieve compactness. GaInAsSb-based p-i-n photodiodes interfaced with SOI waveguides, either through gratings or evanescent coupling, have been demonstrated at 2.29 μm [155,158] (Figure 2c). Similarly, an $InAs_{0.91}Sb$-based p-i-n photodiode has been shown at 3 μm, integrated on the output grating couplers of a spectrometer [159]. In both cases, heterogeneous integration through adhesive bonding was used. Su et al. fabricated an on-chip waveguide integrated device with monolithically integrated PbTe detector film for the detection of methane at 3.3 μm [156] (Figure 2d). For longer wavelength regions, graphene photodetectors integrated with silicon waveguides have been demonstrated [160]. Yazici et al. showed the integration of a MEMS-based broadband infrared thermopile sensor attached through flip flop bonding with an SOI platform integrated with input and output grating couplers [124]. In all these demonstrations, the active area of overlap of the optical mode with the detector element is still small, resulting in lower sensitivity. Highly sensitive detection needs to increase the detector's active area, which, unfortunately, also increases the dark current and thus decreases the signal-to-noise ratio. In order to counter this, ridge waveguide-based detection has been used, where the entire length of the detector element acts as the active material for enhanced detection capability and has particularly been used in context with ICDs and QCDs. In order to improve the performance further, distributed Bragg reflectors (DBR), high reflectivity coatings, or simply etched or cleaved facets that ensure multiple passes through the active region have been proposed for a waveguide-integrated ICD. Such a design simultaneously increases quantum efficiency, as well as reducing dark current [148].

3.1.3. Integration of All Three Components

Among very few works describing prototypes of the on-chip integration of all components, Zhang et al. demonstrated a complete integrated setup involving a fully integrated NIR photonic chip sensor, mounted on a PCB test card, with an on-chip laser, dual photodetectors, reference cell, and an evanescent field-based sensing waveguide on a single silicon substrate. With this device, the authors showed methane sensing with a sub-100 ppmv·Hz$^{-1/2}$ sensitivity [157] (Figure 2e). While this sensitivity still lags behind the state of the art, the fully packaged nature of their demonstration is an important milestone and paves the way for fully integrated devices, particularly in a longer wavelength region. In the MIR, Benedikt Schwarz's work at TU Vienna and Ray Chen's group in UT Austin have shown great progress toward fully developed sensors integrated with QCL and QCD. Schwarz et al. [161,162] demonstrated the integration using QCL technology, relying on the bifunctional functionalities of the active region to work as both laser and detector. Coupled with a dielectric loaded plasmonic waveguide, they exhibited a complete system with liquid sensing capability [153] (Figure 2f). Even though a high-power emission was observed, the detector sensitivity was poor in these bifunctional structures. Later demonstrations separated the functions of the laser and detector, allowing their independent design and optimization [163]. In parallel, in order to extend the lasing wavelength to below 6 μm and simultaneously enable low power consumption, high sensitivity, and sufficient design flexibility, subsequent work focused on ICL technology and integrated setups have been demonstrated for 3.1 μm [164]. On the other front, Ray Chen's group has shown a sensor with QCL and QCD as sources, and detectors integrated with an InGaAs-InP monolithic platform, and gas sensing was demonstrated [165,166] making the pursuit of an on-chip integrated sensor for gas sensing close to reality.

Further advancements have been made through the integration of frequency comb MIR lasers with detectors, which showed ultrafast detection and up to two orders of magnitude lower power consumption, compared to QCLs [167]. In addition, mid-infrared dual-comb spectroscopy is an upcoming area of research, where, through the interference of two mutually coherent mode-locked frequency combs, the absorption spectrum signal can be converted from the optical domain to the radio frequency domain. Dual comb spectroscopy has a fast detection capability with higher resolution and accuracy, making it suitable for gas detection [168,169], and, due to the inherently large bandwidth of the frequency combs, covering even multiple species in parallel [29].

Another strategy to improve the compactness and robustness of the IR spectroscopy setup is to design the sensing functionality within the cavity of the laser, as demonstrated through intra-cavity laser absorption spectroscopy [148,170]. The in situ detection of chemical species within the laser cavity can be monitored directly through the laser's I-V characteristics, which can even eliminate the use of a separate detector [170].

3.2. Applications
3.2.1. Air-Clad

IR absorption spectroscopy on waveguide-based devices has been reported in recent years for various applications in environmental and industrial process monitoring, as well as in the biomedical sector [171–173]. In particular, air-clad waveguides have been proven to be useful for IR absorption spectroscopy, to identify and quantify common gases such as carbon dioxide, acetylene, ammonia and methane, some of them with major implications in global warming. However, most demonstrations are still proof-of-concept experiments, testing the capability of the novel integrated sensors. An overview of the air-clad waveguides used for spectroscopy, together with their principal characteristics, is provided in Table 1, while a more detailed description and a discussion of the achieved results are given in the next paragraphs.

Table 1. Overview of works using WIRAS for gas sensing.

Structure	Cladding	Analyte-LOD	Γ/EFR	Losses -λ	Advantages -Disadvantages	Ref.
Strip Waveguides						
Polysilicon strip waveguides over SiO_2/Si_3N_4	Air	CO_2 500 ppm	14–16% TE (EFR)	3.98–5.6 dB/cm 4.23 μm	Simple with moderate confinement factor. Single wavelength measurement.	[174]
Silicon strip waveguide	Air	CH_4-C_2H_2 <50,000 ppm	13% TE (EFR)	1.74 dB/cm 3–3.3 μm	Simple design and fabrication. Low losses. Moderate confinement factor. High LOD. Wavelength Scanning measurement.	[175]
Silicon strip waveguide	Air	CH_4 < 100 ppm	15% TM (Γ)	Not reported 1.65 μm	Fully integrated chip with a 20-cm-long silicon waveguide. wavelength-scanning measurement.	[157]
Silicon strip waveguides on silica	Air	CH_4 100 ppm	25.5% TM (Γ)	2 dB/cm 1.65 μm	High confinement factor for a strip waveguide. Relatively low losses. Low LOD. Wavelength scanning measurement.	[176]
Germanium on silicon strip waveguides	MS- HDMS- water	Toluene 7 ppm	1% (EFR)	2.5–5 dB/cm 6.5–7.5 μm	100–1000× preconcentration. High LOD. Wavelength Scanning measurement.	[177]
Chalcogenide strip spiral waveguide	Air	CH_4 25000 ppm	8% (EFR)	7 dB/cm 3.28–3.34 μm	Low confinement factor and high losses. Not CMOS compatible. Wavelength scanning measurement.	[178]
Chalcogenide strip waveguide on silica and CaF_2	Water	Phenylethyl amine 1800 ppm (mol/mol) (0.1 mol/L) (12 g/L)	5–15% (EFR)	0.4–1 dB/cm 1.52–1.56 μm	Low losses. Not CMOS compatible, low confinement factor. Single wavelength measurement.	[179]
Chalcogenide strip spiral waveguide	Air	CH_4 10,000 ppm	12.5% (Γ)	8 dB/cm 3.31 μm	Waveguide and detector integrated on the same chip. Single wavelength measurement.	[156]

Table 1. *Cont.*

Structure	Cladding	Analyte-LOD	Γ/EFR	Losses -λ	Advantages -Disadvantages	Ref.
TiO_2 rib porous waveguide on SiO_2	Air	C_2H_2 <100,000 ppm	26% TE (Γ)	2.2–8.5 dB/cm 1.5–1.6 μm	Simple and inexpensive. Highest confinement factor for rib waveguide. Strong fringes. High LOD. Wavelength scanning measurements	[180]
Suspended Waveguides						
Polysilicon-on-Si_3N_4 membrane over Si/SiO_2 walls	Air	CO_2 5000 ppm	19.5% (Γ)	Not reported 4.23 μm	Complicated fabrication, moderate improvement in confinement factor. Single wavelength measurement. 1 cm long.	[84]
Silicon beam on pillars	Air	CO_2 < 1000 ppm	44% (Γ)	3–4 dB/cm 4.24 μm	Sophisticated fabrication and moderate losses. High confinement factor. Few wavelength measurements.	[73]
Suspended tantala rib waveguide	Air	C_2H_2 7 ppm	107% TM (Γ)	6.8 dB/cm 2.55 μm	Highest reported confinement factor. Low fringes. Moderate losses. Low LOD. Wavelength scanning measurements.	[69]
Suspended ring resonator	Air	CO_2 1000 ppm	50% TE (Γ)	Not reported 4.23 μm	High confinement factor. Original but complicated measurement based on dispersion spectroscopy. Wavelength scanning measurements. Ring length935 μm.	[181]
Photonic Crystals						
Photonic crystal	Air	CH_4 > 100 ppm	Not reported $n_g = 30$	Not reported 1660–1670 nm	High losses restrict the length to 300 μm. Wavelength scanning measurements.	[182]
Photonic crystal slot waveguide	Air	TEP 10 ppm	Not reported	Not reported 3.43 μm	No spectroscopic measurements were made. Changes in temperature and refractive index could not be ruled out. 800 μm long.	[183]
SOI holey photonic crystal waveguide	Air	Ethanol 150 ppb	17% (ERF) $n_g = 73$	Not reported 3.4 μm	9 mm long photonic crystal. Due to single wavelength measurement, the results are susceptible to environmental changes.	[184]
Photonic crystal slot waveguide	PDMS-water	Xylene 100 ppb (*v/v*) (86 μg/L)	Not reported $n_g = 20$	Not reported 1.69 μm	Low LOD, small differences between fabrication and design have significant effects. 300 μm long. Spectroscopic measurement.	[185]

Table 1. *Cont.*

Structure	Cladding	Analyte-LOD	Γ/EFR	Losses -λ	Advantages -Disadvantages	Ref.
SOI Photonic crystal waveguide	SU8-water	Xylene 1 ppb Trichloroethane 10 ppb (*v/v*)	Not reported n_g= 23–33	Not reported 1.640–1.680 μm	300μm long. Low LOD. Sigle wavelength measurement for each analyte. The whole device includes a Y-junction combiner, PCW, and MMI.	[186]
Self-standing GaINP Photonic crystal	Air	C_2H_2 < 50,000 ppm	100% TM 31% TE (Γ) n_g = 1.5–6.7	Not reported 1520–1570 μm	High confinement factor. 1.5 mm long photonic crystal waveguide. High LOD. Wavelength scanning measurements.	[80]
InGaAs self-standing holey photonic crystal	Air	NH_3 5 ppm	12% (EFR) n_g = 39.3	39.1 dB/cm 6.15 μm	1 mm long. No spectroscopy measurement was presented. The results are susceptible to environmental changes.	[63]
Subwavelength Grating						
Subwavelengh grating waveguides	Air	NH_3 5 ppm	10% (EFR)	6.15 μm n_g = 14.8 4.1 dB/cm	3 mm long. No spectroscopy results were presented. The results are susceptible to environmental changes.	[63]

The silicon-on-insulator (SOI) platform has been the most popular choice for integrated gas sensor applications in both NIR and MIR. Ranacher et al. [174] demonstrated detection of CO_2 down to a concentration of 500 ppm with polysilicon strip waveguides on silicon dioxide at 4.26 μm. From the measurements, the confinement factor was estimated to be in the range of Γ = 14–16%, and losses down to 3.98 dB/cm were reported. Silicon strip waveguides were also used for the detection of acetylene and methane by Jin et al. [175]. The group fabricated a 1-cm long waveguide with a thickness of 1 μm, which presents a good compromise between coupling efficiency and evanescence field confinement. The losses were determined to be 1.74 dB/cm and the simulated evanescent field ratio (EFR) was around 13% (EFR does not take into account the group index of the mode; so, although related, this should not be taken as a synonym of the confinement factor). Although the limit of detection was not calculated and the lowest concentration measured was 25% for both gases, the experimental results indicate that a concentration down to 5% could be quantified. The SOI platform was also chosen by Tombez et al. [176], with methane gas as the target analyte. They successfully increased the confinement factor by using TM polarization with a simple strip waveguide to 25.5% and achieved losses near to 2 dB/cm when operating at 1650 μm. The results are shown in Figure 3a,b. Two years later, the group took a breakthrough step toward integration, as discussed in the previous section, being able to successfully integrate a 20 cm spiral waveguide with a 15% confinement factor to a source and a detector on a single chip.

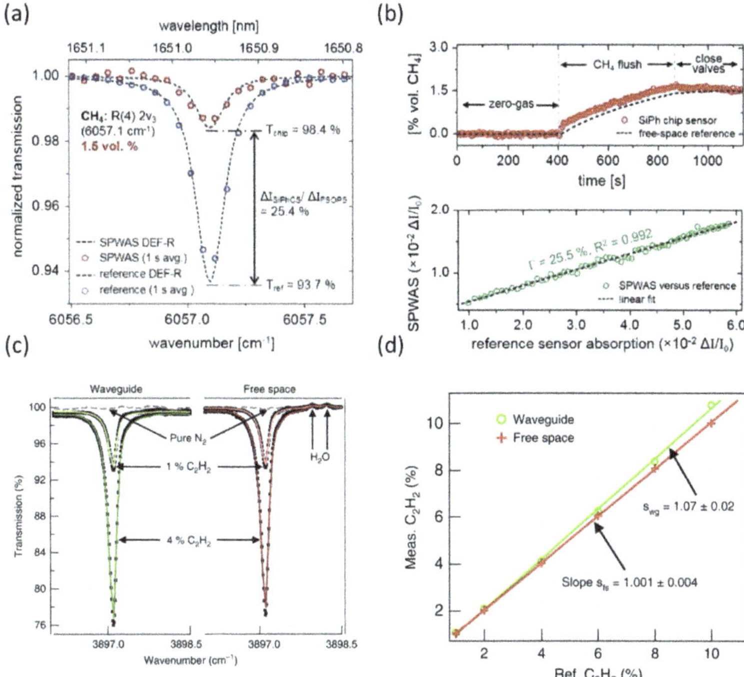

Figure 3. Experimental demonstration of gas detection by WIRAS. (**a**) The spectrum of the methane R(4)2ν_3 line measured by the waveguide-based integrated spectrometer design by Zhang et al. [157]. The Voight spectral fit for 1.5% methane concentration is shown in red, together with the experimental data. (**b**) Upper plot: experimentally measured methane concentration before and after flushing the chamber with methane. Lower plot: correlation between the absorption measured with the waveguide device and a free-space reference beam, indicating $\Gamma = 25.5\%$ in the waveguide (reproduced with permission from reference [157]). (**c**) Comparison of experimental absorption spectra for 4% and 1% acetylene measured using a free-standing tantala waveguide and a free space beam of identical path length, reproduced from Vlk et al. [69]. (**d**) Correlation of the measured concentration to the reference concentration of data in (**c**). The slope gives $\Gamma = 107\%$ (reproduced with permission under a Creative Commons Attribution 4.0 International License. http://creativecommons.org/licenses/by/4.0/ assessed on 9th September 2021).

Due to the light absorption in silicon dioxide bottom cladding at wavelengths over 3.5 µm, silicon-on-nitride (SON) and silicon-on-sapphire (SOS) appeared as an alternative to the commonly used SOI [187,188]. SOS has a transparent window of up to 5.5 µm and a high refractive index contrast between the core and the cladding. Chen and collaborators compared experimentally the performance of photonic crystal waveguides (PCW), slot waveguides and strip waveguides on sapphire. Despite the theoretical 1- to 100-fold slow-light driven improvement in the confinement factor, PC waveguides exhibited only slightly higher light–analyte interaction compared to slot waveguides but they were significantly better than strip waveguides. The same group developed PC waveguides even further [183]. Three PCW designs were fabricated in silicon on sapphire: a regular line-defect PCW (so-called W1 waveguide), a holey PCW (HPCW) with smaller diameter holes etched within the light defect, and a slot PCW, wherein a rectangular slot is etched at the center of the PCW. The designs were optimized for 3.43 µm wavelength to quantify xylene and triethyl phosphate (TEP) vapors. In the case of the slot PCW, the authors simulated that the slow light effect, coupled with the high evanescent field confinement in the slot, should reduce the required absorption path length by a factor of 1000 compared to strip or rib waveguides.

Although the authors observed a detectable signal change when the waveguide was exposed to 10 ppm TEP with an 800 μm long HPCW, the increase in sensitivity due to the slow light effect is difficult to quantify. The measurements only tracked the total power loss at one wavelength instead of spectrally resolved detection; therefore, the band diagram shift due to the refractive index or temperature variations may affect the signal in the sensitive slow-light regime as well.

As an alternative to CMOS-compatible materials, Charrier et al. [179] reported chalcogenide strip waveguides over silica and calcium fluoride. The strip waveguides presented by the group showed losses as low as 0.4 dB/cm at 1.55 μm but the waveguides were tested in solution and not for gas detection. In a similar work, Han and collaborators [178] fabricated a chalcogenide glass ($Ge_{23}Sb_7S_{70}$) strip waveguide over silica. The 2-cm spiral waveguide showed an air confinement factor of 8%, losses of around 7 dB/cm, and a limit of detection of 2.5% for methane at 3.3 μm [143]. However, the comparably high detection limit is due to a broad-band laser source that cannot resolve the narrow methane lines; a better-suited single-mode continuous-wave laser, such as ICL, would enable much better performance. Finally, Agarwal's group used the chalcogenide platform to develop a monolithic-integrated on-chip MIR methane sensor [156]. They demonstrated a 5 mm long spiral chalcogenide strip waveguide ($Ge_{23}Sb_7S_{70}$) capable of sensing methane at 10,000 ppm. The design has a similar configuration to the previous report and provides a 1 cm^2 footprint sensor with both the waveguide and an integrated detector. The losses of the waveguide were measured to be 8 dB/cm, and the air confinement factor, 12.5%, according to simulations.

To further decrease the losses and increase the confinement factor, air-suspended waveguide structures were frequently used over the last few years. Lai et al. [182] proposed photonic crystal slot waveguides for methane detection in the NIR, capable of detecting methane absorption signatures down to several hundred ppm. Dicaire et al. [80] developed a 1.5 mm-long suspended GaInP photonic crystal waveguide and used acetylene to demonstrate its spectroscopic performance. The group indices in the waveguide were 1.5 to 6.7 for the TM and TE modes, while the experimentally obtained confinement factors were 100% and 31%, respectively. The fact that the interaction did not scale with the group index (i.e., the slow-down factor) was due to the considerably larger evanescent field ratio of the TM mode. Based on this result, the authors stress that not only a high group index but also a high evanescent field ratio must be addressed for strong light–analyte interaction, the latter being often neglected in works on photonic crystal waveguides for sensing. The waveguide design also included mode adapters on both end facets to gradually couple into the slow-light mode and thus reduce the Fabry–Perot oscillations. Chen's group [63] designed and fabricated fully suspended InGaAs waveguide devices with holey photonic crystal waveguides and sub-wavelength grating cladding waveguides for the mid-infrared sensing of ammonia at λ = 6.15 μm (Figure 1b,c). The propagation losses for the two waveguide types were 39.1 and 4.1 dB/cm, the light–analyte overlap was calculated to be 12% (TE) and 10% (TM), lengths, 1 and 3 mm, and the group indices, 39 and 15, respectively. Both waveguides were capable of detecting 5 ppm ammonia; nevertheless, no spectroscopy was performed during the measurement. Changes in power were tracked after flushing ammonia at a constant wavelength, leaving the results susceptible to interference from changes in the environment, including refractive index changes or temperature variations. Ranacher and collaborators [84] designed and fabricated a polysilicon waveguide on a silicon nitride membrane suspended over silica walls. They achieved a 19.5% confinement factor at 4.23 μm. The 1 cm-long waveguide was deployed for CO_2 detection, with the lowest detected concentration down to 5000 ppm. However, as in the previous work, only the signal drop at one wavelength was recorded and no spectral scan across a CO_2 absorption line was performed. Gylfason's group [73] also developed a silicon self-standing waveguide, operational at 4.2 μm wavelength, for CO_2 gas sensing. Their waveguide was designed as a Si beam, partially suspended 3 μm above the Si handle substrate and supported by tapered SiO_2 pillars (Figure 1d–f). The waveguide had a large confinement factor of 44%, a total length of 0.5 cm, and losses of 2.9 dB/cm; the CO_2 detection limit was estimated to be 350 ppm. Although the measurement sampled absorption across a spectral line, unfortunately, the spectroscopy data that was provided had very few points to justify the fitting, considering that fringes might be present. The same group was able to measure CO_2 on ring resonators by dispersion spectroscopy [181]. As established by the Kramers–Kronig relationship, a strong variation in absorption implies a sharp change in the real part of the refractive index. The group proved that ring resonators based on suspended rib waveguide, with an air confinement factor of 50%, were able to quantify amounts as low as 1000 ppm of carbon dioxide. In 2021, Vlk et al. [69] reported a thin-film suspended tantala rib waveguide for acetylene detection in the MIR. They simulated and proved experimentally that the waveguide is able to achieve a confinement factor of 107%, and thus surpass free-space light–analyte interaction by the combining of a high evanescent field and a group index larger than unity (Figure 1g,h and Figure 3c,d). The authors also proved that the design can suppress the fringes from facet and defect reflections and improve coupling efficiency. The still relatively high losses of 6.8 dB/cm were mainly attributed to absorption into the tantala film.

To take advantage of the field inside the waveguide in addition to the evanescent field, a mesoporous waveguide for IR spectroscopy was reported by Datta and co-workers [180]. Titania rib waveguides over silica bottom cladding, with 52% field confinement in the waveguide core, were fabricated and tested, using acetylene as a calibration gas. They experimentally confirmed that the mesoporous material enables the gas to rapidly diffuse into the core of the waveguide while maintaining a rather low loss of 2 dB/cm. Nevertheless, the presence of O-H groups on the large surface area of the pore network impairs transparency over time. This problem has been partially overcome by annealing and functionalization [177,189].

3.2.2. Cladding

Top cladding on single-mode waveguides for IR spectroscopy has been only used in aqueous solutions to suppress water background and increase the concentration of the analyte close to the waveguide. Although no reports have been made on the gas phase, the use of cladding to preconcentrate organic compounds from aqueous solution also implies the ability to preconcentrate the same analyte from vapors, as has been described for ATR crystals and QCM sensors [190,191]. Therefore, it is of relevance to cover the topic briefly, as cladding on integrated gas sensors can further push the sensitivity of miniature sensors. Here, we stress that LODs in liquid environments are expected to be much lower than in gasses for the same volumetric unit as the density of molecules per volume is several orders of magnitude higher in the condensed phase.

Polymers and porous silica cladding on integrated waveguides have been reported for the detection of pollutants in water solutions. A PDMS cladding on photonic crystal slot waveguides was tested to detect xylene in water down to 100 ppb *v/v* [185]. The group tested the same device with a 2 μm Su-8 coating to measure xylene and trichloroethylene in water, with a detection limit of 1 and 10 ppb (*v/v*), respectively [186]. Polyisobutylene as a sorbent cladding on chalcogenide strip waveguides was proposed but was not experimentally evaluated [192]. According to their calculations, the cladding can decrease the limit of detection by two orders of magnitude, compared to the waveguide without cladding, due to water background suppression and organic analyte diffusion into the cladding. In another work, germanium rib waveguides were coated with a mesoporous silica coating templated with cetyltrimethylammonium bromide (CTAB) and post-grafted with hexamethyldisilazane [177]. This sensor was used to determine the concentration of toluene in water in the 6.5–7.5 μm wavelength range. It is noteworthy to mention that the use of high refractive-index core material (GOS) led to a very low field confinement factor in the cladding of only around 1% when deposited on a strip waveguide. The resulting sensitivity is consequently considerably lower than with other platforms, e.g., silicon nitride rib waveguides are able to confine light in the mesoporous cladding up to 25% [193] and slot waveguides are able to increase this percentage to 36% [76,194].

Both organic polymers and mesoporous inorganic coatings were previously deposited on ATR crystal for the detection of volatile organic compounds (VOC) from vapors [107,195] and no limitation exists to perform the same experiments on single-mode waveguides. However, the possibility of increasing sensitivity for low-weight gases other than VOC by absorption or preconcentration is dependent on the availability of material designs specific to the task. Typically, these materials hold pockets properly matching the targeted molecule size and functional groups. Examples of such materials are molecularly imprinted polymers or composite materials with cage-like organic molecules, such as cryptophanes, able to match the size of methane and halogenated analogs [196]. Alternatively, reactive centers with high selectivity toward specific reactants such as platinum nanoparticles have been described [99]. Although, still, no reports can be found on specific cladding for integrated infrared spectroscopy for the pre-concentration of gases such as carbon dioxide, methane, or acetylene, some examples can be found in works on ATR crystal [197], quartz crystal microbalance [198,199], and refractive index gas sensors [96,99,200].

4. Waveguide-Enhanced Raman Spectroscopy

4.1. Configuration and Integration

On-chip integration of Raman spectroscopic systems generally requires the enhancement of Raman scattering for greater efficiency, and, at the same time, the scattered light needs to be collected over a small area and a small solid angle (also known as the étendue) to maintain the small size of the device [77]. Single-mode waveguides can provide strong enhancement over a small volume, and therefore constitute an optimal solution for chip-integrated Raman spectroscopic systems. Compared with diffraction-limited systems, waveguide-integrated Raman systems with strong optical field confinement can provide a stronger enhancement of the signal by a few orders of magnitude, allowing for much higher detection sensitivity. Further improvement in signal enhancement can be brought about through the use of nanoplasmonic antennas integrated with waveguides [201].

For Raman spectroscopy, similarly to IR systems, free-space butt coupling through an objective lens, prism-based coupling, and fiber-mediated coupling have been the prominent mechanisms for introducing the light into the waveguides. On the other hand, collection from waveguides can be performed either from the waveguide top surface or from the waveguide facet, which can be in either a back-scattered or forward-scattered configuration (Figure 4a,b). The collection efficiency from the waveguide facet is generally much higher and has been shown to be about 40 times more efficient than that from the surface [202]. As a consequence, this requires less integration time than that from the waveguide surface [28]. However, signals from the surface can provide additional spatially resolved information [52].

While free-space coupling with a high numerical aperture objective provides good coupling efficiency, it also introduces vibrations and critical alignment steps and is unsuitable for use in compact setups. Fiber-mediated coupling solves some of these issues, but it introduces a spurious background signal, including fluorescence and Raman scattering generated from both the input and the output fibers. Kita et al. aimed to eliminate this effect by collecting only the backscattered light. This tactic removes much of the forward-propagating pump beam, which results in a higher signal-to-noise ratio [47]. The collection of the backscattered beam also has the advantage of having virtually no waveguide length limit due to the propagation loss of the waveguide, even though the contribution to the scattered signal for waveguides of lengths longer than $2/\alpha_p$ is negligible, where α_p is the propagation loss of the waveguide. In return, the forward-scattered light collection efficiency generally has a maximum for a particular waveguide length, beyond which the propagation loss dominates, thus reducing the collected signal power (Figure 4c). A different approach to eliminating the influence of the background has been to integrate edge couplers and waveguide filters onto the chips, as demonstrated by Tyndall et al. [204,206]

(Figure 4d). An array of polarization maintaining single-mode fibers is aligned directly to the waveguide facets through edge couplers; the subsequent use of lattice filters helped in separating the background and collecting both the forward- and backward-propagated Raman scattered light at separate outputs. Other on-chip elements, such as a grating-assisted contra-directional coupler, have also been proposed to reject the pump beam by directing it to a separate bus waveguide, resulting in a very high extinction ratio [207]. Nonetheless, the use of dielectric waveguides still introduces some photon background, likely arising from localized thermal fluctuations, which has been difficult to get rid of. The use of a nano-plasmonic slot waveguide, combined with a multi-mode interferometer (MMI) and backward Raman collection, has been shown to mitigate this problem [205] (Figure 4e).

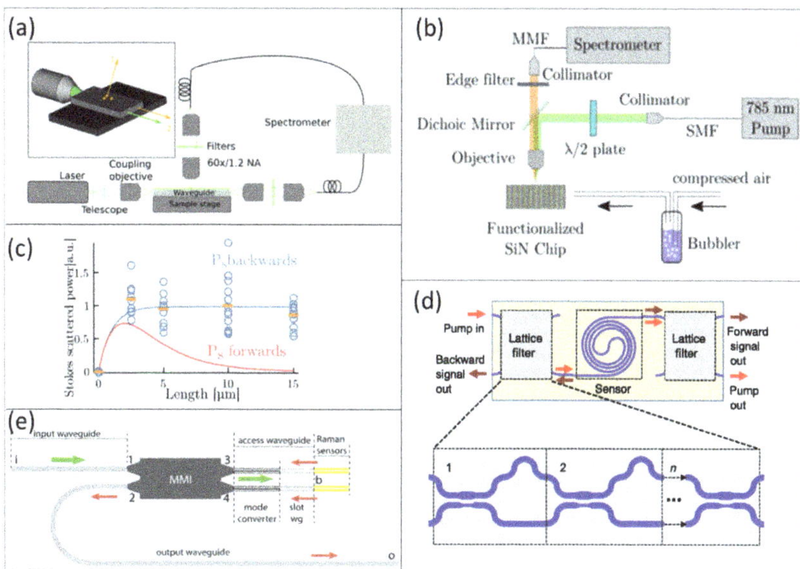

Figure 4. Different configurations as reported for waveguide-enhanced Raman spectroscopy. (**a**) Free space butt coupling of a laser into a waveguide chip for Raman spectroscopy. The figure shows the collection of the scattered signal from both the top surface and from the end facet in a forward configuration (reproduced from [52] licensed under a Creative Commons Attribution 4.0 International License http://creativecommons.org/licenses/by/4.0/ (accessed on 25 October 2021)). (**b**) Backward collection of the Raman scattered signal from a waveguide (reproduced from [203] © 2021, the author(s)). (**c**) Dependence of the scattered signal on the waveguide length for forward- and backward-scattering (reproduced from [48] © 2021, Optical Society of America). (**d**) On-chip scheme to separate the pump beam, the forward- and the backward-scattered signal through the use of lattice filters (reproduced from [204] © the authors). (**e**) WERS setup with a nano-plasmonic waveguide. The design minimizes the background signal, while the use of MMI helps in separating the input pump and the output Raman signal (reproduced from [205] © 2021, Optical Society of America).

Spontaneous Raman scattering signal intensity grows linearly with the average power of a continuous-wave pump laser; this has been a major bottleneck in improving the Raman signal, particularly in a miniaturized system where it is not possible to increase the pump power indefinitely without causing substantial damage. On the other hand, coherent Raman scattering (CRS) is a third-order non-linear phenomenon involving two laser beams, the pump, and the Stokes. When the difference in frequency between both respective lasers equals that of a specific vibrational (or rotational) transition, the probability of this transition is resonantly enhanced. CRS is capable of improving signal by many orders of magnitude and is typically implemented in two configurations, coherent anti-stokes

Raman scattering (CARS), and stimulated Raman scattering (SRS). Of these two techniques, SRS has been shown to be more promising, particularly in the context of waveguide-based Raman sensors. The reasons behind this are a simpler phase-matching relationship between the two lasers, a linear dependence of the Raman spectra on concentration, and an enhancement of the signal due to self-heterodyned detection. However, in spite of these advantages, increased shot noise severely limits the performance, only providing a modest increase of the Raman signal, as demonstrated by Zhao et al. [208]. On the other hand, other techniques such as cavity-enhanced Raman spectroscopy (CERS) and Purcell-enhanced Raman spectroscopy (PERS) have been recently used to resonantly enhance the laser beam power, as well as to improve the rate of Raman scattering, resulting in having higher laser beam–analyte interaction lengths [209–211]. These allow for sufficiently low pump powers and, in combination with their small size, may play a very important role in constructing miniaturized devices in the future.

Despite the above-reviewed efforts, the complete photonic integration of the Raman spectroscopic system is still in its infancy, primarily due to the fact that WERS requires high-power monochromatic light sources at visible or near-infrared wavelengths, high-extinction ratio filters, and sensitive detectors. Beyond this, the suppression of unwanted fluorescence and background from the waveguide material needs to be improved to match the performance of bulk Raman systems [212].

4.2. Applications
4.2.1. Air-Clad

The first steps toward the use of waveguides for Raman spectroscopy came more than 40 years ago, with the use of planar waveguides to characterize polymeric thin films [213]. It was observed that the enhancement of Raman excitation in thin films resulted from maintaining a high excitation intensity over an increased scattering volume of the analyte [214]. Further studies revealed that high-index waveguides yielded the greatest field enhancement for Raman excitation at the surface of an optical waveguide [215]. To better quantify the waveguide performance for Raman sensing, Baets et al. defined the so-called conversion efficiency, a parameter that is dependent on both the waveguide geometry and the material. The group corroborated their finding experimentally with isobutanol at 785 nm [71].

Waveguide-based Raman spectroscopy demonstrations published to date rely on high refractive-index waveguide materials that are transparent in the visible–NIR range, such as silicon nitride and tantalum pentoxide. Similar to IR spectroscopy, the strip, rib and slot waveguides are the most popular designs [65,71,74,216]. Tantala rib waveguides were tested on isopropanol, methanol, and finally, with hemoglobin solutions at physiological concentrations as the first step toward future nanoscopy applications [52]. Silicon nitride strip waveguides were designed in a spiral pattern for WERS at 785 nm and were used to track isopropanol by Baets' group [71]. Soon after, the same group proposed a design based on slot waveguides, which allows for better light–analyte interaction in the slot and, due to lower optical field confinement in the waveguide material, it also mitigates the waveguide Raman background [212]. The introduction of slot waveguides resulted in a 6-fold improvement in performance for silicon nitride waveguides, compared with strip waveguides [51]. Furthermore, Baets and collaborators studied theoretically the influence of the refractive index and polarization on Raman conversion efficiency, concluding that the TE polarization in slot waveguides with a high refractive index contrast presents the highest value from the three designs discussed [217]. Despite the success of Si_3N_4 spiral waveguides, the Raman background and fluorescence signal in the material still pose a serious limitation to detection sensitivity. Therefore, the low density of molecules, such as those found in the gas phase, has not been detected in non-functionalized (or unclad) waveguides so far, and most reports have been on liquid samples.

The use of plasmonics has also been proposed and widely studied, in order to increase the Raman scattering yield. Interaction between nanoplasmonics antennas and waveguides was described both theoretically and experimentally by several authors [218–220] and the implications for sensing have been explored. Although great advances have been made in the field, including integrated plasmonic moieties to the waveguide [48,50], the performance of such devices in comparison to traditional WERS has not shown clear advantages. Plasmonics brings signal enhancement, but it simultaneously increases losses, thus limiting the propagation length to several micrometers in comparison to the centimeter-length scale achieved by common WERS. Therefore, the Raman conversion efficiency results in comparable values, being slightly superior for a 4-cm slot waveguide than 15-μm hybrid plasmonic waveguides. The main advantage of hybrid waveguides has been the reduced spurious Raman background generated from the core of the dielectric waveguide. Nevertheless, neither of these approaches proved sensitive enough for gas sensing.

Stimulated Raman scattering (SRS), as a strategy to increase the Raman signal, was tested experimentally on silicon nitride waveguides by Baets et al. [208]. The signal was enhanced but so was the noise, with a resulting signal-to-noise ratio only slightly improved compared to spontaneous Raman scattering. This mediocre result has been attributed to sub-optimized design and equipment, while more significant improvement with a better-optimized system was not excluded.

4.2.2. Clad/Functionalized

The use of polymer cladding, with the ability to preconcentrate gases and, thus, increase the Raman signal, has been shown recently in several reports. Holmstrom et al. [193] covered Si_3N_4 rib waveguides with a hyperbranched carbosilane fluoroalcohol-based sorbent polymer (HCFSA2). These sensing properties were first tested with ethyl acetate (EA), methyl salicylate (MeS), and dimethyl sulfoxide (DMSO), listed in increasing order of their partitioning coefficient into HCSFA2 (Figure 5a,b). These analytes, exhibiting carboxylate and sulfoxide structures, work as substitutes for the more toxic phosphonate esters, which are well-known warfare agents with a major effect on the nervous system. Later on, the detection of dimethyl methyl phosphonate (DMMP), diethyl methane phosphate (DEMP), trimethyl phosphate (TMP), and triethyl phosphate (TEP) in HCSFA2 over a single-mode waveguide was also studied by Tyndall and collaborators [221]. They reported a limit of detection of approximately 5, 10, 50 and 50 ppb, respectively. The limit of detection was related to the increase in the basicity of the compounds and the affinity to the acid–polymer cladding. The same group recently proposed alternative coatings for the waveguide-enhanced Raman spectroscopy of trace chemical warfare agent simulants. DMMP was chosen to compare the performance of three alternative coatings to HCSFA2: carbosilane chain polymer poly(methyl 2-butanol, 1,1,1-trifluoro-2-(trifluoromethyl)) siloxane (PMBTTS); 2,2-bis(4-hydroxy-3-propyl phenyl) hexafluoropropane (o1pBPAF); and fluoro-polyol (FPOL). Each sorbent is a hydrogen-bond acid designed to target hydrogen-bond basic vapors. The group proved for this particular analyte that o1pBPAF has a better performance over HCSFA2 [222].

Alternatively to polymer cladding, sol-gel-based materials were tested for WERS by Haolan Zhao et al. [194]. Motivated by the two orders of magnitude of improvement for the determination of benzonitrile, valeronitrile and cyclohexanone from water, they tested the performance of a mesoporous silica cladding post-grafted with hexamethyldisilazane against solvent vapors (VOCs) in a gaseous matrix. The resultant hydrophobic coating could absorb VOCs, particularly ethanol, acetone and isopropyl alcohol (Figure 5c,d). This cladding was used to coat a single- and a double-slot silicon nitride waveguide. The confinement factor was estimated to be 36% and 32%, respectively, at a 785 nm wavelength. The single-slot waveguide LOD for isopropanol, ethanol and acetone was 53, 157 and 594 ppm, respectively [194]. The double-slot waveguide device could determine the concentration of isopropanol vapors down to 808 ppm [203]. Table 2 summarizes the most relevant information from clad WERS gas sensors discussed so far.

Despite the undeniable increase in sensitivity enabled by the use of enrichment cladding, several drawbacks can be identified for both polymers and sol-gel layers. VOCs at standard pressure and temperature are liquids according to the phase diagram, readily susceptible to condensation in the cladding, as can be seen in the broad absorption bands analyzed in most of these reports [223]. The broad spectral features are susceptible to cross-sensitivity and limit the sensor selectivity, i.e., the capability to discriminate different compounds in complex matrices [194]. In addition, signal enhancement comparable to VOC cannot be easily achieved for gases with critical temperatures below or near ambient temperature, such as methane, carbon dioxide, or acetylene.

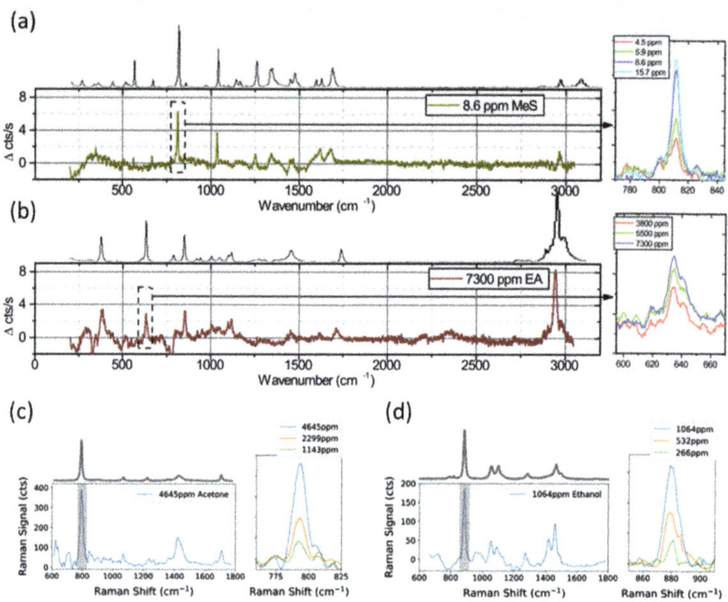

Figure 5. Examples of WERS spectra. Experimental Raman spectrum after background subtraction for vapors absorbed in the cladding (HCSFA2 (**a**,**b**), Ms-HMDS (**c**,**d**)). In each plot, the reference spectrum is placed above the experimental one, and the selected peak used to track different concentrations is shown to the right. Reproduced with permission from reference [193], © 2021, Optical Society of America, and reference [194], © 2021, Optical Society of America.

Table 2. Overview of works using clad waveguides for WERS.

Structure	Cladding	Losses $(\alpha_p)/\lambda$	Analyte	Γ/ Polarization	Length	LOD	Ref.
Si_3N_4 rib waveguide	HCSFA2 FPOL PMBTTS O1pBPAF	1–5 dB/cm 1060–1300 nm	DMMP	Not reported 1064 TM	Spiral length not specified *	3–1000 ppb	[222]
Si_3N_4 rib waveguide	HCSFA2	1–2.5 dB/cm 1060–1200 nm	DMMP DEMP TMP TEP	Not reported 832 nm TM	9.6 mm	5, 10, 50, 50 ppb	[221]
Si_3N_4 rib wavegudies	HCSFA2	2 dB/cm 980–1600 nm	EA MeS DMSO	25% TE 1064 nm	9.6 mm	1.8 ppm 1 ppm 24 ppb	[193]
Si_3N_4 double slot waveguide	MS-HMDS	Not reported	isopropanol	32% TE 785 nm	10 mm	808 ppm	[203]
Si_3N_4 slot waveguide	MS-HMDS	5.6 dB/cm	isopropanol, acetone, ethanol	37% TE 785 nm	8 mm	60, 600, 160 ppm	[194]

* The group reported 3.1–5.7-cm long spirals in another publication [224].

5. Summary of Current Technology–Comparison with Refractive Index Sensing

We summarize the reviewed literature with a technology map (Figure 6). The technology map is a 3D plot, where each data point represents one particular sensing structure, plotted against its LOD, operation wavelength l, and propagation loss, a_{prop}. Different sensing techniques are color-coded to show the general trends for each sensor family. Absorption spectroscopy sensors based on strip or rib waveguides (red) are located in the upper left corner, with LODs above 100 ppm. Lower LODs have been achieved with suspended waveguides (purple) at longer wavelengths, with the most sensitive among them capable of detecting a gas concentration of 7 ppm. Lower LODs are owing to clad systems. Raman measurements (grey) of VOC (liquid at normal standard conditions) present limits of detection two orders of magnitude lower, due to the enrichment properties of the cladding. These devices operate at a shorter wavelength region, closest to the left corner. Subwavelength gratings and photonic crystal-based devices (orange and blue, respectively) were able to detect gases such as NH_3 with a LOD down to 0.150 ppm and are located at the back of the technology map due to high losses (non-reported propagation losses were estimated based on the length of the used waveguide).

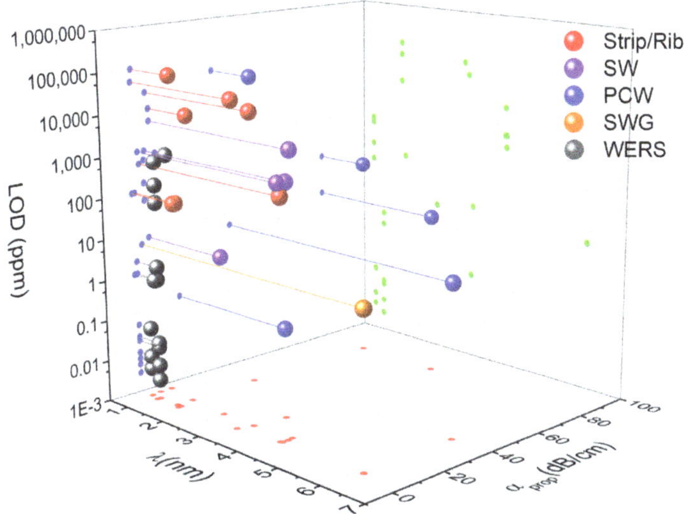

Figure 6. Technology map. A 3D plot of various representative sensing structures, categorized according to their working wavelength l (x-axis), propagation losses α_{prop} (y-axis), and LOD (z-axis). Legend: strip/rib waveguides (strip/rib), suspended waveguides (SW), photonic crystal waveguides (PCW), subwavelength gratings (SWG), clad waveguides used for WERS (WERS).

6. Outlook and Future Perspectives

With this work, we strove to provide a comprehensive review of on-chip waveguide-based IR-absorption and Raman spectroscopy sensors for gas sensing. We discussed the main components, including new and sophisticated waveguide designs, as well as the latest advances in the domain of spectroscopic sensor integration.

From the IR absorption-sensing perspective, traditional simple rib and strip waveguides have evolved in design and processing into self-standing designs, with a corresponding increase in confinement factor from one digit to more than 100%. This dramatic enhancement in confinement factors, together with transition to MIR wavelengths and compatible nanophotonic components, brought the limit of detection down by several orders of magnitude since the first waveguide-based sensor reports. In return, integrated sensors capable of detecting small gas molecules below 10 ppm were reported (see Table 1).

To reduce the limit of detection even further, slotted photonic crystals, capable of reducing the speed of light while maintaining a high air confinement factor, were proposed and fabricated. Nevertheless, the performance is still limited by the high propagation losses of the waveguides, allowing for only centimeter- or, in the case of photonic crystals, millimeter-long pathlengths.

On another front, the use of enrichment cladding that is compatible with integrated waveguide platforms showed promise for enhancing the sensitivity of WIRAS by the selective absorption and up-concentration of volatile analytes. So far, cladding (applied on WIRAS) has been only used to sample solvents in aqueous environments; however, no restrictions exist to use them with gas matrices as long as the cladding remains transparent within the wavelength range of interest. Further development will imply the use of recognition sites in the cladding to boost the specificity and allow for the enrichment of other gases than VOCs. Enrichment remains a challenging topic for small molecules and, particularly, for gases with critical temperatures below room temperature, such as the majority of greenhouse gases including methane and CO_2. Some progress has been made in this direction by developing composite cladding, with cage-like molecules as trapping sites, but the specificity, the transparency, and the processing still need to be matched.

Waveguide-enhanced Raman spectroscopy greatly profits from the high intensity of tightly confined guided modes and from Raman signal collection along the entire waveguide length, increasing both the signal-to-noise ratio and the sensitivity. Nevertheless, the limit of detection of air-clad sensors remains around 50,000 ppm in solution and could not be applied to gases. In a quest to improve this figure, plasmonic structures coupled to waveguides for the surface enhancement of Raman scattering (SERS) were tested, but no great improvement in the performance was reported other than partial background suppression. Only the use of enrichment cladding significantly pushed down the limit of detection. Depending on the Raman cross-section values, the analyte and the cladding, the LOD in a solution can drop down to parts-per-billion.

Comparing WIRAS and WERS, the weak Raman cross-section has limited the WERS sensor performance in comparison with absorption spectroscopy. The absence of direct gas sensing with air-clad waveguides in a Raman configuration and orders of magnitude lower detection limits for absorption spectroscopy, even with the use of cladding, are a direct consequence. The performance of waveguides for both techniques is still limited by losses, while Raman sensors additionally require the careful selection of materials to suppress the spurious Raman background. WERS sensors, however, have a certain advantage over WIRAS for the detection of larger gas molecules, molecules in cladding, or in liquid matrices. Namely, WERS offers the spectral coverage needed to capture the broad spectral features of such analytes and maintain the sensor specificity. Another advantage of WERS is its lower sensitivity to water interference, as Raman spectra typically overlap with a water window spanning from 600 to 2600 cm^{-1}, with a minor band at 1600 cm^{-1} [225]. Moreover, WERS systems operate in VIS-NIR, where the photonic materials and fabrication processes are mature, and traditional cladding materials such as polymers or mesoporous oxide films are transparent. In contrast, in WIRAS, strong MIR water absorption bands must be avoided, e.g., by measurement at low pressure and between water absorption lines, or water needs to be excluded by sample preconditioning or hydrophobic functionalization. WIRAS waveguides are also subject to spurious absorption due to residual OH, NH groups, or organic cladding, which is currently the major factor limiting their performance. Finally, both WIRAS and WERS can handle small sample volumes in combination with microfluidics, which is a major advantage of both techniques compared to the bulk systems.

At present, the miniaturization of WIRAS gas sensors is a large step forward ahead of WERS, with the first systems integrating both laser, waveguide, and detector being successfully demonstrated in both the NIR and MIR. The testing of such sensors in practical applications, followed by commercialization efforts, is underway, and first reports on the deployment of integrated absorption spectroscopic sensors within new platforms such as networks or UAVs will likely emerge within a few years. WERS sensors are, on the other hand, more suitable for large-molecule detection and operation with enrichment cladding. We expect an increasing number of works exploring chemical-spectroscopic detection using such sensors, eventually translating into commercial applications in chemical, biological, and biomedical research.

Author Contributions: All authors contributed to the literature review, writing and editing of the manuscript. All authors have read and agreed to the published version of the manuscript.

Funding: This work was funded by the European Research Council (grant no. 758973) and Tromsø Research Foundation (project ID 17_SG_JJ).

Institutional Review Board Statement: Not applicable.

Informed Consent Statement: Not applicable.

Acknowledgments: The authors would like to thank Olav G. Hellesø for a careful review of the manuscript.

Conflicts of Interest: The authors declare no conflict of interest.

References

1. Breitman, M.; Ruiz-Moreno, S.; Gil, A.L. Experimental Problems in Raman Spectroscopy Applied to Pigment Identification in Mixtures. *Spectrochim. Acta Part A Mol. Biomol. Spectrosc.* **2007**, *68*, 1114–1119. [CrossRef] [PubMed]
2. Keiner, R.; Herrmann, M.; Küsel, K.; Popp, J.; Frosch, T. Rapid Monitoring of Intermediate States and Mass Balance of Nitrogen during Denitrification by Means of Cavity Enhanced Raman Multi-Gas Sensing. *Anal. Chim. Acta* **2015**, *864*, 39–47. [CrossRef] [PubMed]
3. Wagenen, R.A.; Westenskow, D.R.; Benner, R.E.; Gregonis, D.E.; Coleman, D.L. Dedicated Monitoring of Anesthetic and Respiratory Gases by Raman Scattering. *J. Clin. Monit.* **1986**, *2*, 215–222. [CrossRef] [PubMed]
4. Gillibert, R.; Huang, J.Q.; Zhang, Y.; Fu, W.L.; Lamy de la Chapelle, M. Explosive Detection by Surface Enhanced Raman Scattering. *TrAC Trends Anal. Chem.* **2018**, *105*, 166–172. [CrossRef]
5. Tuzson, B.; Graf, M.; Ravelid, J.; Scheidegger, P.; Kupferschmid, A.; Looser, H.; Morales, R.P.; Emmenegger, L. A Compact QCL Spectrometer for Mobile, High-Precision Methane Sensing Aboard Drones. *Atmos. Meas. Tech.* **2020**, *13*, 4715–4726. [CrossRef]
6. Hodgkinson, J.; Tatam, R.P. Optical Gas Sensing: A Review. *Meas. Sci. Technol.* **2013**, *24*, 012004. [CrossRef]
7. Vurgaftman, I.; Weih, R.; Kamp, M.; Meyer, J.R.; Canedy, C.L.; Kim, C.S.; Kim, M.; Bewley, W.W.; Merritt, C.D.; Abell, J.; et al. Interband Cascade Lasers. *J. Phys. D Appl. Phys.* **2015**, *48*, 123001. [CrossRef]
8. Vurgaftman, I.; Bewley, W.W.; Canedy, C.L.; Kim, C.S.; Kim, M.; Merritt, C.D.; Abell, J.; Meyer, J.R. Interband Cascade Lasers with Low Threshold Powers and High Output Powers. *IEEE J. Sel. Top. Quantum Electron.* **2013**, *19*, 1200210. [CrossRef]
9. Razeghi, M.; Lu, Q.Y.; Bandyopadhyay, N.; Zhou, W.; Heydari, D.; Bai, Y.; Slivken, S. Quantum Cascade Lasers: From Tool to Product. *Opt. Express OE* **2015**, *23*, 8462–8475. [CrossRef]
10. Vitiello, M.S.; Scalari, G.; Williams, B.; Natale, P.D. Quantum Cascade Lasers: 20 Years of Challenges. *Opt. Express OE* **2015**, *23*, 5167–5182. [CrossRef]
11. Rahim, M.; Fill, M.; Felder, F.; Chappuis, D.; Corda, M.; Zogg, H. Mid-Infrared PbTe Vertical External Cavity Surface Emitting Laser on Si-Substrate with above 1 W Output Power. *Appl. Phys. Lett.* **2009**, *95*, 241107. [CrossRef]
12. Rahim, M.; Khiar, A.; Fill, M.; Felder, F.; Zogg, H. Continuously Tunable Singlemode VECSEL at 3.3 Mm Wavelength for Spectroscopy. *Electron. Lett.* **2011**, *47*, 1037–1039. [CrossRef]
13. Rey, J.M.; Fill, M.; Felder, F.; Sigrist, M.W. Broadly Tunable Mid-Infrared VECSEL for Multiple Components Hydrocarbon Gas Sensing. *Appl. Phys. B* **2014**, *117*, 935–939. [CrossRef]
14. Picqué, N.; Hänsch, T.W. Frequency Comb Spectroscopy. *Nat. Photonics* **2019**, *13*, 146–157. [CrossRef]
15. Hugi, A.; Villares, G.; Blaser, S.; Liu, H.C.; Faist, J. Mid-Infrared Frequency Comb Based on a Quantum Cascade Laser. *Nature* **2012**, *492*, 229–233. [CrossRef] [PubMed]
16. Spott, A.; Peters, J.; Davenport, M.L.; Stanton, E.J.; Merritt, C.D.; Bewley, W.W.; Vurgaftman, I.; Kim, C.S.; Meyer, J.R.; Kirch, J.; et al. Quantum Cascade Laser on Silicon. *Optica* **2016**, *3*, 545–551. [CrossRef]
17. Zhou, Z.; Yin, B.; Michel, J. On-Chip Light Sources for Silicon Photonics. *Light Sci. Appl.* **2015**, *4*, e358. [CrossRef]
18. Kapsalidis, F.; Shahmohammadi, M.; Süess, M.J.; Wolf, J.M.; Gini, E.; Beck, M.; Hundt, M.; Tuzson, B.; Emmenegger, L.; Faist, J. Dual-Wavelength DFB Quantum Cascade Lasers: Sources for Multi-Species Trace Gas Spectroscopy. *Appl. Phys. B* **2018**, *124*, 107. [CrossRef]
19. Jágerská, J.; Jouy, P.; Hugi, A.; Tuzson, B.; Looser, H.; Mangold, M.; Beck, M.; Emmenegger, L.; Faist, J. Dual-Wavelength Quantum Cascade Laser for Trace Gas Spectroscopy. *Appl. Phys. Lett.* **2014**, *105*, 161109. [CrossRef]
20. Todt, R.; Jacke, T.; Laroy, R.; Morthier, G.; Amann, M.-C. Demonstration of Vernier Effect Tuning in Tunable Twin-Guide Laser Diodes. *IEE Proc. Optoelectron.* **2005**, *152*, 66–71. [CrossRef]

21. Jiang, A.; Jung, S.; Jiang, Y.; Vijayraghavan, K.; Kim, J.H.; Belkin, M.A. Mid-Infrared Quantum Cascade Laser Arrays with Electrical Switching of Emission Frequencies. *AIP Adv.* **2018**, *8*, 085021. [CrossRef]

22. Zéninari, V.; Vallon, R.; Bizet, L.; Jacquemin, C.; Aoust, G.; Maisons, G.; Carras, M.; Parvitte, B. Widely-Tunable Quantum Cascade-Based Sources for the Development of Optical Gas Sensors. *Sensors* **2020**, *20*, 6650. [CrossRef]

23. Barritault, P.; Brun, M.; Labeye, P.; Hartmann, J.-M.; Boulila, F.; Carras, M.; Nicoletti, S. Design, Fabrication and Characterization of an AWG at 4.5 Mm. *Opt. Express OE* **2015**, *23*, 26168–26181. [CrossRef]

24. Granzow, N. Supercontinuum White Light Lasers: A Review on Technology and Applications. In Proceedings of the Photonics and Education in Measurement Science 2019, International Society for Optics and Photonics, Jena, Germany, 17–19 September 2019; Volume 11144, p. 1114408.

25. Montesinos-Ballester, M.; Lafforgue, C.; Frigerio, J.; Ballabio, A.; Vakarin, V.; Liu, Q.; Ramirez, J.M.; Roux, X.L.; Bouville, D.; Barzaghi, A.; et al. On-Chip Mid-Infrared Supercontinuum Generation from 3 to 13 Mm Wavelength. *ACS Photonics* **2020**, *7*, 3423–3429. [CrossRef]

26. Yu, Y.; Gai, X.; Ma, P.; Vu, K.; Yang, Z.; Wang, R.; Choi, D.-Y.; Madden, S.; Luther-Davies, B. Experimental Demonstration of Linearly Polarized 2–10 Mm Supercontinuum Generation in a Chalcogenide Rib Waveguide. *Opt. Lett. OL* **2016**, *41*, 958–961. [CrossRef]

27. Lamont, M.R.E.; Luther-Davies, B.; Choi, D.-Y.; Madden, S.; Eggleton, B.J. Supercontinuum Generation in Dispersion Engineered Highly Nonlinear ($\gamma = 10$ /W/m) As_2S_3 Chalcogenide Planar Waveguide. *Opt. Express OE* **2008**, *16*, 14938–14944. [CrossRef]

28. Du, Q.; Luo, Z.; Zhong, H.; Zhang, Y.; Huang, Y.; Du, T.; Zhang, W.; Gu, T.; Hu, J. Chip-Scale Broadband Spectroscopic Chemical Sensing Using an Integrated Supercontinuum Source in a Chalcogenide Glass Waveguide. *Photon. Res. PRJ* **2018**, *6*, 506–510. [CrossRef]

29. Tagkoudi, E.; Grassani, D.; Grassani, D.; Yang, F.; Herkommer, C.; Kippenberg, T.; Brès, C.-S. Parallel Gas Spectroscopy Using Mid-Infrared Supercontinuum from a Single Si_3N_4 Waveguide. *Opt. Lett. OL* **2020**, *45*, 2195–2198. [CrossRef]

30. Halloran, M.; Traina, N.; Choi, J.; Lee, T.; Yoo, J. Simultaneous Measurements of Light Hydrocarbons Using Supercontinuum Laser Absorption Spectroscopy. *Energy Fuels* **2020**, *34*, 3671–3678. [CrossRef]

31. Cezard, N.; Dobroc, A.; Canat, G.; Duhant, M.; Renard, W.; Alhenc-Gelas, C.; Lefebvre, S.; Fade, J. Supercontinuum Laser Absorption Spectroscopy in the Mid-Infrared Range for Identification and Concentration Estimation of a Multi-Component Atmospheric Gas Mixture. In Proceedings of the Lidar Technologies, Techniques, and Measurements for Atmospheric Remote Sensing VII, International Society for Optics and Photonics, Prague, Czech Republic, 19–22 September 2011; Volume 8182, p. 81820V.

32. Azzam, S.I.; Kildishev, A.V.; Ma, R.-M.; Ning, C.-Z.; Oulton, R.; Shalaev, V.M.; Stockman, M.I.; Xu, J.-L.; Zhang, X. Ten Years of Spasers and Plasmonic Nanolasers. *Light Sci. Appl.* **2020**, *9*, 90. [CrossRef]

33. Wei, J.; Ren, Z.; Lee, C. Metamaterial Technologies for Miniaturized Infrared Spectroscopy: Light Sources, Sensors, Filters, Detectors, and Integration. *J. Appl. Phys.* **2020**, *128*, 240901. [CrossRef]

34. Liu, X.; Tyler, T.; Starr, T.; Starr, A.F.; Jokerst, N.M.; Padilla, W.J. Taming the Blackbody with Infrared Metamaterials as Selective Thermal Emitters. *Phys. Rev. Lett.* **2011**, *107*, 045901. [CrossRef]

35. Liu, B.; Gong, W.; Yu, B.; Li, P.; Shen, S. Perfect Thermal Emission by Nanoscale Transmission Line Resonators. *Nano Lett.* **2017**, *17*, 666–672. [CrossRef]

36. Kuusela, T.; Peura, J.; Matveev, B.A.; Remennyy, M.A.; Stus', N.M. Photoacoustic Gas Detection Using a Cantilever Microphone and III–V Mid-IR LEDs. *Vib. Spectrosc.* **2009**, *51*, 289–293. [CrossRef]

37. Chey, J.W.; Sultan, P.; Gerritsen, H.J. Resonant Photoacoustic Detection of Methane in Nitrogen Using a Room Temperature Infrared Light Emitting Diode. *Appl. Opt. AO* **1987**, *26*, 3192–3194. [CrossRef] [PubMed]

38. Zheng, K.; Zheng, C.; Ma, N.; Liu, Z.; Yang, Y.; Zhang, Y.; Wang, Y.; Tittel, F.K. Near-Infrared Broadband Cavity-Enhanced Spectroscopic Multigas Sensor Using a 1650 Nm Light Emitting Diode. *ACS Sens.* **2019**, *4*, 1899–1908. [CrossRef]

39. Karioja, P.; Alajoki, T.; Cherchi, M.; Ollila, J.; Harjanne, M.; Heinilehto, N.; Suomalainen, S.; Zia, N.; Tuorila, H.; Viheriälä, J.; et al. Integrated Multi-Wavelength Mid-IR Light Source for Gas Sensing. In Proceedings of the Next-Generation Spectroscopic Technologies XI, International Society for Optics and Photonics, Orlando, FL, USA, 15–19 April 2018; Volume 10657, p. 106570A.

40. Popa, D.; Udrea, F. Towards Integrated Mid-Infrared Gas Sensors. *Sensors* **2019**, *19*, 2076. [CrossRef] [PubMed]

41. De Groote, A.; Cardile, P.; Subramanian, A.Z.; Tassaert, M.; Delbeke, D.; Baets, R.; Roelkens, G. A Waveguide Coupled LED on SOI by Heterogeneous Integration of InP-Based Membranes. In Proceedings of the 2015 IEEE 12th International Conference on Group IV Photonics (GFP), Vancouver, BC, Canada, 26–28 August 2015; pp. 31–32.

42. Xie, W.; Zhu, Y.; Aubert, T.; Verstuyft, S.; Hens, Z.; Thourhout, D.V. Low-Loss Silicon Nitride Waveguide Hybridly Integrated with Colloidal Quantum Dots. *Opt. Express OE* **2015**, *23*, 12152–12160. [CrossRef]

43. Lochbaum, A.; Dorodnyy, A.; Koch, U.; Koepfli, S.M.; Volk, S.; Fedoryshyn, Y.; Wood, V.; Leuthold, J. Compact Mid-Infrared Gas Sensing Enabled by an All-Metamaterial Design. *Nano Lett.* **2020**, *20*, 4169–4176. [CrossRef] [PubMed]

44. Pusch, A.; De Luca, A.; Oh, S.S.; Wuestner, S.; Roschuk, T.; Chen, Y.; Boual, S.; Ali, Z.; Phillips, C.C.; Hong, M.; et al. A Highly Efficient CMOS Nanoplasmonic Crystal Enhanced Slow-Wave Thermal Emitter Improves Infrared Gas-Sensing Devices. *Sci. Rep.* **2015**, *5*, 17451. [CrossRef] [PubMed]

45. Li, N.; Yuan, H.; Xu, L.; Tao, J.; Ng, D.K.T.; Lee, L.Y.T.; Cheam, D.D.; Zeng, Y.; Qiang, B.; Wang, Q.; et al. Radiation Enhancement by Graphene Oxide on Microelectromechanical System Emitters for Highly Selective Gas Sensing. *ACS Sens.* **2019**, *4*, 2746–2753. [CrossRef] [PubMed]

46. Wang, Z.; Abbasi, A.; Dave, U.; Groote, A.D.; Kumari, S.; Kunert, B.; Merckling, C.; Pantouvaki, M.; Shi, Y.; Tian, B.; et al. Novel Light Source Integration Approaches for Silicon Photonics. *Laser Photonics Rev.* **2017**, *11*, 1700063. [CrossRef]
47. Kita, D.M.; Michon, J.; Hu, J. A Packaged, Fiber-Coupled Waveguide-Enhanced Raman Spectroscopic Sensor. *Opt. Express OE* **2020**, *28*, 14963–14972. [CrossRef] [PubMed]
48. Wuytens, P.C.; Skirtach, A.G.; Baets, R. On-Chip Surface-Enhanced Raman Spectroscopy Using Nanosphere-Lithography Patterned Antennas on Silicon Nitride Waveguides. *Opt. Express* **2017**, *25*, 12926. [CrossRef]
49. Cao, Q.; Feng, J.; Hongliang, L.; Zhang, H.; Fuling, Z.; Zeng, H. Surface-Enhanced Raman Scattering Using Nanoporous Gold on Suspended Silicon Nitride Waveguides. *Opt. Express* **2018**, *26*, 24614–24620. [CrossRef]
50. Raza, A.; Clemmen, S.; Wuytens, P.; Muneeb, M.; Van Daele, M.; Dendooven, J.; Detavernier, C.; Skirtach, A.; Baets, R. ALD Assisted Nanoplasmonic Slot Waveguide for On-Chip Enhanced Raman Spectroscopy. *APL Photonics* **2018**, *3*, 116105. [CrossRef]
51. Dhakal, A.; Peyskens, F.; Subramanian, A.Z.; Le Thomas, N.; Baets, R. Enhanced Spontaneous Raman Signal Collected Evanescently by Silicon Nitride Slot Waveguides. In Proceedings of the CLEO: Science and Innovations 2015, OSA, San Jose, CA, USA, 10–15 May 2015; p. STh4H.3.
52. Coucheron, D.A.; Wadduwage, D.N.; Murugan, G.S.; So, P.T.C.; Ahluwalia, B.S. Chip-Based Resonance Raman Spectroscopy Using Tantalum Pentoxide Waveguides. *IEEE Photonics Technol. Lett.* **2019**, *31*, 1127–1130. [CrossRef]
53. Atabaki, A.H.; Herrington, W.F.; Burgner, C.; Jayaraman, V.; Ram, R.J. Low-Power Swept-Source Raman Spectroscopy. *Opt. Express OE* **2021**, *29*, 24723–24734. [CrossRef] [PubMed]
54. Haglund, E.; Jahed, M.; Gustavsson, J.S.; Larsson, A.; Goyvaerts, J.; Baets, R.; Roelkens, G.; Rensing, M.; O'Brien, P. High-Power Single Transverse and Polarization Mode VCSEL for Silicon Photonics Integration. *Opt. Express OE* **2019**, *27*, 18892–18899. [CrossRef]
55. Kumari, S.; Haglund, E.P.; Gustavsson, J.S.; Larsson, A.; Roelkens, G.; Baets, R.G. Vertical-Cavity Silicon-Integrated Laser with In-Plane Waveguide Emission at 850 Nm. *Laser Photonics Rev.* **2018**, *12*, 1700206. [CrossRef]
56. Baets, R.; Subramanian, A.Z.; Clemmen, S.; Kuyken, B.; Bienstman, P.; Le Thomas, N.; Roelkens, G.; Van Thourhout, D.; Helin, P.; Severi, S. Silicon Photonics: Silicon Nitride versus Silicon-on-Insulator. In Proceedings of the Optical Fiber Communication Conference, OSA, Anaheim, CA, USA, 20–22 March 2016; p. Th3J.1.
57. Yeniay, A.; Gao, R.; Takayama, K.; Gao, R.; Garito, A.F. Ultra-Low-Loss Polymer Waveguides. *J. Lightwave Technol.* **2004**, *22*, 154–158. [CrossRef]
58. Gutierrez-Arroyo, A.; Baudet, E.; Bodiou, L.; Lemaitre, J.; Hardy, I.; Faijan, F.; Bureau, B.; Nazabal, V.; Charrier, J. Optical Characterization at 77 Mm of an Integrated Platform Based on Chalcogenide Waveguides for Sensing Applications in the Mid-Infrared. *Opt. Express* **2016**, *24*, 23109. [CrossRef]
59. Schmitt, K.; Oehse, K.; Sulz, G.; Hoffmann, C. Evanescent Field Sensors Based on Tantalum Pentoxide Waveguides—A Review. *Sensors* **2008**, *8*, 711–738. [CrossRef]
60. Raza, A.; Clemmen, S.; Wuytens, P.; de Goede, M.; Tong, A.S.K.; Le Thomas, N.; Liu, C.; Suntivich, J.; Skirtach, A.G.; Garcia-Blanco, S.M.; et al. High Index Contrast Photonic Platforms for On-Chip Raman Spectroscopy. *Opt. Express* **2019**, *27*, 23067. [CrossRef] [PubMed]
61. Sipahigil, A.; Evans, R.E.; Sukachev, D.D.; Burek, M.J.; Borregaard, J.; Bhaskar, M.K.; Nguyen, C.T.; Pacheco, J.L.; Atikian, H.A.; Meuwly, C.; et al. An Integrated Diamond Nanophotonics Platform for Quantum-Optical Networks. *Science* **2016**, *354*, 847–850. [CrossRef]
62. Aharonovich, I.; Greentree, A.D.; Prawer, S. Diamond Photonics. *Nat. Photon.* **2011**, *5*, 397–405. [CrossRef]
63. Yoo, K.M.; Midkiff, J.; Rostamian, A.; Chung, C.; Dalir, H.; Chen, R.T. InGaAs Membrane Waveguide: A Promising Platform for Monolithic Integrated Mid-Infrared Optical Gas Sensor. *ACS Sens.* **2020**, *5*, 861–869. [CrossRef]
64. Yadav, A. Integrated Photonic Materials for the Mid-Infrared. *Int. J. Appl. Glass Sci.* **2020**, *11*, 491–510. [CrossRef]
65. Lin, H.; Luo, Z.; Gu, T.; Kimerling, L.C.; Wada, K.; Agarwal, A. Mid-Infrared Integrated Photonics on Silicon: A Perspective. *Nanophotonics* **2018**, *7*, 393–420. [CrossRef]
66. Wu, J.; Yue, G.; Chen, W.; Xing, Z.; Wang, J.; Wong, W.R.; Cheng, Z.; Set, S.Y.; Senthil Murugan, G.; Wang, X.; et al. On-Chip Optical Gas Sensors Based on Group-IV Materials. *ACS Photonics* **2020**, *7*, 2923–2940. [CrossRef]
67. Mi, S.; Kiss, M.; Graziosi, T.; Quack, N. Integrated Photonic Devices in Single Crystal Diamond. *J. Phys. Photonics* **2020**, *2*, 042001. [CrossRef]
68. Williams, K.R.; Member, S.; Gupta, K.; Member, S.; Wasilik, M. Etch Rates for Micromachining Processing—Part II. *J. Microelectromechanical Syst.* **2003**, *12*, 761–778. [CrossRef]
69. Vlk, M.; Datta, A.; Alberti, S.; Yallew, H.D.; Mittal, V.; Murugan, G.S.; Jágerská, J. Extraordinary Evanescent Field Confinement Waveguide Sensor for Mid-Infrared Trace Gas Spectroscopy. *Light Sci. Appl.* **2021**, *10*, 26. [CrossRef]
70. Kita, D.M.; Michon, J.; Johnson, S.G.; Hu, J. Are Slot and Sub-Wavelength Grating Waveguides Better than Strip Waveguides for Sensing? *Optica* **2018**, *5*, 1046. [CrossRef]
71. Dhakal, A.; Subramanian, A.Z.; Wuytens, P.; Peyskens, F.; Le Thomas, N.; Baets, R. Evanescent Excitation and Collection of Spontaneous Raman Spectra Using Silicon Nitride Nanophotonic Waveguides. *Opt. Lett.* **2014**, *39*, 4025. [CrossRef]
72. Milvich, J.; Kohler, D.; Freude, W.; Koos, C. Surface Sensing with Integrated Optical Waveguides: A Design Guideline. *Opt. Express* **2018**, *26*, 19885. [CrossRef] [PubMed]

73. Ottonello-Briano, F.; Errando-Herranz, C.; Rödjegård, H.; Martin, H.; Sohlström, H.; Gylfason, K.B. Carbon Dioxide Absorption Spectroscopy with a Mid-Infrared Silicon Photonic Waveguide. *Opt. Lett.* **2020**, *45*, 109. [CrossRef]
74. Subramanian, A.Z.; Ryckeboer, E.; Dhakal, A.; Peyskens, F.; Malik, A.; Kuyken, B.; Zhao, H.; Pathak, S.; Ruocco, A.; De Groote, A.; et al. Silicon and Silicon Nitride Photonic Circuits for Spectroscopic Sensing On-a-Chip. *Photon. Res.* **2015**, *3*. [CrossRef]
75. Olio, F.D.; Passaro, V.M.N. Optical Sensing by Optimized Silicon Slot Waveguides. *Opt. Express* **2007**, *15*, 4977–4993. [CrossRef] [PubMed]
76. Liu, Z.; Zhao, H.; Baumgartner, B.; Lendl, B.; Stassen, A.; Skirtach, A.; Le Thomas, N.; Baets, R. Ultra-Sensitive Slot-Waveguide-Enhanced Raman Spectroscopy for Aqueous Solutions of Non-Polar Compounds Using a Functionalized Silicon Nitride Photonic Integrated Circuit. *Opt. Lett.* **2021**, *46*, 1153. [CrossRef] [PubMed]
77. Dhakal, A.; Peyskens, F.; Clemmen, S.; Raza, A.; Wuytens, P.; Zhao, H.; Le Thomas, N.; Baets, R. Single Mode Waveguide Platform for Spontaneous and Surface-Enhanced on-Chip Raman Spectroscopy. *Interface Focus.* **2016**, *6*, 20160015. [CrossRef] [PubMed]
78. Soler Penades, J.; Khokhar, A.; Nedeljkovic, M.; Mashanovich, G. Low Loss Mid-Infrared SOI Slot Waveguides. *IEEE Photonics Technol. Lett.* **2015**, *27*, 1197–1199. [CrossRef]
79. Chen, L.R.; Wang, J.; Naghdi, B.; Glesk, I. Subwavelength Grating Waveguide Devices for Telecommunications Applications. *IEEE J. Select. Top. Quantum Electron.* **2019**, *25*, 1–11. [CrossRef]
80. Dicaire, I.; De Rossi, A.; Combrié, S.; Thévenaz, L. Probing Molecular Absorption under Slow-Light Propagation Using a Photonic Crystal Waveguide. *Opt. Lett.* **2012**, *37*, 4934. [CrossRef]
81. Reimer, C.; Nedeljkovic, M.; Stothard, D.J.M.; Esnault, M.O.S.; Reardon, C.; O'Faolain, L.; Dunn, M.; Mashanovich, G.Z.; Krauss, T.F. Mid-Infrared Photonic Crystal Waveguides in Silicon. *Opt. Express* **2012**, *20*, 29361. [CrossRef]
82. Shankar, R.; Leijssen, R.; Bulu, I.; Lončar, M. Mid-Infrared Photonic Crystal Cavities in Silicon. *Opt. Express* **2011**, *19*, 5579. [CrossRef]
83. Lin, P.T.; Singh, V.; Hu, J.; Richardson, K.; Musgraves, J.D.; Luzinov, I.; Hensley, J.; Kimerling, L.C.; Agarwal, A. Chip-Scale Mid-Infrared Chemical Sensors Using Air-Clad Pedestal Silicon Waveguides. *Lab Chip* **2013**, *13*, 2161. [CrossRef]
84. Ranacher, C.; Consani, C.; Tortschanoff, A.; Jannesari, R.; Bergmeister, M.; Grille, T.; Jakoby, B. Mid-Infrared Absorption Gas Sensing Using a Silicon Strip Waveguide. *Sens. Actuators A Phys.* **2018**, *277*, 117–123. [CrossRef]
85. Penades, J.S.; Ortega-Moñux, A.; Nedeljkovic, M.; Wangüemert-Pérez, J.G.; Halir, R.; Khokhar, A.Z.; Alonso-Ramos, C.; Qu, Z.; Molina-Fernández, I.; Cheben, P.; et al. Suspended Silicon Mid-Infrared Waveguide Devices with Subwavelength Grating Metamaterial Cladding. *Opt. Express* **2016**, *24*, 22908. [CrossRef]
86. Stievater, T.H.; Pruessner, M.W.; Rabinovich, W.S.; Park, D.; Mahon, R.; Kozak, D.A.; Bradley Boos, J.; Holmstrom, S.A.; Khurgin, J.B. Suspended Photonic Waveguide Devices. *Appl. Opt.* **2015**, *54*, F164. [CrossRef]
87. Yamada, M.; Ohmori, Y.; Takada, K.; Kobayashi, M. Evaluation of Antireflection Coatings for Optical Waveguides. *Appl. Opt. AO* **1991**, *30*, 682–688. [CrossRef] [PubMed]
88. Schmid, J.H.; Cheben, P.; Janz, S.; Lapointe, J.; Post, E.; Xu, D.-X. Gradient-Index Antireflective Subwavelength Structures for Planar Waveguide Facets. *Opt. Lett. OL* **2007**, *32*, 1794–1796. [CrossRef] [PubMed]
89. Zhang, E.J.; Tombez, L.; Teng, C.C.; Wysocki, G.; Green, W.M.J. Adaptive Etalon Suppression Technique for Long-Term Stability Improvement in High Index Contrast Waveguide-Based Laser Absorption Spectrometers. *Electron. Lett.* **2019**, *55*, 851–853. [CrossRef]
90. Demtröder, W. Widths and Profiles of Spectral Lines. In *Laser Spectroscopy*; Springer: Berlin/Heidelberg, Germany, 1981; Volume 5, pp. 78–114. ISBN 978-3-662-08259-1.
91. Martínez-Máñez, R.; Sancenón, F.; Biyikal, M.; Hecht, M.; Rurack, K. Mimicking Tricks from Nature with Sensory Organic–Inorganic Hybrid Materials. *J. Mater. Chem.* **2011**, *21*, 12588. [CrossRef]
92. Boulart, C.; Mowlem, M.C.; Connelly, D.P.; Dutasta, J.-P.; German, C.R. A Novel, Low-Cost, High Performance Dissolved Methane Sensor for Aqueous Environments. *Opt. Express* **2008**, *16*, 12607. [CrossRef] [PubMed]
93. Mateescu, A.; Wang, Y.; Dostalek, J.; Jonas, U. Thin Hydrogel Films for Optical Biosensor Applications. *Membranes* **2012**, *2*, 40–69. [CrossRef]
94. Bliem, C.; Piccinini, E.; Knoll, W.; Azzaroni, O. Enzyme Multilayers on Graphene-Based FETs for Biosensing Applications. In *Methods in Enzymology*; Elsevier: Amsterdam, The Netherlands, 2018; Volume 609, pp. 23–46. ISBN 978-0-12-815240-9.
95. Benéitez, N.T.; Missinne, J.; Shi, Y.; Chiesura, G.; Luyckx, G.; Degrieck, J.; Van Steenberge, G. Highly Sensitive Waveguide Bragg Grating Temperature Sensor Using Hybrid Polymers. *IEEE Photonics Technol. Lett.* **2016**, *28*, 1150–1153. [CrossRef]
96. Dullo, F.T.; Lindecrantz, S.; Jágerská, J.; Hansen, J.H.; Engqvist, M.; Solbø, S.A.; Hellesø, O.G. Sensitive On-Chip Methane Detection with a Cryptophane-A Cladded Mach-Zehnder Interferometer. *Opt. Express* **2015**, *23*, 31564. [CrossRef]
97. Antonacci, G.; Goyvaerts, J.; Zhao, H.; Baumgartner, B.; Lendl, B.; Baets, R. Ultra-Sensitive Refractive Index Gas Sensor with Functionalized Silicon Nitride Photonic Circuits. *APL Photonics* **2020**, *5*. [CrossRef]
98. Sulabh; Singh, L.; Jain, S.; Kumar, M. Optical Slot Waveguide with Grating-Loaded Cladding of Silicon and Titanium Dioxide for Label-Free Bio-Sensing. *IEEE Sens. J.* **2019**, *19*, 6126–6133. [CrossRef]
99. Yebo, N.A.; Taillaert, D.; Roels, J.; Lahem, D.; Debliquy, M.; Van Thourhout, D.; Baets, R. Silicon-on-Insulator (SOI) Ring Resonator-Based Integrated Optical Hydrogen Sensor. *IEEE Photonics Technol. Lett.* **2009**, *21*, 960–962. [CrossRef]
100. Pang, F.; Han, X.; Chu, F.; Geng, J.; Cai, H.; Qu, R.; Fang, Z. Sensitivity to Alcohols of a Planar Waveguide Ring Resonator Fabricated by a Sol–Gel Method. *Sens. Actuators B Chem.* **2007**, *120*, 610–614. [CrossRef]

101. Stach, R.; Pejcic, B.; Crooke, E.; Myers, M.; Mizaikoff, B. Mid-Infrared Spectroscopic Method for the Identification and Quantification of Dissolved Oil Components in Marine Environments. *Anal. Chem.* **2015**, *87*, 12306–12312. [CrossRef]

102. Howley, R.; MacCraith, B.D.; O'Dwyer, K.; Kirwan, P.; McLoughlin, P. A Study of the Factors Affecting the Diffusion of Chlorinated Hydrocarbons into Polyisobutylene and Polyethylene-Co-Propylene for Evanescent Wave Sensing. *Vib. Spectrosc.* **2003**, *31*, 271–278. [CrossRef]

103. Göbel, R.; Seitz, R.W.; Tomellini, S.A.; Krska, R.; Kellner, R. Infrared Attenuated Total Reflection Spectroscopic Investigations of the Diffusion Behaviour of Chlorinated Hydrocarbons into Polymer Membranes. *Vib. Spectrosc.* **1995**, *8*, 141–149. [CrossRef]

104. Mizaikoff, B.; Göbel, R.; Krska, R.; Taga, K.; Kellner, R.; Tacke, M.; Katzir, A. Infrared Fiber-Optical Chemical Sensors with Reactive Surface Coatings. *Sens. Actuators B Chem.* **1995**, *29*, 58–63. [CrossRef]

105. Murphy, B.; Mcloughlin, P. Determination of Chlorinated Hydrocarbon Species in Aqueous Solution Using Teflon Coated ATR Waveguide/FTIR Spectroscopy. *Int. J. Environ. Anal. Chem.* **2003**, *83*, 653–662. [CrossRef]

106. Howley, R.; MacCraith, B.D.; O'Dwyer, K.; Masterson, H.; Kirwan, P.; McLoughlin, P. Determination of Hydrocarbons Using Sapphire Fibers Coated with Poly(Dimethylsiloxane). *Appl. Spectrosc.* **2003**, *57*, 400–406. [CrossRef] [PubMed]

107. Flavin, K.; Hughes, H.; Dobbyn, V.; Kirwan, P.; Murphy, K.; Steiner, H.; Mizaikoff, B.; Mcloughlin, P. A Comparison of Polymeric Materials as Pre-Concentrating Media for Use with ATR/FTIR Sensing. *Int. J. Environ. Anal. Chem.* **2006**, *86*, 401–415. [CrossRef]

108. Regan, F.; Meaney, M.; Vos, J.G.; MacCraith, B.D.; Walsh, J.E. Determination of Pesticides in Water Using ATR-FTIR Spectroscopy on PVC/Chloroparaffin Coatings. *Anal. Chim. Acta* **1996**, *334*, 85–92. [CrossRef]

109. McKelvy, M.L.; Britt, T.R.; Davis, B.L.; Gillie, J.K.; Lentz, L.A.; Leugers, A.; Nyquist, R.A.; Putzig, C.L. Infrared Spectroscopy. *Anal. Chem.* **1996**, *68*, 93–160. [CrossRef]

110. Stach, R.; Pejcic, B.; Heath, C.; Myers, M.; Mizaikoff, B. Mid-Infrared Sensor for Hydrocarbon Monitoring: The Influence of Salinity, Matrix and Aging on Hydrocarbon–Polymer Partitioning. *Anal. Methods* **2018**, *10*, 1516–1522. [CrossRef]

111. Alberti, S.; Jágerská, J. Sol-Gel Thin Film Processing for Integrated Waveguide Sensors. *Front. Mater.* **2021**, *8*, 629822. [CrossRef]

112. Scott, B.J.; Wirnsberger, G.; Stucky, G.D. Mesoporous and Mesostructured Materials for Optical Applications. *Chem. Mater.* **2001**, *13*, 3140–3150. [CrossRef]

113. Lionello, D.F.; Steinberg, P.Y.; Zalduendo, M.M.; Soler-Illia, G.J.A.A.; Angelomé, P.C.; Fuertes, M.C. Structural and Mechanical Evolution of Mesoporous Films with Thermal Treatment: The Case of Brij 58 Templated Titania. *J. Phys. Chem. C* **2017**, *121*, 22576–22586. [CrossRef]

114. Bhatia, S.K.; Jepps, O.G.; Nicholson, D. Adsorbate Transport in Nanopores. *Adsorption* **2005**, *11*, 443–447. [CrossRef]

115. Zelcer, A.; Saleh Medina, L.M.; Hoijemberg, P.A.; Fuertes, M.C. Optical Quality Mesoporous Alumina Thin Films. *Microporous Mesoporous Mater.* **2019**, *287*, 211–219. [CrossRef]

116. Deckoff-Jones, S.; Lin, H.; Kita, D.; Zheng, H.; Li, D.; Zhang, W.; Hu, J. Chalcogenide Glass Waveguide-Integrated Black Phosphorus Mid-Infrared Photodetectors. *J. Opt.* **2018**, *20*, 044004. [CrossRef]

117. Huang, L.; Dong, B.; Guo, X.; Chang, Y.; Chen, N.; Huang, X.; Liao, W.; Zhu, C.; Wang, H.; Lee, C.; et al. Waveguide-Integrated Black Phosphorus Photodetector for Mid-Infrared Applications. *ACS Nano* **2019**, *13*, 913–921. [CrossRef]

118. Ma, Y.; Dong, B.; Wei, J.; Chang, Y.; Huang, L.; Ang, K.-W.; Lee, C. High-Responsivity Mid-Infrared Black Phosphorus Slow Light Waveguide Photodetector. *Adv. Opt. Mater.* **2020**, *8*, 2000337. [CrossRef]

119. Youngblood, N.; Chen, C.; Koester, S.J.; Li, M. Waveguide-Integrated Black Phosphorus Photodetector with High Responsivity and Low Dark Current. *Nat. Photonics* **2015**, *9*, 247–252. [CrossRef]

120. Liu, J.; Xia, F.; Xiao, D.; García de Abajo, F.J.; Sun, D. Semimetals for High-Performance Photodetection. *Nat. Mater.* **2020**, *19*, 830–837. [CrossRef] [PubMed]

121. Li, J.V.; Yang, R.Q.; Hill, C.J.; Chuang, S.L. Interband Cascade Detectors with Room Temperature Photovoltaic Operation. *Appl. Phys. Lett.* **2005**, *86*, 101102. [CrossRef]

122. Gendron, L.; Carras, M.; Huynh, A.; Ortiz, V.; Koeniguer, C.; Berger, V. Quantum Cascade Photodetector. *Appl. Phys. Lett.* **2004**, *85*, 2824–2826. [CrossRef]

123. Giorgetta, F.R.; Baumann, E.; Graf, M.; Yang, Q.; Manz, C.; Kohler, K.; Beere, H.E.; Ritchie, D.A.; Linfield, E.; Davies, A.G.; et al. Quantum Cascade Detectors. *IEEE J. Quantum Electron.* **2009**, *45*, 1039–1052. [CrossRef]

124. Yazici, M.S.; Dong, B.; Hasan, D.; Sun, F.; Lee, C. Integration of MEMS IR Detectors with MIR Waveguides for Sensing Applications. *Opt. Express* **2020**, *28*, 11524–11537. [CrossRef]

125. Ng, D.K.T.; Ho, C.-P.; Xu, L.; Zhang, T.; Siow, L.-Y.; Ng, E.J.; Cai, H.; Zhang, Q.; Lee, L.Y.T. Cmos-Mems SC0.12AL0.88N-Based Pyroelectric Infared Detector with CO_2 Gas Sensing. In Proceedings of the 2021 IEEE 34th International Conference on Micro Electro Mechanical Systems (MEMS), Online. 25–29 January 2021; pp. 852–855.

126. Ng, D.K.T.; Wu, G.; Zhang, T.-T.; Xu, L.; Sun, J.; Chung, W.-W.; Cai, H.; Zhang, Q.; Singh, N. Considerations for an 8-Inch Wafer-Level CMOS Compatible AlN Pyroelectric 5–14 Mm Wavelength IR Detector Towards Miniature Integrated Photonics Gas Sensors. *J. Microelectromechanical Syst.* **2020**, *29*, 1199–1207. [CrossRef]

127. Yang, Z.; Albrow-Owen, T.; Cai, W.; Hasan, T. Miniaturization of Optical Spectrometers. *Science* **2021**, *371*. [CrossRef]

128. Fathy, A.; Sabry, Y.M.; Nazeer, S.; Bourouina, T.; Khalil, D.A. On-Chip Parallel Fourier Transform Spectrometer for Broadband Selective Infrared Spectral Sensing. *Microsyst. Nanoeng.* **2020**, *6*, 10. [CrossRef]

129. Kim, J.; Deutsch, E.R. A Monolithic MEMS Michelson Interferometer for Ftir Spectroscopy. In Proceedings of the 2011 16th International Solid-State Sensors, Actuators and Microsystems Conference, Beijing, China, 5–9 June 2011; pp. 1524–1526.

130. Rissanen, A.; Mannila, R.; Tuohiniemi, M.; Akujärvi, A.; Antila, J. Tunable MOEMS Fabry-Perot Interferometer for Miniaturized Spectral Sensing in near-Infrared. In Proceedings of the MOEMS and Miniaturized Systems XIII, International Society for Optics and Photonics, San Francisco, CA, USA, 1–6 February 2014; Volume 8977, p. 89770X.

131. Huang, J.; Wen, Q.; Nie, Q.; Chang, F.; Zhou, Y.; Wen, Z. Miniaturized NIR Spectrometer Based on Novel MOEMS Scanning Tilted Grating. *Micromachines* **2018**, *9*, 478. [CrossRef]

132. Beć, K.B.; Grabska, J.; Huck, C.W. Principles and Applications of Miniaturized Near-Infrared (NIR) Spectrometers. *Chemistry* **2021**, *27*, 1514–1532. [CrossRef] [PubMed]

133. Tittl, A.; Leitis, A.; Liu, M.; Yesilkoy, F.; Choi, D.-Y.; Neshev, D.N.; Kivshar, Y.S.; Altug, H. Imaging-Based Molecular Barcoding with Pixelated Dielectric Metasurfaces. *Science* **2018**, *360*, 1105–1109. [CrossRef]

134. Li, E.; Chong, X.; Ren, F.; Wang, A.X. Broadband On-Chip near-Infrared Spectroscopy Based on a Plasmonic Grating Filter Array. *Opt. Lett. OL* **2016**, *41*, 1913–1916. [CrossRef]

135. Nitkowski, A.; Chen, L.; Lipson, M. Cavity-Enhanced on-Chip Absorption Spectroscopy Using Microring Resonators. *Opt. Express OE* **2008**, *16*, 11930–11936. [CrossRef] [PubMed]

136. Alshamrani, N.; Alshamrani, N.; Alshamrani, N.; Grieco, A.; Grieco, A.; Hong, B.; Fainman, Y. Miniaturized Integrated Spectrometer Using a Silicon Ring-Grating Design. *Opt. Express OE* **2021**, *29*, 15279–15287. [CrossRef]

137. Hu, T.; Zhang, X.; Zhang, M.; Yan, X. A High-Resolution Miniaturized Ultraviolet Spectrometer Based on Arrayed Waveguide Grating and Microring Cascade Structures. *Opt. Commun.* **2021**, *482*, 126591. [CrossRef]

138. Hartmann, W.; Varytis, P.; Gehring, H.; Walter, N.; Beutel, F.; Busch, K.; Pernice, W. Waveguide-Integrated Broadband Spectrometer Based on Tailored Disorder. *Adv. Opt. Mater.* **2020**, *8*, 1901602. [CrossRef]

139. Florjańczyk, M.; Cheben, P.; Janz, S.; Scott, A.; Solheim, B.; Xu, D.-X. Planar Waveguide Spatial Heterodyne Spectrometer. In Proceedings of the Photonics North 2007, International Society for Optics and Photonics, Ottawa, ON, Canada, 4–6 June 2007; Volume 6796, p. 67963J.

140. Dinh, T.T.D.; González-Andrade, D.; Montesinos-Ballester, M.; Deniel, L.; Szelag, B.; Roux, X.L.; Cassan, E.; Marris-Morini, D.; Vivien, L.; Cheben, P.; et al. Silicon Photonic On-Chip Spatial Heterodyne Fourier Transform Spectrometer Exploiting the Jacquinot's Advantage. *Opt. Lett. OL* **2021**, *46*, 1341–1344. [CrossRef]

141. Velasco, A.V.; Cheben, P.; Bock, P.J.; Delâge, A.; Schmid, J.H.; Lapointe, J.; Janz, S.; Calvo, M.L.; Xu, D.-X.; Florjańczyk, M.; et al. High-Resolution Fourier-Transform Spectrometer Chip with Microphotonic Silicon Spiral Waveguides. *Opt. Lett. OL* **2013**, *38*, 706–708. [CrossRef]

142. Nedeljkovic, M.; Velasco, A.V.; Khokhar, A.Z.; Delage, A.; Cheben, P.; Mashanovich, G.Z. Mid-Infrared Silicon-on-Insulator Fourier-Transform Spectrometer Chip. *IEEE Photonics Technol. Lett.* **2016**, *28*, 528–531. [CrossRef]

143. Podmore, H.; Scott, A.; Cheben, P.; Velasco, A.V.; Schmid, J.H.; Vachon, M.; Lee, R. Demonstration of a Compressive-Sensing Fourier-Transform on-Chip Spectrometer. *Opt. Lett. OL* **2017**, *42*, 1440–1443. [CrossRef] [PubMed]

144. Montesinos-Ballester, M.; Liu, Q.; Vakarin, V.; Ramirez, J.M.; Alonso-Ramos, C.; Roux, X.L.; Frigerio, J.; Ballabio, A.; Talamas, E.; Vivien, L.; et al. On-Chip Fourier-Transform Spectrometer Based on Spatial Heterodyning Tuned by Thermo-Optic Effect. *Sci. Rep.* **2019**, *9*, 14633. [CrossRef]

145. Le Coarer, E.; Blaize, S.; Benech, P.; Stefanon, I.; Morand, A.; Lérondel, G.; Leblond, G.; Kern, P.; Fedeli, J.M.; Royer, P. Wavelength-Scale Stationary-Wave Integrated Fourier-Transform Spectrometry. *Nat. Photonics* **2007**, *1*, 473–478. [CrossRef]

146. Nie, X.; Ryckeboer, E.; Roelkens, G.; Baets, R. CMOS-Compatible Broadband Co-Propagative Stationary Fourier Transform Spectrometer Integrated on a Silicon Nitride Photonics Platform. *Opt. Express OE* **2017**, *25*, A409–A418. [CrossRef] [PubMed]

147. Hatori, N.; Shimizu, T.; Okano, M.; Ishizaka, M.; Yamamoto, T.; Urino, Y.; Mori, M.; Nakamura, T.; Arakawa, Y. A Hybrid Integrated Light Source on a Silicon Platform Using a Trident Spot-Size Converter. *J. Lightwave Technol.* **2014**, *32*, 1329–1336. [CrossRef]

148. Meyer, J.R.; Kim, C.S.; Kim, M.; Canedy, C.L.; Merritt, C.D.; Bewley, W.W.; Vurgaftman, I. Interband Cascade Photonic Integrated Circuits on Native III-V Chip. *Sensors* **2021**, *21*, 599. [CrossRef] [PubMed]

149. Spott, A.; Stanton, E.J.; Volet, N.; Peters, J.D.; Meyer, J.R.; Bowers, J.E. Heterogeneous Integration for Mid-Infrared Silicon Photonics. *IEEE J. Sel. Top. Quantum Electron.* **2017**, *23*, 1–10. [CrossRef]

150. Komljenovic, T.; Davenport, M.; Hulme, J.; Liu, A.Y.; Santis, C.T.; Spott, A.; Srinivasan, S.; Stanton, E.J.; Zhang, C.; Bowers, J.E. Heterogeneous Silicon Photonic Integrated Circuits. *J. Lightwave Technol.* **2016**, *34*, 20–35. [CrossRef]

151. Jung, S.; Palaferri, D.; Zhang, K.; Xie, F.; Okuno, Y.; Pinzone, C.; Lascola, K.; Belkin, M.A. Homogeneous Photonic Integration of Mid-Infrared Quantum Cascade Lasers with Low-Loss Passive Waveguides on an InP Platform. *Optica* **2019**, *6*, 1023. [CrossRef]

152. Crosnier, G.; Sanchez, D.; Bouchoule, S.; Monnier, P.; Beaudoin, G.; Sagnes, I.; Raj, R.; Raineri, F. Hybrid Indium Phosphide-on-Silicon Nanolaser Diode. *Nat. Photon.* **2017**, *11*, 297–300. [CrossRef]

153. Schwarz, B.; Reininger, P.; Ristanić, D.; Detz, H.; Andrews, A.M.; Schrenk, W.; Strasser, G. Monolithically Integrated Mid-Infrared Lab-on-a-Chip Using Plasmonics and Quantum Cascade Structures. *Nat. Commun.* **2014**, *5*, 4085. [CrossRef] [PubMed]

154. Consani, C.; Ranacher, C.; Tortschanoff, A.; Grille, T.; Irsigler, P.; Jakoby, B. Mid-Infrared Photonic Gas Sensing Using a Silicon Waveguide and an Integrated Emitter. *Sens. Actuators B Chem.* **2018**, *274*, 60–65. [CrossRef]

155. Gassenq, A.; Hattasan, N.; Cerutti, L.; Rodriguez, J.B.; Tournié, E.; Roelkens, G. Study of Evanescently-Coupled and Grating-Assisted GaInAsSb Photodiodes Integrated on a Silicon Photonic Chip. *Opt. Express OE* **2012**, *20*, 11665–11672. [CrossRef] [PubMed]

156. Su, P.; Han, Z.; Kita, D.; Becla, P.; Lin, H.; Deckoff-Jones, S.; Richardson, K.; Kimerling, L.C.; Hu, J.; Agarwal, A. Monolithic On-Chip Mid-IR Methane Gas Sensor with Waveguide-Integrated Detector. *Appl. Phys. Lett.* **2019**, *114*, 051103. [CrossRef]

157. Zhang, E.J.; Martin, Y.; Orcutt, J.S.; Xiong, C.; Glodde, M.; Barwicz, T.; Schares, L.; Duch, E.A.; Marchack, N.; Teng, C.C.; et al. Trace-Gas Spectroscopy of Methane Using a Monolithically Integrated Silicon Photonic Chip Sensor. In Proceedings of the Conference on Lasers and Electro-Optics, OSA, San Jose, CA, USA, 5–10 May 2019; p. STh1F.2.

158. Hattasan, N.; Gassenq, A.; Cerutti, L.; Rodriguez, J.-B.; Tournie, E.; Roelkens, G. Heterogeneous Integration of GaInAsSb P-i-n Photodiodes on a Silicon-on-Insulator Waveguide Circuit. *IEEE Photonics Technol. Lett.* **2011**, *23*, 1760–1762. [CrossRef]

159. Muneeb, M.; Vasiliev, A.; Ruocco, A.; Malik, A.; Chen, H.; Nedeljkovic, M.; Penades, J.S.; Cerutti, L.; Rodriguez, J.B.; Mashanovich, G.Z.; et al. III-V-on-Silicon Integrated Micro—Spectrometer for the 3μm Wavelength Range. *Opt. Express OE* **2016**, *24*, 9465–9472. [CrossRef]

160. Ma, Y.; Chang, Y.; Dong, B.; Wei, J.; Liu, W.; Lee, C. Heterogeneously Integrated Graphene/Silicon/Halide Waveguide Photodetectors toward Chip-Scale Zero-Bias Long-Wave Infrared Spectroscopic Sensing. *ACS Nano* **2021**, *15*, 10084–10094. [CrossRef]

161. Schwarz, B.; Reininger, P.; Detz, H.; Zederbauer, T.; Maxwell Andrews, A.; Kalchmair, S.; Schrenk, W.; Baumgartner, O.; Kosina, H.; Strasser, G. A Bi-Functional Quantum Cascade Device for Same-Frequency Lasing and Detection. *Appl. Phys. Lett.* **2012**, *101*, 191109. [CrossRef]

162. Schwarz, B.; Reininger, P.; Detz, H.; Zederbauer, T.; Andrews, A.M.; Schrenk, W.; Strasser, G. Monolithically Integrated Mid-Infrared Quantum Cascade Laser and Detector. *Sensors* **2013**, *13*, 2196–2205. [CrossRef]

163. Hitaka, M.; Dougakiuchi, T.; Ito, A.; Fujita, K.; Edamura, T. Stacked Quantum Cascade Laser and Detector Structure for a Monolithic Mid-Infrared Sensing Device. *Appl. Phys. Lett.* **2019**, *115*, 161102. [CrossRef]

164. Lotfi, H.; Li, L.; Shazzad Rassel, S.M.; Yang, R.Q.; Corrége, C.J.; Johnson, M.B.; Larson, P.R.; Gupta, J.A. Monolithically Integrated Mid-IR Interband Cascade Laser and Photodetector Operating at Room Temperature. *Appl. Phys. Lett.* **2016**, *109*, 151111. [CrossRef]

165. Chakravarty, S.; Midkiff, J.; Yoo, K.; Rostamian, A.; Chen, R.T. Monolithic Integration of Quantum Cascade Laser, Quantum Cascade Detector, and Subwavelength Waveguides for Mid-Infrared Integrated Gas Sensing. In Proceedings of the Quantum Sensing and Nano Electronics and Photonics XVI, International Society for Optics and Photonics, San Francisco, CA, USA, 2–7 February 2019; Volume 10926, p. 109261V.

166. Midkiff, J.; Yoo, K.M.; Dalir, H.; Chen, R.T. Monolithic Integration of Quantum Cascade Laser, Quantum Cascade Detector, and Passive Components for Absorption Sensing at [Lambda] = 4.6 Mm. In Proceedings of the Quantum Sensing and Nano Electronics and Photonics XVII, International Society for Optics and Photonics, San Francisco, CA, USA, 1–6 February 2020; Volume 11288, p. 112882F.

167. Schwarz, B.; Hillbrand, J.; Beiser, M.; Andrews, A.M.; Strasser, G.; Detz, H.; Schade, A.; Weih, R.; Höfling, S. Monolithic Frequency Comb Platform Based on Interband Cascade Lasers and Detectors. *Optica* **2019**, *6*, 890. [CrossRef]

168. Yu, M.; Okawachi, Y.; Griffith, A.G.; Picqué, N.; Lipson, M.; Gaeta, A.L. Silicon-Chip-Based Mid-Infrared Dual-Comb Spectroscopy. *Nat. Commun.* **2018**, *9*, 1869. [CrossRef]

169. Zhang, Z.; Gardiner, T.; Reid, D.T. Mid-Infrared Dual-Comb Spectroscopy with an Optical Parametric Oscillator. *Opt. Lett. OL* **2013**, *38*, 3148–3150. [CrossRef]

170. Medhi, G.; Muravjov, A.V.; Saxena, H.; Fredricksen, C.J.; Brusentsova, T.; Peale, R.E.; Edwards, O. Intracavity Laser Absorption Spectroscopy Using Mid-IR Quantum Cascade Laser. In Proceedings of the Next-Generation Spectroscopic Technologies IV, International Society for Optics and Photonics, Orlando, FL, USA, 25–29 April 2011; Volume 8032, p. 80320E.

171. Ryckeboer, E.; Bockstaele, R.; Vanslembrouck, M.; Baets, R. Glucose Sensing by Waveguide-Based Absorption Spectroscopy on a Silicon Chip. *Biomed. Opt. Express* **2014**, *5*, 1636. [CrossRef]

172. Lin, P.T. Mid-Infrared Photonic Chip for Label-Free Glucose Sensing. In Proceedings of the Biophotonics Congress: Biomedical Optics Congress 2018 (Microscopy/Translational/Brain/OTS), OSA, Washington, DC, USA, 3–6 April 2018; p. JW3A.11.

173. Jin, T.; Li, L.; Zhang, B.; Lin, H.-Y.G.; Wang, H.; Lin, P.T. Real-Time and Label-Free Chemical Sensor-on-a-Chip Using Monolithic Si-on-BaTiO3 Mid-Infrared Waveguides. *Sci. Rep.* **2017**, *7*, 5836. [CrossRef] [PubMed]

174. Ranacher, C.; Consani, C.; Vollert, N.; Tortschanoff, A.; Bergmeister, M.; Grille, T.; Jakoby, B. Characterization of Evanescent Field Gas Sensor Structures Based on Silicon Photonics. *IEEE Photonics J.* **2018**, *10*, 1–14. [CrossRef]

175. Jin, T.; Zhou, J.; Lin, P.T. Real-Time and Non-Destructive Hydrocarbon Gas Sensing Using Mid-Infrared Integrated Photonic Circuits. *RSC Adv.* **2020**, *10*, 7452–7459. [CrossRef]

176. Tombez, L.; Zhang, E.J.; Orcutt, J.S.; Kamlapurkar, S.; Green, W.M.J. Methane Absorption Spectroscopy on a Silicon Photonic Chip. *Optica* **2017**, *4*, 1322–1325. [CrossRef]

177. Benéitez, N.T.; Baumgartner, B.; Missinne, J.; Radosavljevic, S.; Wacht, D.; Hugger, S.; Leszcz, P.; Lendl, B.; Roelkens, G. Mid-IR Sensing Platform for Trace Analysis in Aqueous Solutions Based on a Germanium-on-Silicon Waveguide Chip with a Mesoporous Silica Coating for Analyte Enrichment. *Opt. Express* **2020**, *28*, 27013. [CrossRef]

178. Han, Z.; Lin, P.; Singh, V.; Kimerling, L.; Hu, J.; Richardson, K.; Agarwal, A.; Tan, D.T.H. On-Chip Mid-Infrared Gas Detection Using Chalcogenide Glass Waveguide. *Appl. Phys. Lett.* **2016**, *108*, 141106. [CrossRef]

179. Charrier, J.; Brandily, M.-L.; Lhermite, H.; Michel, K.; Bureau, B.; Verger, F.; Nazabal, V. Evanescent Wave Optical Micro-Sensor Based on Chalcogenide Glass. *Sens. Actuators B Chem.* **2012**, *173*, 468–476. [CrossRef]

180. Datta, A.; Alberti, S.; Vlk, M.; Jágerská, J. Spectroscopic Gas Detection Using Thin-Film Mesoporous Waveguides. In Proceedings of the 2021 Conference on Lasers and Electro-Optics Europe & European Quantum Electronics Conference (CLEO/Europe-EQEC), Munich, Germany, 21–25 June 2021.

181. Briano, F.O.; Errando-Herranz, C.; Gylfason, K.B. On-Chip Dispersion Spectroscopy of CO_2 Using a Mid-Infrared Microring Resonator. *Opt. InfoBase Conf. Pap.* **2020**, 2–5. [CrossRef]

182. Lai, W.-C.; Chakravarty, S.; Wang, X.; Lin, C.; Chen, R.T. On-Chip Methane Sensing by near-IR Absorption Signatures in a Photonic Crystal Slot Waveguide. *Opt. Lett.* **2011**, *36*, 984. [CrossRef] [PubMed]

183. Zou, Y.; Wray, P.; Chakravarty, S.; Chen, R.T. Silicon on Sapphire Chip Based Photonic Crystal Waveguides for Detection of Chemical Warfare Simulants and Volatile Organic Compound. In Proceedings of the CLEO: Applications and Technology 2015, OSA, San Jose, CA, USA, 10–15 May 2015; p. AF2J.1.

184. Rostamian, A.; Madadi-Kandjani, E.; Dalir, H.; Sorger, V.J.; Chen, R.T. Towards Lab-on-Chip Ultrasensitive Ethanol Detection Using Photonic Crystal Waveguide Operating in the Mid-Infrared. *Nanophotonics* **2021**, *10*, 1675–1682. [CrossRef]

185. Lai, W.; Chakravarty, S.; Wang, X.; Lin, C.; Chen, R.T. Photonic Crystal Slot Waveguide Absorption Spectrometer for On-Chip near-Infrared Spectroscopy of Xylene in Water. *Appl. Phys. Lett.* **2011**, *98*, 023304. [CrossRef]

186. Lai, W.-C.; Chakravarty, S.; Zou, Y.; Chen, R.T. Multiplexed Detection of Xylene and Trichloroethylene in Water by Photonic Crystal Absorption Spectroscopy. *Opt. Lett.* **2013**, *38*, 3799. [CrossRef]

187. Ranacher, C.; Consani, C.; Maier, F.J.; Hedenig, U.; Jannesari, R.; Lavchiev, V.; Tortschanoff, A.; Grille, T.; Jakoby, B. Spectroscopic Gas Sensing Using a Silicon Slab Waveguide. *Procedia Eng.* **2016**, *168*, 1265–1269. [CrossRef]

188. Kumari, B.; Barh, A.; Varshney, R.K.; Pal, B.P. Mid-IR Evanescent Field Gas Sensor Based on Silicon-on-Nitride Slot Waveguide. In Proceedings of the 12th International Conference on Fiber Optics and Photonics, OSA, Kharagpur, India, 13–16 December 2014; p. M4A.12.

189. Kanta, A.; Sedev, R.; Ralston, J. Thermally- and Photoinduced Changes in the Water Wettability of Low-Surface-Area Silica and Titania. *Langmuir* **2005**, *21*, 2400–2407. [CrossRef] [PubMed]

190. Baumgartner, B.; Freitag, S.; Gasser, C.; Lendl, B. A Pocket-Sized 3D-Printed Attenuated Total Reflection-Infrared Filtometer Combined with Functionalized Silica Films for Nitrate Sensing in Water. *Sens. Actuators B Chem.* **2020**, *310*, 127847. [CrossRef]

191. Matsuguchi, M.; Uno, T. Molecular Imprinting Strategy for Solvent Molecules and Its Application for QCM-Based VOC Vapor Sensing. *Sens. Actuators B Chem.* **2006**, *113*, 94–99. [CrossRef]

192. Baudet, E.; Gutierrez-Arroyo, A.; Baillieul, M.; Charrier, J.; Němec, P.; Bodiou, L.; Lemaitre, J.; Rinnert, E.; Michel, K.; Bureau, B.; et al. Development of an Evanescent Optical Integrated Sensor in the Mid-Infrared for Detection of Pollution in Groundwater or Seawater. *Adv. Device Mater.* **2017**, *3*, 23–29. [CrossRef]

193. Holmstrom, S.A.; Stievater, T.H.; Kozak, D.A.; Pruessner, M.W.; Tyndall, N.; Rabinovich, W.S.; Andrew McGill, R.; Khurgin, J.B. Trace Gas Raman Spectroscopy Using Functionalized Waveguides. *Optica* **2016**, *3*, 891. [CrossRef]

194. Zhao, H.; Baumgartner, B.; Raza, A.; Skirtach, A.; Lendl, B.; Baets, R. Multiplex Volatile Organic Compound Raman Sensing with Nanophotonic Slot Waveguides Functionalized with a Mesoporous Enrichment Layer. *Opt. Lett.* **2020**, *45*, 447. [CrossRef]

195. Janotta, M.; Karlowatz, M.; Vogt, F.; Mizaikoff, B. Sol–Gel Based Mid-Infrared Evanescent Wave Sensors for Detection of Organophosphate Pesticides in Aqueous Solution. *Anal. Chim. Acta* **2003**, *496*, 339–348. [CrossRef]

196. Takacs, Z.; Soltesova, M.; Kowalewski, J.; Lang, J.; Brotin, T.; Dutasta, J.-P. Host-Guest Complexes between Cryptophane-C and Chloromethanes Revisited: Host-Guest Complexes between Cryptophane-C and Chloromethanes. *Magn. Reson. Chem.* **2013**, *51*, 19–31. [CrossRef]

197. Schädle, T.; Pejcic, B.; Mizaikoff, B. Monitoring Dissolved Carbon Dioxide and Methane in Brine Environments at High Pressure Using IR-ATR Spectroscopy. *Anal. Methods* **2016**, *8*, 756–762. [CrossRef]

198. Tsuge, Y.; Moriyama, Y.; Tokura, Y.; Shiratori, S. Silver Ion Polyelectrolyte Container as a Sensitive Quartz Crystal Microbalance Gas Detector. *Anal. Chem.* **2016**, *88*, 10744–10750. [CrossRef]

199. Procek, M.; Stolarczyk, A.; Pustelny, T.; Maciak, E. A Study of a QCM Sensor Based on TiO2 Nanostructures for the Detection of NO2 and Explosives Vapours in Air. *Sensors* **2015**, *15*, 9563–9581. [CrossRef]

200. Yebo, N.A.; Sree, S.P.; Levrau, E.; Detavernier, C.; Hens, Z.; Martens, J.A.; Baets, R. Selective and Reversible Ammonia Gas Detection with Nanoporous Film Functionalized Silicon Photonic Micro-Ring Resonator. *Opt. Express* **2012**, *20*, 11855. [CrossRef] [PubMed]

201. Peyskens, F.; Dhakal, A.; Van Dorpe, P.; Le Thomas, N.; Baets, R. Surface Enhanced Raman Spectroscopy Using a Single Mode Nanophotonic-Plasmonic Platform. *ACS Photonics* **2016**, *3*, 102–108. [CrossRef]

202. Wang, Z.; Zervas, M.N.; Bartlett, P.N.; Wilkinson, J.S. Surface and Waveguide Collection of Raman Emission in Waveguide-Enhanced Raman Spectroscopy. *Opt. Lett. OL* **2016**, *41*, 4146–4149. [CrossRef] [PubMed]

203. Zhao, H.; Raza, A.; Baumgartner, B.; Clemmen, S.; Lendl, B.; Skirtach, A.; Baets, R. Waveguide-Enhanced Raman Spectroscopy Using a Mesoporous Silica Sorbent Layer for Volatile Organic Compound (VOC) Sensing. In Proceedings of the Conference on Lasers and Electro-Optics, OSA, San Jose, CA, USA, 5–10 May 2019; p. STh1F.7.

204. Tyndall, N.F.; Stievater, T.H.; Kozak, D.A.; Pruessner, M.W.; Holmstrom, S.A.; Rabinovich, W.S. Ultrabroadband Lattice Filters for Integrated Photonic Spectroscopy and Sensing. *Opt. Eng.* **2018**, *57*, 127103. [CrossRef]

205. Reynkens, K.; Clemmen, S.; Raza, A.; Zhao, H.; Santo-Domingo Penaranda, J.; Detavernier, C.; Baets, R. Mitigation of Photon Background in Nanoplasmonic All-on-Chip Raman Sensors. *Opt. Express OE* **2020**, *28*, 33564–33572. [CrossRef] [PubMed]

206. Tyndall, N.F.; Stievater, T.H.; Kozak, D.A.; Pruessner, M.W.; Rabinovich, W.S. Passive Photonic Integration of Lattice Filters for Waveguide-Enhanced Raman Spectroscopy. *Opt. Express OE* **2020**, *28*, 34927–34934. [CrossRef] [PubMed]

207. Nie, X.; Turk, N.; Li, Y.; Liu, Z.; Baets, R. High Extinction Ratio On-Chip Pump-Rejection Filter Based on Cascaded Grating-Assisted Contra-Directional Couplers in Silicon Nitride Rib Waveguides. *Opt. Lett. OL* **2019**, *44*, 2310–2313. [CrossRef]

208. Zhao, H.; Clemmen, S.; Raza, A.; Baets, R. Stimulated Raman Spectroscopy of Analytes Evanescently Probed by a Silicon Nitride Photonic Integrated Waveguide. *Opt. Lett. OL* **2018**, *43*, 1403–1406. [CrossRef]

209. Niklas, C.; Wackerbarth, H.; Ctistis, G. A Short Review of Cavity-Enhanced Raman Spectroscopy for Gas Analysis. *Sensors* **2021**, *21*, 1698. [CrossRef] [PubMed]

210. Petrak, B.; Cooper, J.; Konthasinghe, K.; Peiris, M.; Djeu, N.; Hopkins, A.J.; Muller, A. Isotopic Gas Analysis through Purcell Cavity Enhanced Raman Scattering. *Appl. Phys. Lett.* **2016**, *108*, 091107. [CrossRef]

211. Petrak, B.; Djeu, N.; Muller, A. Purcell-Enhanced Raman Scattering from Atmospheric Gases in a High-Finesse Microcavity. *Phys. Rev. A* **2014**, *89*, 023811. [CrossRef]

212. Dhakal, A.; Wuytens, P.; Raza, A.; Le Thomas, N.; Baets, R. Silicon Nitride Background in Nanophotonic Waveguide Enhanced Raman Spectroscopy. *Materials* **2017**, *10*, 140. [CrossRef]

213. Levy, Y.; Imbert, C.; Cipriani, J.; Racine, S.; Dupeyrat, R. Raman Scattering of Thin Films as a Waveguide. *Opt. Commun.* **1974**, *11*, 66–69. [CrossRef]

214. Rabolt, J.F.; Santo, R.; Swalen, J.D. Raman Spectroscopy of Thin Polymer Films Using Integrated Optical Techniques. *Appl. Spectrosc.* **1979**, *33*, 549–551. [CrossRef]

215. Kanger, J.S.; Otto, C.; Slotboom, M.; Greve, J. Waveguide Raman Spectroscopy of Thin Polymer Layers and Monolayers of Biomolecules Using High Refractive Index Waveguides. *J. Phys. Chem.* **1996**, *100*, 3288–3292. [CrossRef]

216. Beshkov, G.; Lei, S.; Lazarova, V.; Nedev, N.; Georgiev, S.S. IR and Raman Absorption Spectroscopic Studies of APCVD, LPCVD and PECVD Thin SiN Films. *Vacuum* **2003**, *69*, 301–305. [CrossRef]

217. Dhakal, A.; Raza, A.; Peyskens, F.; Subramanian, A.Z.; Clemmen, S.; Le Thomas, N.; Baets, R. Efficiency of Evanescent Excitation and Collection of Spontaneous Raman Scattering near High Index Contrast Channel Waveguides. *Opt. Express* **2015**, *23*, 27391. [CrossRef] [PubMed]

218. Tang, F.; Adam, P.M.; Boutami, S. Theoretical Investigation of SERS Nanosensors Based on Hybrid Waveguides Made of Metallic Slots and Dielectric Strips. *Opt. Express* **2016**, *24*, 21244–21255. [CrossRef] [PubMed]

219. Raza, A.; Peyskens, F.; Clemmen, S.; Baets, R. Towards Single Antenna On-Chip Surface Enhanced Raman Spectroscopy: Arch Dipole Antenna. In Proceedings of the META'16, the 7th International Conference on Metamaterials, Photonic Crystals and Plasmonics, Malaga, Spain, 25–28 July 2016.

220. Singh, G.; Bi, R.; Dinish, U.S.; Olivo, M. Generating Localized Plasmonic Fields on an Integrated Photonic Platform Using Tapered Couplers for Biosensing Applications. *Sci. Rep.* **2017**, *7*, 15587. [CrossRef] [PubMed]

221. Tyndall, N.F.; Stievater, T.H.; Kozak, D.A.; Koo, K.; McGill, R.A.; Pruessner, M.W.; Rabinovich, W.S.; Holmstrom, S.A. Waveguide-Enhanced Raman Spectroscopy of Trace Chemical Warfare Agent Simulants. *Opt. Lett.* **2018**, *43*, 4803. [CrossRef]

222. Tyndall, N.F.; Stievater, T.H.; Kozak, D.A.; Pruessner, M.W.; Roxworthy, B.J.; Rabinovich, W.S.; Roberts, C.A.; McGill, R.A.; Miller, B.L.; Luta, E.; et al. Figure-of-Merit Characterization of Hydrogen-Bond Acidic Sorbents for Waveguide-Enhanced Raman Spectroscopy. *ACS Sens.* **2020**, *5*, 831–836. [CrossRef]

223. Yampolskii, Y. *Materials Science of Membranes for Gas and Vapor Separation*; Wiley: Hoboken, NJ, USA, 2006; ISBN 9780470853450.

224. Stievater, T.H.; Tyndall, N.F.; Kozak, D.A.; McGill, R.A.; Holmstrom, S.A.; Koo, K.; Goetz, P.G.; Pruessner, M.W. Chemical Sensors Fabricated by a Photonic Integrated Circuit Foundry. In *Frontiers in Biological Detection: From Nanosensors to Systems X*; Miller, B.L., Weiss, S.M., Danielli, A., Eds.; SPIE: San Francisco, CA, USA, 2018; p. 17.

225. Ikeda, T. Infrared Absorption and Raman Scattering Spectra of Water under Pressure via First Principles Molecular Dynamics. *J. Chem. Phys.* **2014**, *141*, 044501. [CrossRef] [PubMed]

sensors

Article

Fabrication of Microchannels in a Nodeless Antiresonant Hollow-Core Fiber Using Femtosecond Laser Pulses

Paweł Kozioł [1,*], Piotr Jaworski [1], Karol Krzempek [1], Viktoria Hoppe [2], Grzegorz Dudzik [1], Fei Yu [3,4], Dakun Wu [3,4], Meisong Liao [3], Jonathan Knight [5] and Krzysztof Abramski [1]

[1] Laser & Fiber Electronics Group, Faculty of Electronics, Photonics and Microsystems, Wroclaw University of Science and Technology, 50-370 Wroclaw, Poland; piotr.jaworski@pwr.edu.pl (P.J.); karol.krzempek@pwr.edu.pl (K.K.); grzegorz.dudzik@pwr.edu.pl (G.D.); krzysztof.abramski@pwr.edu.pl (K.A.)

[2] Centre for Advanced Manufacturing Technologies (CAMT), Faculty of Mechanical Engineering, Wroclaw University of Science and Technology, 50-370 Wroclaw, Poland; viktoria.hoppe@pwr.edu.pl

[3] Key Laboratory of Materials for High Power Laser, Shanghai Institute of Optics and Fine Mechanics, Chinese Academy of Sciences, Shanghai 201800, China; yufei@siom.ac.cn (F.Y.); wudakun@ucas.ac.cn (D.W.); liaomeisong@siom.ac.cn (M.L.)

[4] Hangzhou Institute for Advanced Study, University of Chinese Academy of Sciences, Hangzhou 310024, China

[5] Centre for Photonics and Photonic Materials, Department of Physics, University of Bath, Claverton Down, Bath BA2 7AY, UK; j.c.knight@bath.ac.uk

* Correspondence: pawel.koziol@pwr.edu.pl

Citation: Kozioł, P.; Jaworski, P.; Krzempek, K.; Hoppe, V.; Dudzik, G.; Yu, F.; Wu, D.; Liao, M.; Knight, J.; Abramski, K. Fabrication of Microchannels in a Nodeless Antiresonant Hollow-Core Fiber Using Femtosecond Laser Pulses. *Sensors* **2021**, *21*, 7591. https://doi.org/10.3390/s21227591

Academic Editor: Vittorio M.N. Passaro

Received: 27 October 2021
Accepted: 14 November 2021
Published: 16 November 2021

Publisher's Note: MDPI stays neutral with regard to jurisdictional claims in published maps and institutional affiliations.

Abstract: In this work, we present femtosecond laser cutting of microchannels in a nodeless antiresonant hollow-core fiber (ARHCF). Due to its ability to guide light in an air core combined with exceptional light-guiding properties, an ARHCF with a relatively non-complex structure has a high application potential for laser-based gas detection. To improve the gas flow into the fiber core, a series of 250×30 μm microchannels were reproducibly fabricated in the outer cladding of the ARHCF directly above the gap between the cladding capillaries using a femtosecond laser. The execution time of a single lateral cut for optimal process parameters was 7 min. It has been experimentally shown that the implementation of 25 microchannels introduces low transmission losses of 0.17 dB (<0.01 dB per single microchannel). The flexibility of the process in terms of the length of the performed microchannel was experimentally demonstrated, which confirms the usefulness of the proposed method. Furthermore, the performed experiments have indicated that the maximum bending radius for the ARHCF, with the processed 100 μm long microchannel that did not introduce its breaking, is 15 cm.

Keywords: antiresonant hollow core fibers; femtosecond laser micromachining; microchannel fabrication; microstructured fibers

1. Introduction

The main goal of laser-based gas spectroscopy research is the development of sensors with high sensitivity and selectivity, preferably maintaining a non-complex design of their setup [1–3]. According to the Beer–Lambert law, the detection ability of such sensors depends mainly on the length of the interaction path between the light beam and the measured gas molecules [4]. Hence, long optical paths implemented within the sensors setup are highly desired, especially in the case of low-concentration (typically single parts-per-million (ppm) or below—parts-per-billion (ppb) levels) gas molecules detection. Usually, long optical paths are realized by bulk-optics-based multipass cells (MPCs) that indeed fulfill this crucial requirement, however, at the expense of reduced stability of the sensor due to possible opto-mechanical drifts and optical fringes contributing to the overall noise level. Recent reports demonstrate the use of hollow-core fibers (HCFs), especially

63

the relatively new type of such fibers, ARHCFs, as an alternative to sensors based on MPCs [5–8]. A gas-filled ARHCF can form a versatile absorption cell characterized by a small volume and can be tailored to a desired length with respect to the specific application. The filling process of the ARHCF core with the target gas mixture can be realized by either gas diffusion through both fiber ends or via specially designed gas cells located at both fiber ends, through which the gas is delivered to the hollow core using overpressure. The gas filling time can vary from several hours per meter for free gas diffusion [9,10] to a few seconds for pressure-supported methods [11–13]. A purely diffusion-based gas exchange approach is less complex; however, it significantly increases the response time of an ARHCF-based gas sensor. On the other hand, the use of overpressure filling via additional gas cells increases the sensor complexity and can negatively affect the long-term stability of the sensor. Hence, an alternative solution that will enable noninvasive, simple, and efficient gas filling of the ARHCF core is highly desired.

An alternative method of filling the ARHCF core with the target gas can be realized via side incisions that allow direct access to the fiber air core along the entire length of the fiber. This solution allows non-complex gas exchange without interfering with the optical coupling at the fiber end facets [14]. Due to the fragility of the material, noncontact processing methods are most often used. The most popular method is based on the usage of a femtosecond (fs) laser. For example, a V-shaped hole was made in an HCF [15,16] using the layer-by-layer scanning method. In [17], the researchers presented fabrication of a side hole in an HCF using an fs laser, where the process was carried out in a fluid with a matching refractive index to reduce the debris that is generated during the laser ablation process. The fabrication of microchannels with the use of femtosecond laser pulses introduces certain transmission losses, which mostly result from local damage of the fiber structure or the presence of debris generated during laser ablation.

Another noncontact method used for the production of microchannels is based on the use of a focused ion beam (FIB) as presented in [18,19] for which, due to lack of destructive factors, no losses resulting from the fabrication of the microchannel were recorded. The disadvantage of this method is the time required to process a single hole (~30 min for a single hole with a diameter of 10 µm [19]). Furthermore, because of the limited space in the FIB-based setup, it is impossible to process long (i.e., several meters) fibers in multiple places, which is required for providing optimum gas filling conditions within the long optical path provided by the fiber. For this reason, an attempt was made to find a way to improve the quality of microchannels produced using laser micromachining.

The methods of microchannel realization using an fs laser presented in the literature so far have introduced relatively high losses at the level of 0.45 dB per hole [20], and more. The lowest loss of 0.35 dB per hole was recorded for a single microchannel made in an HCF assisted by liquid [17]. These results are not satisfactory from the potential applications viewpoint since in the case where a higher number of holes is required, the total losses will effectively limit light propagation in such fiber. The most common cause of losses are local modifications of the fiber structure due to laser ablation, leading to an increased attenuation of the propagating light. An additional factor that increases losses is the debris that is formed during the process and is deposited inside the fiber. Both of these factors are the result of the non-optimized laser micromachining process used for modifying the HCF structure.

In this work, we propose a new approach based on the use of the fs laser micromachining process, enabling damage-free and efficient processing of the outer cladding of nodeless ARHCFs. To the best of our knowledge, this is the first attempt to utilize fs laser pulses to access the air core of nodeless ARHCFs via side-drilled microchannels. The developed process does not affect the cladding capillaries that define the low-loss guidance properties of this particular fiber. Such a modification in the outer structure of the ARHFC fiber, as shown in the literature [21–23], should not introduce additional transmission losses since the outer cladding has a negligible impact on the light confinement in the air core. With a properly optimized laser-micromachining process, the gap between the capillaries can be used to directly access the hollow region of the microstructured fiber. The key element of the proposed method is the precision of laser processing and proper selection of process parameters in order to significantly reduce the formation of the debris that can accumulate in the air core of the fiber. Minimization of the debris is crucial as it significantly influences the guidance properties of ARHCFs. Here, we will compare two approaches of fabricating lateral channels in ARHCFs in terms of amount of debris, channel surface quality, and processing time. The influence of the microchannel dimensions and bending of the processed fiber will be experimentally verified.

2. Materials and Methods

2.1. Antiresonant Hollow-Core Fiber

The ARHCF used for the experiments was fabricated from high purity fused silica glass (Suprasil F300) using the stack and draw technique [24]. The outer diameter of the ARHCF was D_a ~310 µm, and the thickness of the solid outer cladding was t_s ~54 µm. The hollow core of the ARHCF with a diameter of D ~87.6 µm was defined by the fiber cladding consisting of seven non-touching, circular capillaries with a diameter of d ~54 µm. The distance between the capillaries was between 10 and 12 µm, due to irregularities during the drawing process. The ARHCF was covered with an acrylate polymer layer with a thickness of ~120 µm to minimize mechanical damage factors. Figure 1 shows a cross-section of the fiber used in the experiments (the polymer coating was removed for the fiber cutting process).

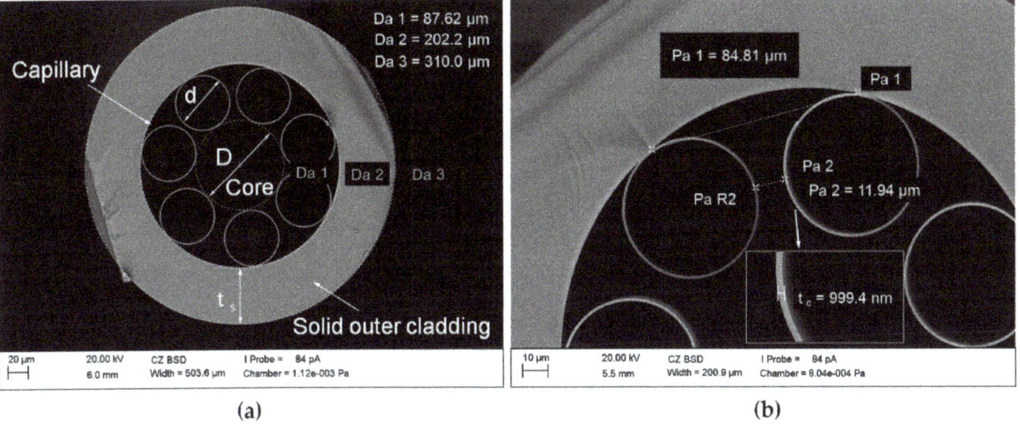

Figure 1. Scanning Electron Micrograph (SEM) cross-sectional image of the ARHCF: (**a**) view of the ARHCF with an air core of 87 µm and (**b**) view of the capillaries forming the fiber cladding.

According to the ARROW (Antiresonant Reflecting Optical Waveguiding) model [25], the guided wavelength range in an ARHCF is mainly defined by the thickness of the core-wall (capillary). For the fabricated fiber, the thickness of the core-wall was t_c ~1 µm, which enabled low-loss transmission of light in both near- (NIR) and mid-infrared (MIR) spectral bands, in the vicinity of 1.5 and 3.34 µm, respectively [12].

2.2. Methodology

The method of fabricating the microchannels in the outer cladding of the fiber is shown in Figure 2. The developed laser micromachining process allows for non-contact and damage-free processing of the fragile fiber structure with the required high accuracy and precision. At the point of impact of a high-energy fs laser beam, the material evaporates, leading to the formation of a crater, the depth of which depends on the selected process parameters, i.e., pulse duration, repetition rate, scan speed, pulse energy, and the number of consecutive scans across the processed area. The key element of the method was the precise positioning of the fiber in such a way that the laser beam moved in the area between the inner capillaries, which in our case were 10–12 μm apart. For this reason, it was necessary to control the position of the fiber, which was realized using a CMOS (Complementary Metal-Oxide-Semiconductor) camera and precise motorized XYZ translation stages and theta rotation.

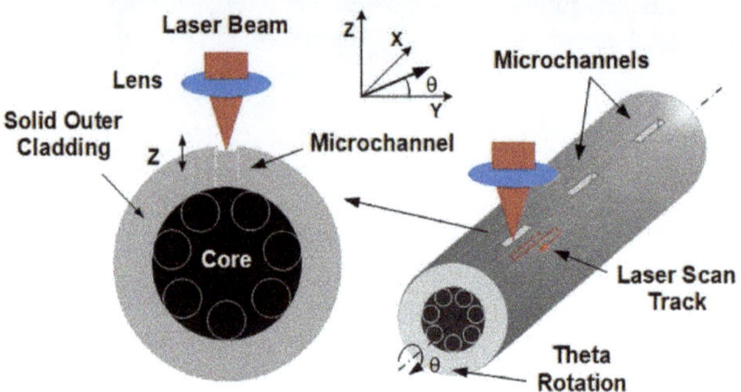

Figure 2. Methods used to process the lateral channels in the outer cladding of the ARHCF.

The critical point in the laser processing of such fibers is obtaining a contamination-free microchannel in the outer cladding of the fiber. There is a risk of contamination of the air core by debris generated during the process, which is composed of the glass material that did not evaporate during the process. Any damage to the inner capillaries should also be avoided, as it directly leads to local changes in the resonance properties of this fiber. For this reason, the cutting process should be performed layer by layer so that debris does not penetrate the core and no damage to the capillaries is present.

2.3. Experimental Setup and Procedure

The experiments were carried out using a TruMicro 2030 laser (Trumpf Laser GmbH, Schramberg, Germany), generating fs pulses with adjustable duration from 270 fs to 20 ps, at a wavelength of 1030 nm, an average power of 20 W, and adjustable repetition rate from 200 kHz (100 μJ pulse energy) to 2 MHz (10 μJ pulse energy). Furthermore, the laser has a built-in "pulse picker" that allows operation at selected frequencies (>1 kHz). In order to develop the most effective way of making the lateral cuts in terms of quality and time of implementation, two configurations of delivering the laser beam to the material surface were investigated. In the first approach, shown in Figure 3a, the laser beam, after passing through the beam expander, was directed to the biaxial galvanometer scanners (model IntelliSCAN 14, ScanLab, Munich, Germany) controlled by the TruTops PFO software (Trumpf Laser GmbH, Schramberg, Germany), which was used for fast and precise positioning of the mirrors to deflect the laser beam. The beam ($1/e^2$ diameter of 10 mm) leaving the scanner was focused on the sample using an F-theta lens with a focal length of f = 56 mm, producing a focused spot with $1/e^2$ diameter of ~12 μm. Galvanometer scanners in combination with the F-theta lens allowed for rapid scanning of the surface of the processed material (with speeds of m/s or greater), which significantly reduced the time of the process. To ensure precise positioning of the ARHCF with respect to the incident beam, the fiber was placed in a rotating holder (model CRM1, Thorlabs GmbH, Bergkirchen, Germany) mounted on biaxial motorized stages (model 8MTF-102, Standa Inc., Vilnius, Lithuania). Visual control of the process was achieved using a CMOS camera (ALPHA1080 Series, Hangzhou, China).

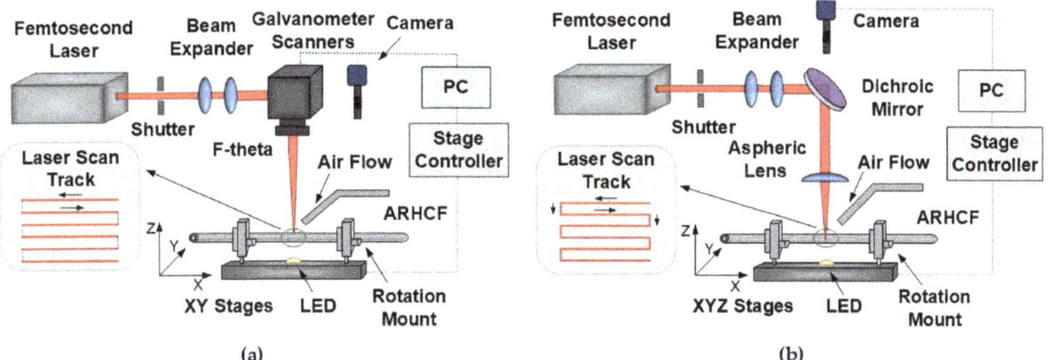

Figure 3. Schematic of the experimental setup for: (**a**) a system with galvanometer scanners and an F-theta lens and (**b**) a moving field and an aspherical lens.

The second variant of laser micromachining is a system with a moving field of work as depicted in Figure 3b, where the beam, after passing through the beam expander, is focused on the fiber using an aspherical lens (model AL1210-B, Thorlabs GmbH, Bergkirchen, Germany) with a focal length of 10 mm and NA = 0.55, providing a focused spot diameter. ~2 μm ($1/e^2$). The fiber placed in a rotating holder (model CRM1, Thorlabs GmbH, Bergkirchen, Germany) embedded in the platform was moved in relation to the incident laser beam using precise motorized XYZ stages at a speed of ~1 mm/s. All three stages were driven by linear motors on the X-Y (model 8MTF-102, Standa Inc., Vilnius, Lithuania), Z axis (model 8MT175-50, Standa Inc., Vilnius, Lithuania), and the positioning resolution for each axis was 0.31 μm. The advantage of this system is the possibility of using the CMOS camera on the same axis as the laser beam, which enables real-time monitoring of the laser ablation processes.

To effectively reduce the occurrence of debris in the laser treated area for both configurations, experiments were conducted in a natural environment without the use of shielding gases except for the transverse airflow used to remove ablation residues.

The research work began with the precise alignment of the fiber in relation to the incident laser beam. For this purpose, a CMOS camera with a source of light transmitted through the fiber, situated perpendicularly to the top layer of the fiber, and a rotating holder were used. By focusing the camera on the surface of the inner capillaries and rotating the fiber with the rotating holder, the tangent points of the inner capillaries to the outer cladding were controlled so that their positions were on the same level. Thus, the space between the capillary pair was centrally located just below the focus area of the laser beam, as shown in Figure 2. In the next step, the location of the focus of the laser beam in relation to the fiber surface was determined. This was achieved by intentionally positioning the fiber far below the focal length of the lens and subsequently moving the fiber mounting platform on the Z axis upward in 2.5 μm steps. During this process a single line scan across the ARHCF was performed at low pulse energy, until the effects of laser ablation were observed (polymer coating was locally mechanically removed using a scalpel before the laser processing). After the location of the focal point was defined with respect to the position of the fiber, subsequent experiments were conducted. The laser cutting is a complex process dependent on several variable parameters related to both the laser source (i.e., pulse energy, repetition rate, average power) and the process itself (e.g., scanning speed, distance between lines—hatching, multiplicity of the process, or a change of the lens focus position in relation to the material). We chose manufactured microchannels with dimensions 250 × 30 μm, which we found to be optimal for the ARHCF used in the experiments (due to the distance between the capillaries of ~12 μm as shown in Figure 1).

In order to define the optimal parameters for the process of cutting a microchannel and the pattern of scanning the surface, a series of systematic experiments were carried out for each setup configuration. A summary of the individual parameters used during the experiments is presented in Table 1. Additionally, we investigated the possibility to remove the protective polymer layer from the fiber with the use of fs laser pulses for both techniques.

Table 1. Laser and process parameters used for optimization of laser cutting of the microchannel.

Parameter	F-theta Lens	Aspheric Lens
Wavelength (nm)	1030	
Pulse duration (fs)	270	
Spot size ($1/e^2$) (μm)	~12	~2
Pulse Energy (μJ)	6–14	1–6
Repetition ratio (kHz)	20–1000	1–200
Speed (mm/s)	50–1000	0.5–5
Z move step (μm)	2.5–20	0.5–10
Number of cycles N	1 per step × 10–× 50	1 per step -
Hatching line (μm)	1–10	0.5–2.5

After laser treatment, the macroscale analysis of the quality of the microchannels was performed using a digital microscope (model VHX 5000 Keyence, Osaka, Japan) with transmitted light illumination of the sample. Detailed analysis was carried out using the scanning electron microscope (SEM, model EVO MA 25, Zeiss, Oberkochen, Germany). To minimize the charging effect, the fiber was coated with a ~5 nm gold layer using a sputter coater (model Quorum Q150R ES, Lewes, UK). Moreover, the fiber was placed on top of a conductive carbon strip attached to a grounded sample holder.

To investigate the influence of the fabricated microchannels on transmission losses we used a custom-built difference frequency generation-based optical frequency comb source (DFG COMB) operating within the fundamental transmission band of the fiber (3.3–3.4 μm). The broadband comb source was coupled into the ARHCF, and the losses were estimated by measuring the difference in the exiting optical power before and after laser processing of the microchannels. Measurements were taken with the aid of a digital optical power meter (PM100D with high-resolution thermal power sensors S401C, Thorlabs GmbH, Bergkirchen, Germany). This setup was also used to measure the ARHCF bending losses (see Section 3.3 for further details).

3. Results

Figure 4 presents the selected examples of laser cutting of microchannels in the solid outer cladding of the ARHCF. Photographs of the fabricated microchannels were compared in five sections according to different process parameters used during the laser ablation process. The main criterion for assessing the laser cutting process was the quality of the cut and the efficiency of the process. Qualitative analysis was carried out on the basis of the number and size of microchannels, the occurrence of microcracks in the processing region of the material and the amount of debris generated during the process. Comparison of the quality of the laser-processed microchannels has indicated that the structures made with the use of the F-theta lens are characterized by greater debris in the area of laser treatment, both on the surface and in the air core of the ARHCF. On the other hand, the aspherical lens-aided process introduced significantly less debris and provided a higher quality of the machined features in the fiber structure. In addition, incisions made with the F-theta lens exhibited microcracks that resulted from the generation of high stresses due to the interaction of the high-energy laser beam over a much larger area than in the process where an aspherical lens was used. When analyzing the execution time for the microchannels fabricated by using the F-theta lens, significantly shorter time was achieved (less than one minute) as compared to the process completed using an aspherical lens, where the average time of a laser cutting was in the range between 4 and 15 min per microchannel (depending on the parameters used). Such large differences in the processing time resulted from different diameters of the laser beam and the speed of the processes used.

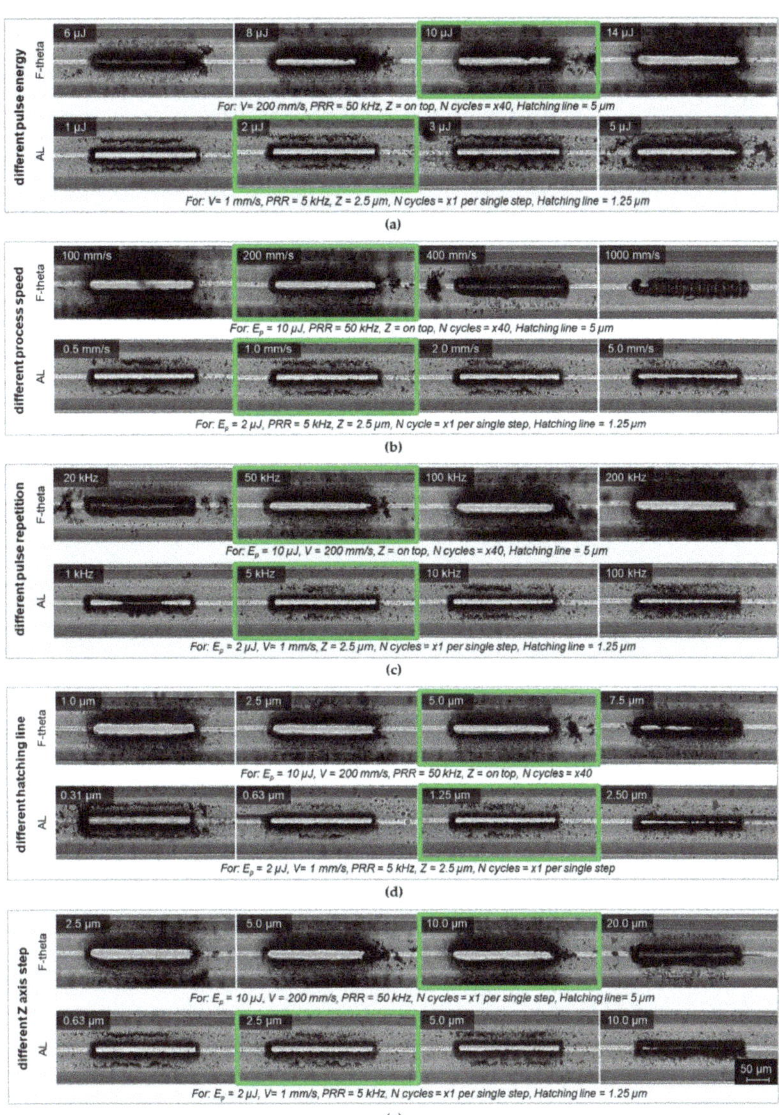

Figure 4. Microscope photographs of microchannels fabricated with the aid of a laser using a galvanometer scanner with an F-theta lens—F-theta and a setup based on a moving field and an aspherical lens, AL. Results are presented for varying parameters: (**a**) pulse energy, (**b**) scanning speed, (**c**) pulse repetition frequency, (**d**) different hatching lines, and (**e**) the Z-axis step of the beam focus position.

On the basis of the microscope photographs analysis, the process parameters giving the best results are indicated in green. For the AL-based configuration the optimal parameters were: pulse energy 2 µJ, speed 1 mm/s, repetition frequency 5 kHz, distance between lines 1.25 µm (hatching) and increment along the Z axis with a 2.5 µm step. With these parameters, a single microchannel was fabricated with dimensions equal to 250 × 30 µm in 7 min. For the F-theta configuration, the following parameters were found to be optimal: pulse energy 10 µJ, speed 200 mm/s, repetition ratio 50 kHz, distance between lines 5 µm (hatching) and increment in the Z axis with a step of 10 µm. The optimal parameters result in a total processing time equal to 30 s. After optimization of the processing parameters, both methods of fabricating the microchannels resulted in acceptable quality. The air core of the ARHFC was exposed with the desired microchannel cross-section, no capillaries were damaged, no microcracks in the outer cladding were present, and low accumulation of debris was observed inside the fiber.

Figure 5 shows the results of the fabrication of the microchannels in the ARHCF using both the AL and F-theta setup for the optimized parameters. In both cases, unobstructed microchannels were created along their entire length. The SEM images shown in Figure 5b clearly show that the AL approach delivers a superior overall quality of the fabricated microchannels compared to the F-theta lens-based approach. It can be seen that the amount of debris accumulated in the laser cutting area is negligible, and the edge of each microchannel is very sharp with no visible chipping of the glass material. This is due to the fact that the processing of materials using a small diameter laser beam (~2 µm $1/e^2$) allows the material to be removed in a more controlled manner—layer by layer with a small increment in the Z axis, however, at a cost of increased process duration. A microscope photography of the fabricated microchannel taken with visible light illumination clearly shows the differences in the quality of the fabricated features.

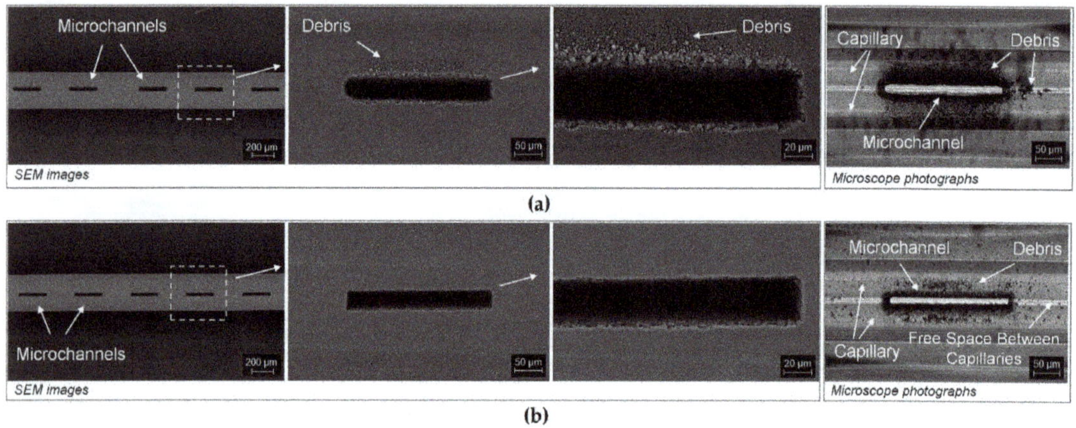

Figure 5. SEM images and microscope photographs of fabricated microchannels using fs laser pulses: (**a**) for the configuration with the F-theta lens and (**b**) for the AL configuration.

Figure 6 shows the results of the experiments during which the microchannels were fabricated without prior removal of the outer polymer layer. This approach eliminates the requirement of stripping the polymer coating of the fiber using, e.g., a scalpel blade, which is not straightforward and can easily damage the fiber, especially in the case when several tens of meters long fiber need to be laser-processed in the middle of its length. For both processing setups, the polymer layer was successfully laser-removed during the process before ablating the glass-based outer clad of the fiber. This method of selective laser removal of the polymer layer is superior when compared to preprocessing mechanical removal from a larger, neighboring area. The laser-based method is ablating the polymer directly above the microchannel area, which does not weaken the overall structure of the fiber as much as pre-processing removal of the fiber. The parameters used for polymer removal with the fs laser were identical as those for the microchannel cutting, except that for the AL method the number of steps on the Z-axis was increased to fully remove the polymer layer. Analyzing the results obtained for the F-theta lens, a much more efficient removal of the polymer was observed (Figure 6a) than for the AL system. This is a direct result of using a larger beam, which resulted in evaporation of the material from a larger surface for a single laser pulse. Moreover, it was observed that the exposure of the outer optical cladding in a small area causes disturbance of the air flow used to remove the free fractions of the debris from the site of interaction of the laser beam with the fiber (polymer layer thickness ~120 μm), which is visible in the case of the AL technique (Figure 6b). As a result, the debris that arose during the polymer coating removal was deposited in the crater formed during laser cutting, effectively blocking the laser ablation process and directly leading to the partial free-flow fabricated microchannel. The use of vertical air flow in this case would mitigate the problem, but there is a risk of blowing the generated debris into the ARHCF core, which would have a negative impact on the transmission properties of the fiber, in the worst case causing severe losses. Therefore, the proposed solution to this problem is based on material removal from a larger area to ensure improved external air flow at the bottom of the microchannel being processed.

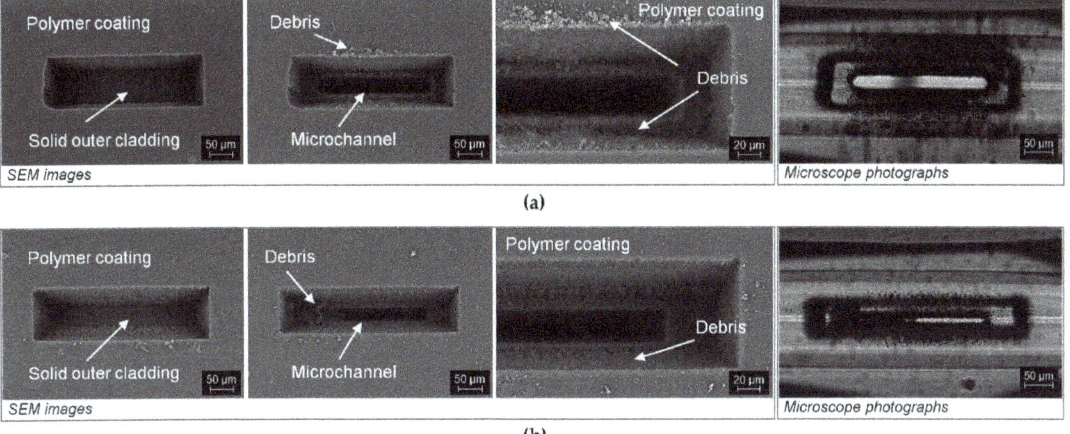

Figure 6. SEM images and microscope photographs of the polymer coating removal process and the laser microcutting of the microchannel realized in one process: (**a**) with the F-theta configuration and (**b**) with the AL configuration.

3.1. Influence of Laser Microchannel Processing on the ARHCF Transmission Characteristic

To verify to what extent the fabricated microchannels affect the transmission properties of the ARHCF, an experiment was performed in which a set of microchannels were manufactured with selected processing parameters. In this stage, a 30 m long ARHCF was used and 5 sections with 5 microchannels each (250 × 30 μm) were manufactured. The distance between the microchannels was 1 mm and the separation between the sections was equal to 10 cm. A diagram of the experimental setup is presented in Figure 7a. A custom-built DFG COMB was used as a broadband light source operating in the 3.3–3.4 μm wavelength region, matching a part of the low-loss transmission band of the ARHCF. The loss introduced by the fiber modification was calculated by comparing the optical power delivered by the unprocessed fiber and the fiber with laser-machined microchannels. First, the fiber-delivered power was measured for the ARHCF with 5 sections of microchannels processed at its end. Subsequently, one section of the microchannels was cut off from the fiber using a ceramic blade, and the measurement of the delivered optical power was performed again. This procedure was repeated for each section until the entire processed part of the fiber was removed. Note that during the measurements the input light coupling conditions were maintained the same for each step, so the average optical power of the DFG COMB radiation coupled into the ARHCF core was constant. The results obtained are presented in Figure 7b.

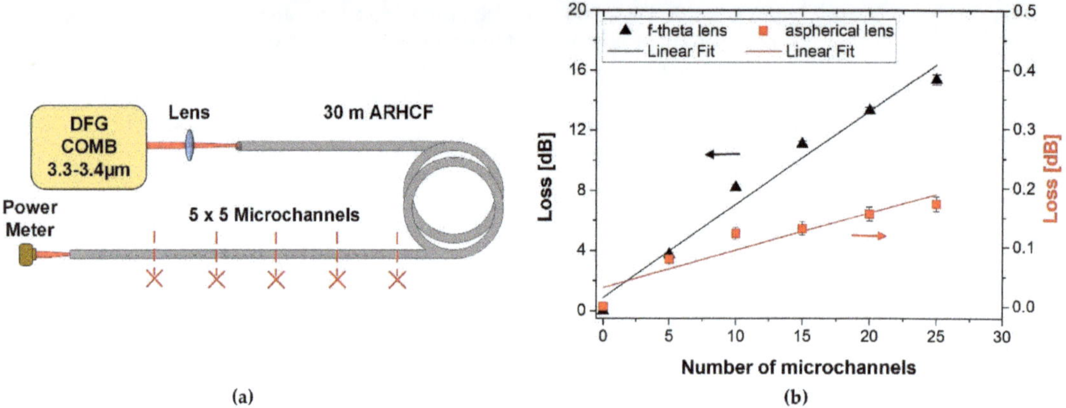

Figure 7. Influence of microchannel fabrication on transmission loss in the ARHCF for wavelengths 3.3–3.4 µm: (**a**) experimental setup and (**b**) loss measurement calculated as a function of the number of microchannels.

The microchannels produced with the F-theta lens introduced significantly greater transmission losses (black triangle) compared to the lateral cutting made with the aspherical lens (red square). For example, for the 25 microchannels made with the F-theta lens, the loss was 15.49 dB, while for the aspherical lens it was only 0.17 dB. Based on theoretical calculations presented in [21–23], the introduction of lateral cuts in the outer cladding of the nodeless ARHCF should not lead to significant transmission deterioration since the outer cladding does not participate directly in guiding the light inside the core. To investigate the cause of the increase in loss, a detailed analysis of the fabricated microchannels was carried out.

Figure 8 shows photographs of randomly selected microchannels made with the aid of the aspherical lens (Figure 8a,b) and the F-theta lens (Figure 8c,d). In both cases, the microchannels were made in a region exactly above the free space between the inner capillaries, without damaging their structure. The photographs taken with visible light illumination show the presence of impurities inside the fiber within the manufactured microchannels only in the case of using the F-theta lens approach. By analyzing the SEM images of the fiber cross-sections (Figure 8b,d), we can conclude that for the microchannels made with the use of an aspherical lens, only a low degree of impurities occurs in the space between the outer cladding and the capillaries. For microchannels made with the F-theta lens, the amount of impurities inside the fiber is much greater (Figure 8c,d). The formation of large particles, which settle on the walls of the capillaries forming the core, is visible not only in close proximity to the lateral cut, but they also displace into the inner section of the air core, for example due to the pressure difference, which is shown in Figure 8c. The presence of such large fractions leads to deterioration of transmission in the fiber, due to scattering effects, attenuation of the glass particles at wavelengths longer than 2.5 µm and as a result of influencing the ARHCF guiding mechanism, which is directly dependent on the thickness of the capillaries. The formation of debris is directly related to the laser ablation process, during which the material is removed by evaporation. The condensation of vapors induced by the collision with cooler gas molecules from the environment leads to the deposition of the debris in the vicinity of the laser beam or on the walls of the formed crater. This leads to a deterioration in the effectiveness of laser ablation in the next cycle of the process [26–28]. Therefore, the increased amount of debris in the case of laser processing with the F-theta lens results mainly from the processed surface area (due to the relatively large focused spot size) from which the material is evaporated in a short time. With this in mind, further experiments were performed using an aspherical lens for which; due to almost an order of magnitude smaller focused beam diameter on the sample and the speed of the process, the amount of debris was significantly reduced.

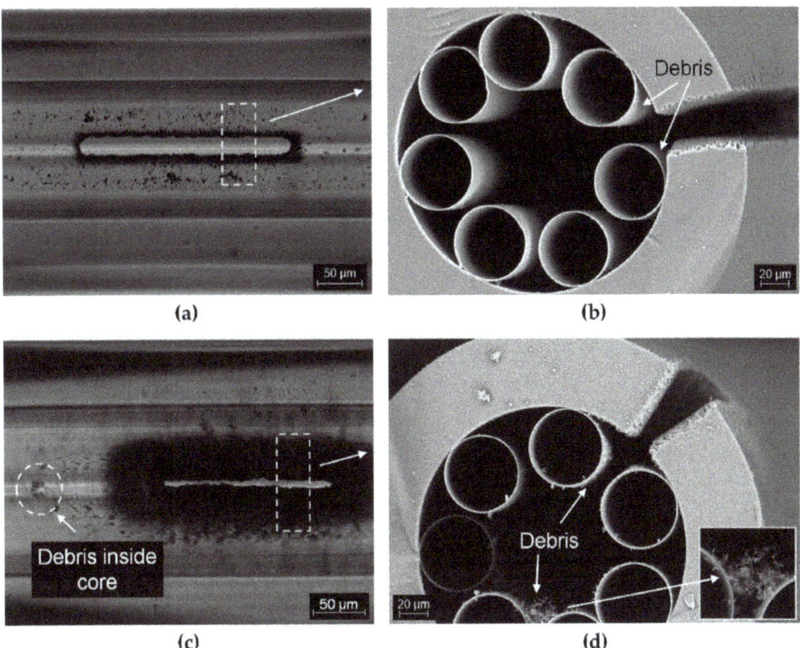

Figure 8. Images of microchannels processed in the ARHCF: (**a**) using an aspherical lens, (**b**) SEM cross-sectional image with debris in the vicinity of the microchannel made using an aspherical lens, (**c**) using the F-theta lens, and (**d**) SEM cross-sectional images with debris in the vicinity of the microchannel made with the F-theta lens.

3.2. Influence of Microchannel Length on the ARHCF Transmission

To determine the usefulness of the proposed method for modifying the ARHCF structure with the aid of the ultrafast laser micromachining process, we have investigated the possibility of cutting long microchannels with a length of up to ~2 mm. Here, 35 microchannels were made at the output end of a 15 m-long ARHCF (the same fiber as in the previous experiment) in 7 sections, each containing 5 microchannels with different lengths and fixed widths equal to 30 µm. The distance between the sections was 10 cm and the microchannels were separated by 1 mm. The polymer coating was removed locally, mechanically using a scalpel prior to laser processing. All microchannels were made using an aspherical lens for optimal process parameters, as defined at the beginning of this section. Figure 9a,b shows fabricated microchannels with lengths of 0.1, 0.25, 0.5, 0.75, 1, 1.5 and 2 mm. The quality of the fabricated microchannels was at the same level regardless of their length. In-depth analysis of the fabricated structures has shown that no damage occurred to the internal capillaries that form the actual fiber cladding. The impact of a microchannel length on the transmission properties of the ARHCF was determined using a similar approach to that previously described in Section 3.1. The optical power at the fiber output was measured after individual sections of the microchannels were cut off. The results are plotted in Figure 9c. It can be seen that microchannels with a length below

0.5 mm introduce very low transmission loss, not exceeding 0.1 dB. For microchannels with lengths greater than 0.5 mm, an increase in light attenuation was observed. The higher loss level observed in the case of longer microchannels was caused by the accumulation of debris in the vicinity of the processing area, which was pushed into the core by the air flow used to clean the processed region.

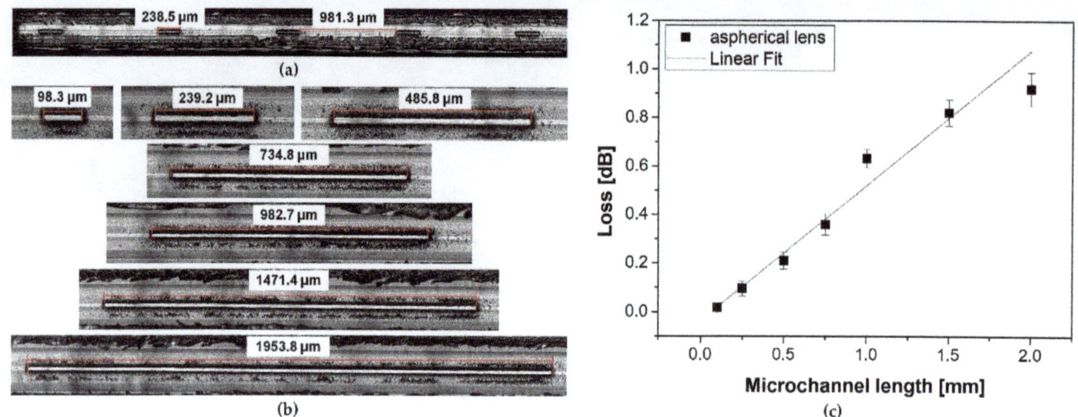

Figure 9. Additional loss of the ARHFC plotted as a function of the length of the microchannel: (**a**) view of a single section containing 5 microchannels with a length of 0.25 mm, (**b**) microchannels with lengths equal to 0.1, 0.25, 0.5, 0.75, 1, 1.5, 2 mm, respectively, and (**c**) loss vs. microchannel length.

3.3. Bending Performance of the ARHCF with Manufactured Microchannels

From an application point of view, an interesting aspect is determining to what extent the fabrication of lateral microchannels will affect the strength of the ARHCF and its transmission. For this purpose, an ARHCF with a length of 1 m was laser processed in its midsection to fabricate 5 microchannels with two different lengths of 100 and 250 µm and a width of 30 µm. Microchannels were produced with the use of an AL configuration and using the parameters defined at the beginning of Section 3. Figure 10a shows a schematic of the measurement system, where the fiber was bent at radii equal to 10, 15, 20 and 25 cm. Microchannels did not change their location relative to the plane during the shift of bending radius. The additional loss introduced by the modification of the fiber structure was defined by comparing the optical power transmitted through an unprocessed fiber and a fiber with the fabricated microchannels with different bending radiuses. The results are shown in Figure 10b.

The performed experiments indicated that for a 100 µm long microchannel (blue triangles), the bending-related loss is comparable to an unprocessed fiber for bending radius in the range between 15 and 25 cm. When the bending radius was reduced to 10 cm, a significant increase in loss (~9 dB) was observed, which was caused by a crack in the outer optical cladding within one of the microchannels. For 250 µm long microchannels (red dots), low-loss transmission was observed for a bending radius in the range of 20 to 25 cm. The lower bending radius resulted in permanent damage (fracture) to the fiber directly in one of the fabricated microchannels. This can be attributed to a significant weakening of the ARHCF structure in the processing region.

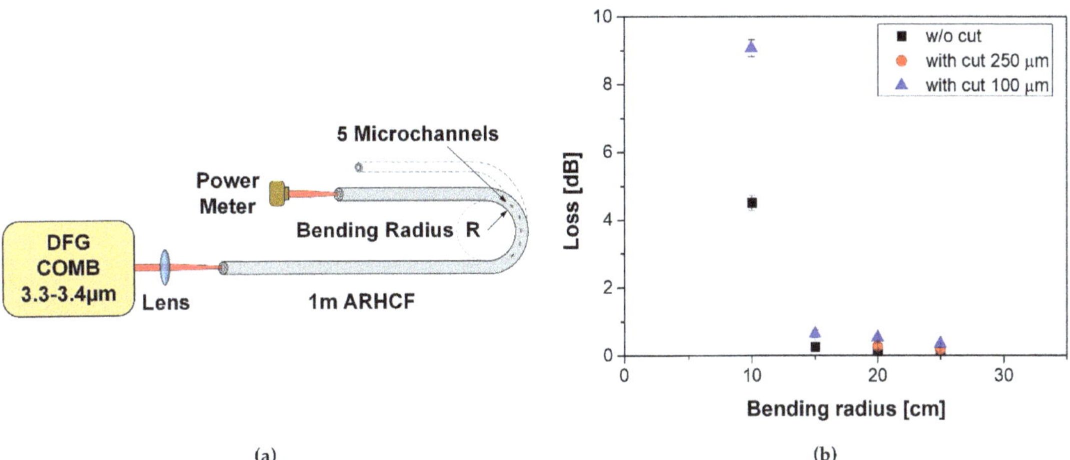

(a) (b)

Figure 10. Bending loss of the ARHCF with microchannels: (**a**) schematic of the measurement setup and (**b**) results of measurements plotted as a function of the bending radius.

4. Discussion

In this work, we have demonstrated the method of fabricating microchannels in an ARHCF based on the laser ablation process. Conducting detailed research on the strategy and process parameters allowed for reproducible fabrication of microchannels in the ARHCF, which introduced relatively low losses (<0.01 dB for a single microchannel). So far, in the literature, the lowest loss for a single microchannel made with an fs laser on an HCF fiber was 0.35 dB [17]. In comparison, this result is over 35 times worse than the result obtained using the method developed within our research. In another article, a 20 μm microchannel made by laser ablation introduced 1 dB loss [29]. In both cited cases, the holes were made in the hollow-core photonic bandgap fiber (HC-PBG). Due to the complex structure and the light-guiding mechanism in HC-PBG, during the fabrication of microchannels interference in the structure of the fiber is inevitable through microdefects that locally disturb the periodicity and symmetry of the fiber, leading to a deformation of the PBG structure, and hence increase in the transmission loss of the fiber. Recently, there was a report about the creation of a 150 μm long channel using an fs laser in a hollow core negative curvature fiber (HC-NCF), which in its structure is similar to the fiber used by us, however, its cladding structure is formed by a set of non-gapless capillaries. The channel fabricated in the HC-NCF introduced a 0.45 dB loss, which mainly resulted from the microdestruction of the cladding capillaries, which are responsible for confining the light into the air core of the fiber. Table 2 summarizes the best results obtained for single microchannels made in the laser ablation process in HCFs.

The fabrication of a microchannel that showed a low loss (<0.01 dB) partly results from the relatively non-complex structure of the fiber used by us. The mechanism of light guidance in ARHCF takes place within the air core limited by internal capillaries, and interference with the outer cladding by the fabrication of microchannels of different lengths does not have a significant impact on the optical performance of the fiber, which has already been described in detail in [30]. In our case, the gap between the capillaries constituted a natural part of the channel, for which it was not necessary to introduce changes to the fiber structure. Therefore, the key element of the process was the precision of guiding the laser beam so that, in the final stage of the implementation of the microchannel in the outer cladding, the internal capillaries would not be accidentally damaged. The second important factor was the amount of debris formed during the laser ablation process, which, after the process was conducted, was found both outside and inside the ARHCF. It was possible to reduce the problem related to formed debris to some extent by means of an appropriate correlation of the applied process parameters: scanning speed, radiation power and pulse repetition rate, or spot size of the laser beam. An additional advantageous element was the use of air blown parallel to the fiber plane during the process. In this way, the debris generated during laser ablation was to some extent blown off the surface of the treated fiber, without causing excessive accumulation and priming of the laser beam. Obtaining low losses for single microchannels fabricated in the ARHCF allows us in future research to use them in a greater number to significantly accelerate the diffusion of gas into its interior.

Table 2. Comparison of the losses induced by microchannels in HCFs.

Type of Fiber	Microchannel Type	Loss (Per Single Channel)	Ref.
HC-PBGF	V-shaped microchannel 20 µm diameter (on top cladding)	1 dB	[29]
SC-HF	V-shaped microchannel 20 µm diameter (on top cladding)	0.5 dB	[16]
HC-PBGF	Hole diameter 1.5 µm (on top cladding)	0.35 dB	[17]
HC-NCF	Slot x-axis—150 µm, y-axis across the fiber	0.45 dB	[20]
ARHCF	Microchannel size x-axis—250 µm, y-axis—30 µm	0.007 dB	[this work]

In most of the works, the fabricated microchannels were realized for relatively short sections of the fiber (up to several centimeters) and their analysis was limited only to the losses they introduced without addressing the issue of weakening the fiber structure and lowering mechanical resistance. In this work, it was shown that while maintaining precision and limiting the amount of generated debris, it is possible to fabricate microchannels from single µm to several mm. The height of the microchannels in this case was limited to the free space between the capillaries, which was 10–12 µm at the narrowest point. In addition, studies were carried out to determine to what extent the fabrication of microchannels affects the strength of the fiber and bending resistance, which for potential applications in the construction of long absorption cells or sensors is important in their miniaturization. It has been proven that when bending is required, it will be more advantageous to fabricate microchannels having a relatively short length, since they exhibit greater bending resistance due to the shorter length of the fiber subjected to modification.

5. Conclusions

We have demonstrated an efficient method of producing microchannels in a nodeless ARHCF using an fs laser. The development of optimal process parameters and a laser cutting strategy allowed the fabrication of microchannels in the outer cladding of the ARHCF without damaging the internal capillaries forming the antiresonant structure. The key element of the proposed technology was to carry out the process in a way that allowed us to minimize the formation of the debris that occurred during laser ablation. For a single microchannel, there was a slight increase in transmission losses (<0.01 dB), which is more than 35 times lower than reported so far [13]. Moreover, the implementation of microchannels of different lengths did not cause a significant increase in losses, which proves that in the case of ARHCF it is possible to make microchannels of any length, provided that the fiber is precisely positioned in relation to the laser beam. Additionally, it has been shown that making lateral incisions in the ARHCF weakens its mechanical resistance to bending. Coiling of the fiber in a small diameter is possible only in the case of short microchannels <0.1 mm, since for longer microchannels the bending forces can fracture the ARHCF in the area subjected to laser modification. The presented work shows that the proposed micromachining process can be used to manufacture microchannels in ARHCFs, thus allowing the construction of low-loss fiber-based absorption gas cells for applications in laser-based gas sensing.

Author Contributions: Conceptualization, P.K., P.J. and K.K.; methodology, P.K. and P.J.; formal analysis, P.K.; investigation, P.K., P.J., K.K. and G.D.; resources, F.Y. and D.W.; data curation, P.K.; writing—original draft preparation, P.K.; writing—review and editing, P.J. and K.K.; visualization, V.H.; supervision, M.L., J.K. and K.A.; project administration, M.L. and K.A.; funding acquisition, K.A. All authors have read and agreed to the published version of the manuscript.

Funding: This research was funded by the National Science Centre (NCN), grant number UMO-2018/30/Q/ST3/00809, the Polish National Agency for Academic Exchange, grant number PPI/APM/2018/1/00031/U/001, the National Natural Science Foundation of China (61935002, 61961136003), the Key Research Program of Frontier Sciences, Chinese Academy of Sciences (ZDBS-LY- JSC020) and the CAS Pioneer Hundred Talents Program; the Open Fund of the Guangdong Provincial Key Laboratory of Fiber Laser Materials and Applied Techniques (South China University of Technology). The laser source for the MIR range was supported by the Foundation for Polish Science within the First TEAM program co-financed by the European Union under the European Regional Development Fund (contract no. POIR.04.04.00-00-434D/17-00).

Institutional Review Board Statement: Not applicable.

Informed Consent Statement: Not applicable.

Data Availability Statement: Not applicable.

Acknowledgments: The Authors would like to thank A. Antończak for providing a system for laser micromachining and Aleksander Budnicki from Trumpf Laser GmbH for lending a femtosecond laser source.

Conflicts of Interest: The authors declare no conflict of interest.

References

1. Cui, R.; Dong, L.; Wu, H.; Li, S.; Zhang, L.; Ma, W.; Yin, W.; Xiao, L.; Jia, S.; Tittel, F.K. Highly Sensitive and Selective CO Sensor Using a 2.33 μm Diode Laser and Wavelength Modulation Spectroscopy. *Opt. Express* **2018**, *26*, 24318. [CrossRef] [PubMed]
2. Wan, F.; Zhou, F.; Hu, J.; Wang, P.; Wang, J.; Chen, W.; Zhu, C.; Liu, Y. Highly Sensitive and Precise Analysis of SF6 Decomposition Component CO by Multi-Comb Optical-Feedback Cavity Enhanced Absorption Spectroscopy with a 2.3 μm Diode Laser. *Sci. Rep.* **2019**, *9*, 9690. [CrossRef] [PubMed]
3. Dong, L.; Yao, Y.; Zhao, Y.; Cui, N.; Liu, S.; Zhang, H.; Zhang, Y.; Xu, Y. A High-Sensitive Methane Measuring Telemetry System Based on a Direct Optical Absorption Technique. *Laser Phys.* **2020**, *30*, 126201. [CrossRef]
4. Swinehart, D.F. The Beer-Lambert Law. *J. Chem. Educ.* **1962**, *39*, 333–335. [CrossRef]
5. Yu, R.; Chen, Y.; Shui, L.; Xiao, L. Hollow-Core Photonic Crystal Fiber Gas Sensing. *Sensors* **2020**, *20*, 2996. [CrossRef]
6. Krzempek, K.; Abramski, K.; Nikodem, M. Kagome Hollow Core Fiber-Based Mid-Infrared Dispersion Spectroscopy of Methane at Sub-Ppm Levels. *Sensors* **2019**, *19*, 3352. [CrossRef]
7. Jaworski, P.; Krzempek, K.; Dudzik, G.; Sazio, P.J.; Belardi, W. Nitrous Oxide Detection at 5.26 Mm with a Compound Glass Antiresonant Hollow-Core Optical Fiber. *Opt. Lett. OL* **2020**, *45*, 1326–1329. [CrossRef]
8. Li, J.; Yan, H.; Dang, H.; Meng, F. Structure Design and Application of Hollow Core Microstructured Optical Fiber Gas Sensor: A Review. *Opt. Laser Technol.* **2021**, *135*, 106658. [CrossRef]
9. Hoo, Y.L.; Jin, W.; Ho, H.L.; Ju, J.; Wang, D.N. Gas Diffusion Measurement Using Hollow-Core Photonic Bandgap Fiber. *Sens. Actuators B Chem.* **2005**, *105*, 183–186. [CrossRef]
10. Hoo, Y.L.; Jin, W.; Shi, C.; Ho, H.L.; Wang, D.N.; Ruan, S.C. Design and Modeling of a Photonic Crystal Fiber Gas Sensor. *Appl. Opt.* **2003**, *42*, 3509–3515. [CrossRef]
11. Ritari, T.; Tuominen, J.; Ludvigsen, H.; Petersen, J.C.; Sørensen, T.; Hansen, T.P.; Simonsen, H.R. Gas Sensing Using Air-Guiding Photonic Bandgap Fibers. *Opt. Express OE* **2004**, *12*, 4080–4087. [CrossRef]
12. Jaworski, P.; Kozioł, P.; Krzempek, K.; Wu, D.; Yu, F.; Bojęś, P.; Dudzik, G.; Liao, M.; Abramski, K.; Knight, J. Antiresonant Hollow-Core Fiber-Based Dual Gas Sensor for Detection of Methane and Carbon Dioxide in the Near- and Mid-Infrared Regions. *Sensors* **2020**, *20*, 3813. [CrossRef]
13. Challener, W.A.; Kasten, A.M.; Yu, F.; Puc, G.; Mangan, B.J. Dynamics of Trace Methane Diffusion/Flow Into Hollow Core Fiber Using Laser Absorption Spectroscopy. *IEEE Sens. J.* **2021**, *21*, 6287–6292. [CrossRef]
14. Wynne, R.M.; Barabadi, B.; Creedon, K.J.; Ortega, A. Sub-Minute Response Time of a Hollow-Core Photonic Bandgap Fiber Gas Sensor. *J. Lightwave Technol.* **2009**, *27*, 1590–1596. [CrossRef]

15. Karp, J.; Challener, W.; Kasten, M.; Choudhury, N.; Palit, S.; Pickrell, G.; Homa, D.; Floyd, A.; Cheng, Y.; Yu, F.; et al. Fugitive Methane Leak Detection Using Mid-Infrared Hollow-Core Photonic Crystal Fiber Containing Ultrafast Laser Drilled Side-Holes. In *Proceedings of the Fiber Optic Sensors and Applications XIII*; International Society for Optics and Photonics: Baltimore, MD, USA, 2016; Volume 9852, p. 985210.

16. van Brakel, A.; Grivas, C.; Petrovich, M.N.; Richardson, D.J. Micro-Channels Machined in Microstructured Optical Fibers by Femtosecond Laser. *Opt. Express OE* **2007**, *15*, 8731–8736. [CrossRef] [PubMed]

17. Hensley, C.J.; Broaddus, D.H.; Schaffer, C.B.; Gaeta, A.L. Photonic Band-Gap Fiber Gas Cell Fabricated Using Femtosecond Micromachining. *Opt. Express OE* **2007**, *15*, 6690–6695. [CrossRef] [PubMed]

18. Martelli, C.; Olivero, P.; Canning, J.; Groothoff, N.; Gibson, B.; Huntington, S. Micromachining Structured Optical Fibers Using Focused Ion Beam Milling. *Opt. Lett. OL* **2007**, *32*, 1575–1577. [CrossRef]

19. Amanzadeh, M.; Aminossadati, S.M. A New Approach for Drilling Lateral Microchannels in Photonic Crystal Fibres. In Proceedings of the 2016 IEEE Photonics Conference (IPC), Waikoloa, HI, USA, 2–6 October 2016; pp. 779–780.

20. Novo, C.C.; Choudhury, D.; Siwicki, B.; Thomson, R.R.; Shephard, J.D. Femtosecond Laser Machining of Hollow-Core Negative Curvature Fibres. *Opt. Express OE* **2020**, *28*, 25491–25501. [CrossRef] [PubMed]

21. Belardi, W. Design and Properties of Hollow Antiresonant Fibers for the Visible and Near Infrared Spectral Range. *J. Lightwave Technol.* **2015**, *33*, 4497–4503. [CrossRef]

22. Adamu, A.I.; Wang, Y.; Correa, R.A.; Bang, O.; Bang, O.; Bang, O.; Markos, C.; Markos, C.; Markos, C. Low-Loss Micro-Machining of Anti-Resonant Hollow-Core Fiber with Focused Ion Beam for Optofluidic Application. *Opt. Mater. Express OME* **2021**, *11*, 338–344. [CrossRef]

23. Hao, Y.; Xiao, L.; Benabid, F. Optimized Design of Unsymmetrical Gap Nodeless Hollow Core Fibers for Optofluidic Applications. *J. Lightwave Technol.* **2018**, *36*, 3162–3168. [CrossRef]

24. Yu, F.; Song, P.; Wu, D.; Birks, T.; Bird, D.; Knight, J. Attenuation Limit of Silica-Based Hollow-Core Fiber at Mid-IR Wavelengths. *APL Photonics* **2019**, *4*, 080803. [CrossRef]

25. Litchinitser, N.M.; Abeeluck, A.K.; Headley, C.; Eggleton, B.J. Antiresonant Reflecting Photonic Crystal Optical Waveguides. *Opt. Lett. OL* **2002**, *27*, 1592–1594. [CrossRef]

26. Matsumura, T.; Kazama, A.; Yagi, T. Generation of Debris in the Femtosecond Laser Machining of a Silicon Substrate. *Appl. Phys. A* **2005**, *81*, 1393–1398. [CrossRef]

27. Odachi, G.; Sakamoto, R.; Hara, K.; Yagi, T. Effect of Air on Debris Formation in Femtosecond Laser Ablation of Crystalline Si. *Appl. Surf. Sci.* **2013**, *282*, 525–530. [CrossRef]

28. Bulgakova, N.M.; Zhukov, V.P.; Collins, A.R.; Rostohar, D.; Derrien, T.J.-Y.; Mocek, T. How to Optimize Ultrashort Pulse Laser Interaction with Glass Surfaces in Cutting Regimes? *Appl. Surf. Sci.* **2015**, *336*, 364–374. [CrossRef]

29. Hou, M.; Zhu, F.; Wang, Y.; Wang, Y.; Liao, C.; Liu, S.; Lu, P. Antiresonant Reflecting Guidance Mechanism in Hollow-Core Fiber for Gas Pressure Sensing. *Opt. Express* **2016**, *24*, 27890. [CrossRef]

30. Song, P.; Phoong, K.Y.; Bird, D. Quantitative Analysis of Anti-Resonance in Single-Ring, Hollow-Core Fibres. *Opt. Express* **2019**, *27*, 27745. [CrossRef]

Article

Application of High-Speed Quantum Cascade Detectors for Mid-Infrared, Broadband, High-Resolution Spectroscopy

Tatsuo Dougakiuchi [1,*] **and Naota Akikusa** [2]

[1] Central Research Laboratory, Hamamatsu Photonics K.K., 5000 Hirakuchi, Hamakita-ku, Hamamatsu City 434-8601, Japan

[2] Laser Promotion Division, Hamamatsu Photonics K.K., 5000 Hirakuchi, Hamakita-ku, Hamamatsu City 434-8601, Japan; aki@lpd.hpk.co.jp

* Correspondence: tatsuo.dogakiuchi@crl.hpk.co.jp; Tel.: +81-53-586-7111

Abstract: Broadband, high-resolution, heterodyne, mid-infrared absorption spectroscopy was performed with a high-speed quantum cascade (QC) detector. By strictly reducing the device capacitance and inductance via air-bridge wiring and a small mesa structure, a 3-dB frequency response over 20 GHz was obtained for the QC detector, which had a 4.6-μm peak wavelength response. In addition to the high-speed, it exhibited low noise characteristics limited only by Johnson–Nyquist noise, bias-free operation without cooling, and photoresponse linearity over a wide dynamic range. In the detector characterization, the noise-equivalent power was 7.7×10^{-11} W/Hz$^{1/2}$ at 4.6 μm, and it had good photoresponse linearity up to 250 mW, with respect to the input light power. Broadband and high-accuracy molecular spectroscopy based on heterodyne detection was demonstrated by means of two distributed-feedback 4.5-μm QC lasers. Specifically, several nitrous oxide absorption lines were acquired over a wavelength range of 0.8 cm^{-1} with the wide-band QC detector.

Keywords: mid-infrared; quantum cascade detector; high-speed operation; heterodyne detection; high-resolution spectroscopy

Citation: Dougakiuchi, T.; Akikusa, N. Application of High-Speed Quantum Cascade Detectors for Mid-Infrared, Broadband, High-Resolution Spectroscopy. *Sensors* **2021**, *21*, 5706. https://doi.org/10.3390/s21175706

Academic Editors: Krzysztof M. Abramski and Piotr Jaworski

Received: 30 July 2021
Accepted: 18 August 2021
Published: 24 August 2021

Publisher's Note: MDPI stays neutral with regard to jurisdictional claims in published maps and institutional affiliations.

1. Introduction

Quantum cascade (QC) photovoltaic infrared photodetectors are based on intersubband transitions of electrons [1,2]. Unlike quantum-well infrared photodetectors [3], the active regions of QC detectors can be engineered with highly flexibility, as shown for QC lasers [4]. Consequently, a variety of active region designs have been reported that expand the operational wavelength range or improve the responsivity [5–7]. Hence, photoresponses in QC detectors have been demonstrated from the near- to the far-infrared wavelengths [8,9]. QC detectors also exhibit low-noise and high-speed. Due to their bias-free operation, dark currents induced by external voltages are absent, which is important for high detectivity without an elaborate cooling system. Regarding high-speed operation, an electron transit time of less than 1 ps was substantiated in a near-infrared QC detector by a time-resolved pump–probe measurement [10]. More recently, frequency responses of several tens of gigahertz and picosecond response times have been demonstrated in the mid-infrared (MIR) QC detectors [11,12]. The intrinsic short response times are determined by high-speed electron transport via sub-picosecond intersubband scattering processes. Low-noise, high-speed QC detectors could be key devices for high-speed MIR applications.

The MIR is a molecular finger print region and, thus, is very important in fundamental science, medicine, and industry. A large number of unique and strong absorption lines that correspond to fundamental vibrational modes for many molecules are in the MIR, and they can be used to identify and quantify specific molecules. In particular, laser absorption spectroscopy (LAS) in the MIR is a powerful tool for high-precision and high-sensitivity molecular sensing because of the strong absorption of narrow-linewidth laser light [13–16]. Its most prominent application is the detection and identification of gas-phase molecules, because significantly high sensitivity can be obtained by long-range propagation of a collimated beam in a gas. The atmospheric windows in the MIR enable environmental measurements in free space [17,18]. As multiple absorption lines of several gases can overlap in a specific spectral region, high-resolution, high-sensitivity broadband is always required in LAS.

Heterodyne detection is an established technique to obtain high-resolution in LAS [19], and by using a high-speed detector, broadband measurements can be achieved simultaneously. In the MIR application field, HgCdTe (MCT) detectors are the most widely used photodetectors because of their high-responsivity and broad responsive wavelength covering a few micrometers [20], unlike that of QC detectors. However, in MCT detectors, a high-speed operation of several tens gigahertz is difficult in principle and not adequate for broadband heterodyne detection. On the other hand, the frequency range of a heterodyne beat signal up to 20–30 GHz, the upper limit of practical processers, corresponds to the wavelength tuning range of ~1 cm^{-1}. In this point of view, the narrow response spectrum of QC detectors is not concern for heterodyne spectroscopy and is a preferable property to avoid background noises. Here, we demonstrate broadband heterodyne LAS by using QC lasers and a QC detector over a spectral range centered at 4.5 μm. To enhance the high speed of the QC detector, we reduced the parasitic capacitance and inductance with air-bridge wiring and a small mesa structure for the thick active region constructed from 90 cascade modules. A 3-dB cutoff frequency was measured for over 20 GHz, and the wide-band frequency response guaranteed a 0.8 cm^{-1} broadband spectral range for heterodyne spectroscopy. Several absorption lines of nitrous oxide (N_2O) were observed over the range 2220.59–2219.76 cm^{-1} with 5 MHz resolution.

2. Characterization of a Quantum Cascade Detector

A schematic of the conduction band of the QC detector with a coupled-quantum-well design is shown in Figure 1a [7,11,21]. The response wavelength was determined by the energy separation between levels 7 and 1 (E_{71} = 289 meV). An incident light that has an energy corresponding to E_{71} associated with the electron excitation was absorbed. Due to the asymmetric conduction-band potential, the excited electrons were transferred in a preferential direction in line with the step-like energy levels formed in the sequential quantum wells. In the coupled-quantum-well design, the center of the wavefunction for the upper absorption level 7 was slightly shifted from that of level 1 to the thin well side. Consequently, longer longitudinal optical phonon scattering times (τ_{71}~3 ps) were obtained, while the dipole length (d_{71}~1.1 nm) remained almost unchanged. Additionally, reverse currents caused by electron transitions from levels 6–4 to 1 were prevented by the spatial separation via the thin well between the absorption and transport regions. As the level 7 wavefunction extended to the transport region and overlapped with that of level 6, photoexcited electrons could escape from the absorption region across the thin well and thick barrier.

All of the layer structures consisting of 90 cascade modules were grown on a semi-insulating InP substrate via metal–organic vapor-phase epitaxy. The wafer was processed into a 25-μm-wide mesa stripe and cleaved to a 100-μm length. The cleaved facet was used as the acceptance surface for strong absorption of the incident light propagating along the stripe direction. Both the thick active region of the 90 cascade modules and the narrow 25 μm × 100 μm mesa were essential to reducing the parasitic capacitance. Furthermore, to cut the device inductance, air-bridge wiring was used for electrical connection to the signal-output electrode. The device capacitance was 0.19 pF, as determined with a C-meter (4280A, Hewlett-Packard, Palo Alto, CA, USA), and the inductance was 0.21 nH, as calculated from the geometry of the air-bridge wiring [11]. The 3-dB cutoff frequency was estimated to be ~23 GHz, based on an equivalent circuit model, including the 1-ps ultimate electron transition time across one cascade module [11]. The experimental confirmation of the frequency response of the QC detector is shown later.

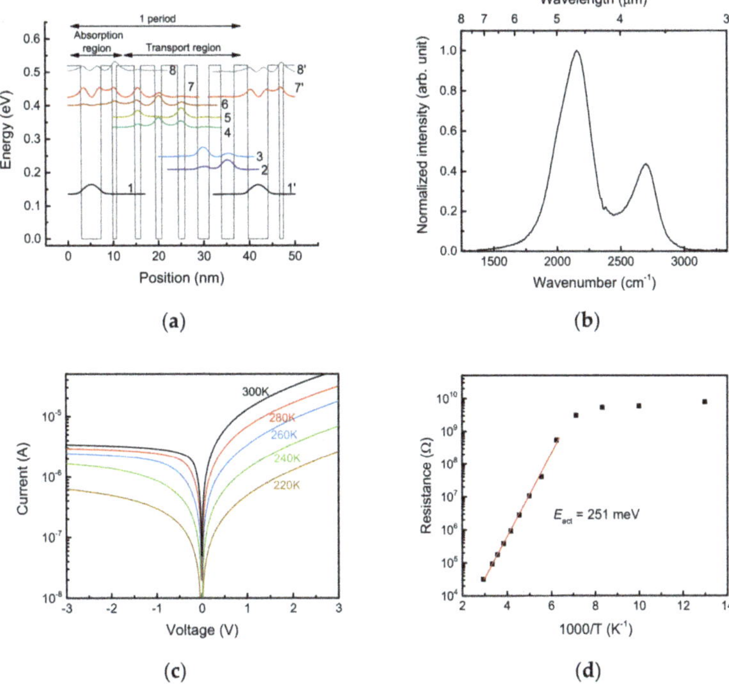

(a)

(b)

(c)

(d)

Figure 1. (a) Schematic of the conduction band and moduli squared of the relevant wave-functions of the quantum cascade detector. The $In_{0.53}Ga_{0.47}As/In_{0.52}Al_{0.48}As$ layer sequence of one period of the active regions, in angstroms, starting from the absorption well, is **44**/25/**9**/40/**13**/33/**14**/36/**15**/28/**25**/27/**28**/30, where InAlAs barrier layers are in bold, InGaAs quantum-well layers are in roman type, and the doped layer (Si, 4×10^{17} cm^{-3}) is underlined. (b) Photoresponse spectrum of the device measured without cooling. (c) Dark current–voltage characteristics over the temperature range 220–300 K. (d) Arrhenius plot of the measured differential resistance. The red line is the fit of the Arrhenius model, $L = A\exp[E_{act}/k_B T]$, where L is lifetime, A is a constant, E_{act} is the activation energy, k_B is Boltzmann's constant, and T is the device temperature.

Figure 1b shows a response spectrum of the uncooled QC detector obtained with a Fourier-transform infrared spectrometer (Nicolet 8700, Thermo Fisher Scientific, Waltham, MA, USA). The peak response wavelength was 4.6 μm (2160 cm^{-1}, 267 meV), and E_{71} was less than the calculated value of 287 meV. The reason for this difference was because of the insufficient band offset for the higher 7 and 8 energy levels. The energy difference for the second peak, which appeared at a shorter wavelength region, was ~30 meV because of weaker quantum confinement near the top of the barrier height. This difference between design and experiment could be reduced by applying a strain-compensated condition in the InGaAs/InAlAs or by using other wide-bandgap materials, as in shorter-wavelength QC detectors [8,10].

The dark voltage–current characteristics measured at various temperatures over 220–300 K, with 20 K intervals, are shown in Figure 1c. An asymmetric behavior originating from the band structure was observed. Figure 1d is an Arrhenius plot of the differential resistances at zero bias. The estimated activation energy of the device was 251 meV, which corresponded to the transition energy between E_{61} (264 meV) and E_{51} (228 meV). This indicated that the dark current induced by the transition from the ground level 1 to the 5–2 levels in Figure 1a was suppressed in the coupled-quantum-well design.

Figure 2a plots the output photocurrent as a function of the incident light power. The incident light was the 2220 cm^{-1} continuous wave (CW) distributed-feedback QC laser described below. The photoresponse exhibits good linearity with the incident light power up to 250 mW; the slope derived from the linear fit was estimated to be 4.7 mA/W for the specific wavelength of 2220 cm^{-1} (without compensation for the coupling losses from the focusing lens, surface reflections of the optics, and the cleaved facet acceptance area of the QC detector). Simultaneously, the peak responsivity at 2160 cm^{-1} was 5.7 mA/W, determined from the ratio of the signal intensities between the two wavelengths in the response spectrum in Figure 1b. Figure 2b plots the current noise power spectrum density of the QC detector at room temperature, obtained with a low-noise current amplifier (LCA-40K-100M, FEMTO, Berlin, Germany) and an audio analyzer (SR1, Stanford Research Systems, Sunnyvale, CA, USA). At frequencies higher than 100 Hz, the measured flat noise level matched the calculated Johnson–Nyquist noise level for an 89-kΩ device resistance. Due to the excellent low noise in bias-free operation, the detectivity was improved despite the low responsivity relative to other MIR detectors. The calculated noise-equivalent power was 7.7×10^{-11} W/Hz$^{1/2}$, with a peak responsivity and flat noise level of 4.4×10^{-13} A/Hz$^{1/2}$.

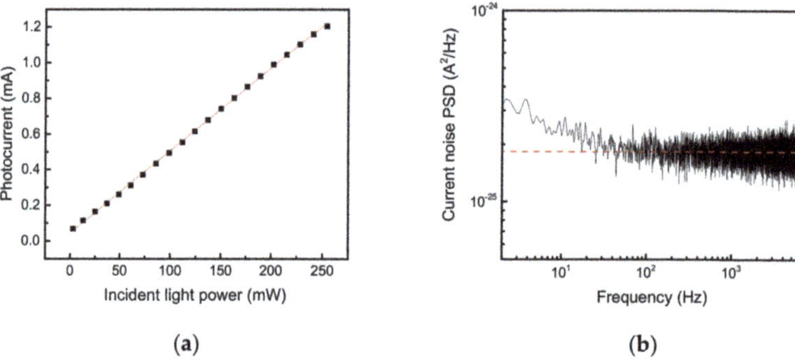

(a)   (b)

Figure 2. (a) Measured photocurrent vs. incident light power of the 2220-cm^{-1} distributed-feedback quantum cascade (QC) laser. The measured results are plotted as square black dots, and the red line is a linear fit. (b) Current noise power spectrum density of the QC detector at room temperature (no cooling). The red dashed line indicates the Johnson–Nyquist noise level calculated from $4k_{B}T/R$, where k_{B} is Boltzmann's constant, T is the device temperature, and R is the device resistance.

3. Broadband Heterodyne Spectroscopy

The high-speed QC detector was used to acquire broadband heterodyne absorption spectra of N_2O with the optical setup depicted in Figure 3. A heterodyne beat signal was generated with two identical 4.5 μm distributed-feedback QC lasers (L12004-2209H-C, HAMAMATSU PHOTONICS, Hamamatsu, Japan) [11,21,22]. The emissions from the lasers were collimated with aspheric lenses and combined in a beam splitter for collinear propagation and to focus on the acceptance surface of the QC detector via an optical isolator (MESOS optical isolator, Electro-Optics Technology, Traverse City, MI, USA). For use as a local oscillator, the wavelength of one QC laser (Fixed QC laser in Figure 3) was stabilized at a locked heatsink temperature and a fixed injection current supplied by a low-noise current driver (C16174-01, HAMAMATSU PHOTONICS, Hamamatsu, Japan). The emission wavelength of the "Tuned-QC laser" (Figure 3) was modulated with a ramp wave controlled by a function generator (FGX-2220, TEXIO, Yokohama, Japan). To observe narrow N_2O absorption lines, a multi-pass cell (2.4-PA, Infrared Analysis, Anaheim, CA, USA) with a 2.4 m optical path length was used in the beam path of the tuned QC laser. The pressure of the N_2O enclosed in the multi-pass cell was controlled with a vacuum gauge (not shown in Figure 3). The scanning beat signal was detected by the QC detector and accumulated in the spectrum analyzer. The N_2O absorption lines were observed as extinctions of the signal intensity associated with the wavelength modulation.

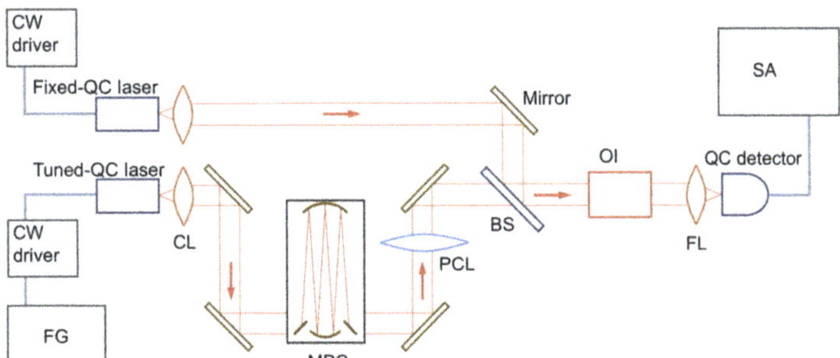

Figure 3. Schematic of heterodyne spectroscopy. FG: function generator, MPC: multi-pass cell, CL: collimating lens (ZnSe, aspheric, working distance of 3 mm), PCL: plano-convex lens (CaF$_2$, focal length of 300 mm), FL: focusing lens (ZnSe, aspheric, working distance of 1 mm), BS: beam splitter, OI: optical isolator, and SA: spectrum analyzer.

3.1. Control and Measurement of Beat Signals

Figure 4 shows the CW current–voltage–light output characteristics of the fixed and tuned QC lasers measured at locked heatsink temperatures of 28 °C and 27 °C, respectively. The temperatures were carefully determined for an expedient scanning range of the heterodyne beat signal that included several N$_2$O absorption lines, as described in the next section.

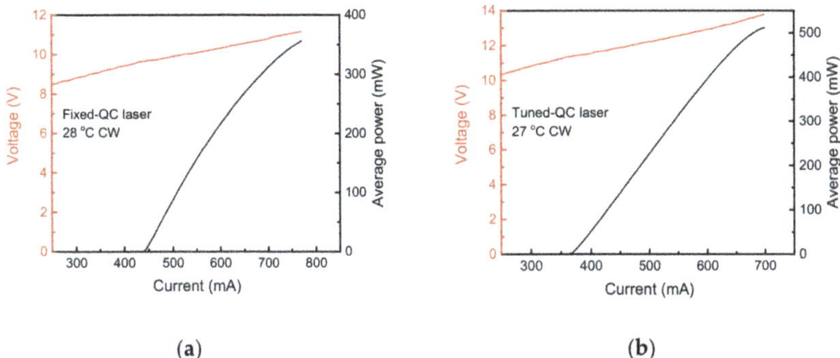

(a) (b)

Figure 4. Continuous-wave current–voltage–light output characteristics of (**a**) the fixed quantum cascade (QC) laser operated at the heatsink temperature of 28 °C and (**b**) the tuned QC laser operated at the heatsink temperature of 27 °C.

The injection current of the fixed QC laser was set to 625 mA to maintain the 2220.6 cm^{-1} emission wavelength, as shown in Figure 5a, where the Fourier-transform infrared spectrum was calibrated with a N$_2$O absorption line. To generate a wide-band beat signal, the injection current of the tuned QC laser was varied in a mode-hop-free manner, as exhibited in Figure 5a,b. The frequency of the heterodyne beat signal was tuned from 40 MHz to 26 GHz by varying the injection current of the tuned QC laser over the range 407–494 mA. The frequency of the heterodyne beat signal could be calculated from the difference in wavelengths of the fixed and tuned QC lasers. In Figure 5b, the beat frequency was thus calculated from the fixed and the varied wavelengths for a certain injection current, with good agreement with the measured beat frequencies. Hence, broadband heterodyne spectroscopy could be performed with the well-controlled beat signal over 25 GHz.

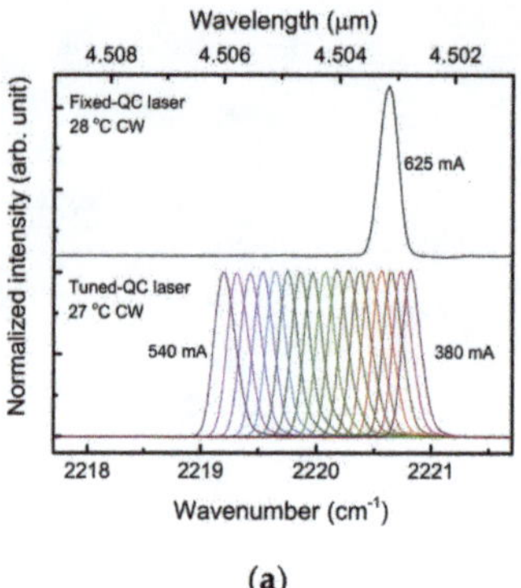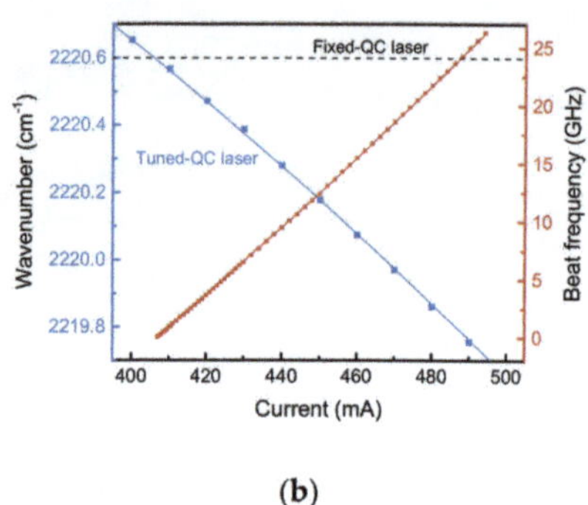

(a)

(b)

Figure 5. Emission wavelengths of the quantum cascade (QC) lasers and the frequency of the generated beat signal. (**a**) The spectra of the QC lasers depend on the injection current. The upper spectrum corresponds to the fixed QC laser with a fixed current of 625 mA, and the lower spectra correspond to the current-tuned QC laser. (**b**) The beat frequency (red square dots) measured with a QC detector and spectrum analyzer. The red line is a prediction of the beat frequency calculated from the wavelength difference of the fixed (dashed line) and tuned QC lasers (blue square dots). The blue line is the fit.

The frequency response of the QC detector shown in Figure 6 was measured with the heterodyne setup (Figure 3), without the multi-pass cell, and recorded in the max-hold trace mode of the spectrum analyzer (N9000B, KEYSIGHT TECHNOLOGIES, Santa Rosa, CA, USA). The result was normalized at 0 dB by using the average of the data below 3 GHz, and the signal level was maintained over 35 dB for the entire the frequency range. The theoretical curve based on an equivalent circuit model [11] is also exhibited in Figure 6. Due to the reductions in the parasitic capacitance and inductance, a 3-dB cutoff frequency over 20 GHz was confirmed. The experimental data differed with the theoretical curve at around 25 GHz because of a small impedance mismatch between the device and the measurement system in the high-frequency range. Such a behavior of the frequency response was possible to clear up with measurements over 30 GHz; however, this was limited by our instrument. The optimization of a QC detector for radio-frequency operation was reported in Ref. [12], where a well-designed coplanar waveguide was used to match the 50 Ω impedance, and no noticeable artifacts appeared in the frequency response up to 50 GHz. However, for broadband heterodyne spectroscopy, the variation of the beat signal intensity up to 25 GHz can be regarded as within the 3-dB cutoff.

3.2. Observation of N₂O Absorption Lines

Broadband heterodyne absorption spectroscopy of N_2O was demonstrated by adding 50 Pa of N_2O to the multi-pass cell. To reduce the data volume, the spectral acquisition was performed over three separate frequency regions that could be covered with the tuned QC laser while the wavelength of the fixed QC laser was constant. Each measurement region had a 2 GHz width and a 5 MHz resolution bandwidth (they are arbitrarily conditions specified with the spectrum analyzer), and the absorption spectra were recorded for four minutes in the max-hold trace mode of the spectrum analyzer. During data accumulation, the wavelength of the tuned QC laser was repeatedly modulated with the ramp current pulse at a 1 kHz repetition rate and an amplitude of 4 mA for the wavelength sweep.

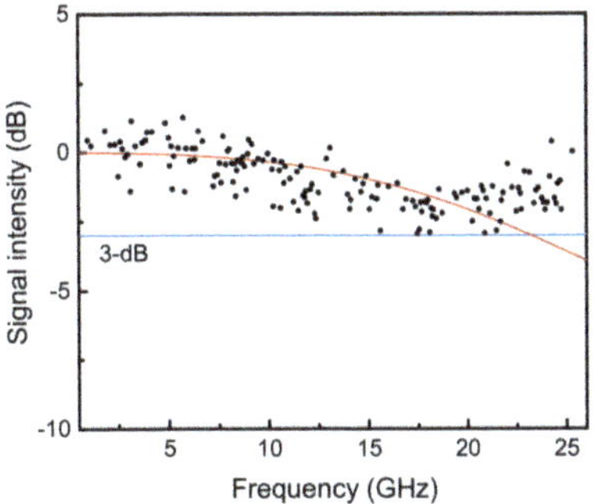

Figure 6. Frequency response of the quantum cascade (QC) detector observed in heterodyne beat measurement. The black dots are experimental data obtained by using the max-hold trace mode of the spectrum analyzer. The red curve is a theoretical prediction of the frequency response using the measured capacitance of 0.19 pF, the calculated inductance of 0.21 nH, and the input impedance of 50 Ω.

Figure 7 plots the heterodyne spectroscopy of the N_2O absorption lines. Figure 7a shows the entire measurement range, while Figure 7b–d were enlarged spectra. Each absorption signal was extracted by using the background obtained with the empty multi-pass cell. The black dots and red lines correspond to the experimental and calculated results, respectively (SpectraPlot [23,24]), where both were normalized by maximum values around 2220.4 cm^{-1}. The experimental absorption strengths, peak positions, and spectral widths in the three spectra agree well with the calculations in the spectral range of ~0.8 cm^{-1}.

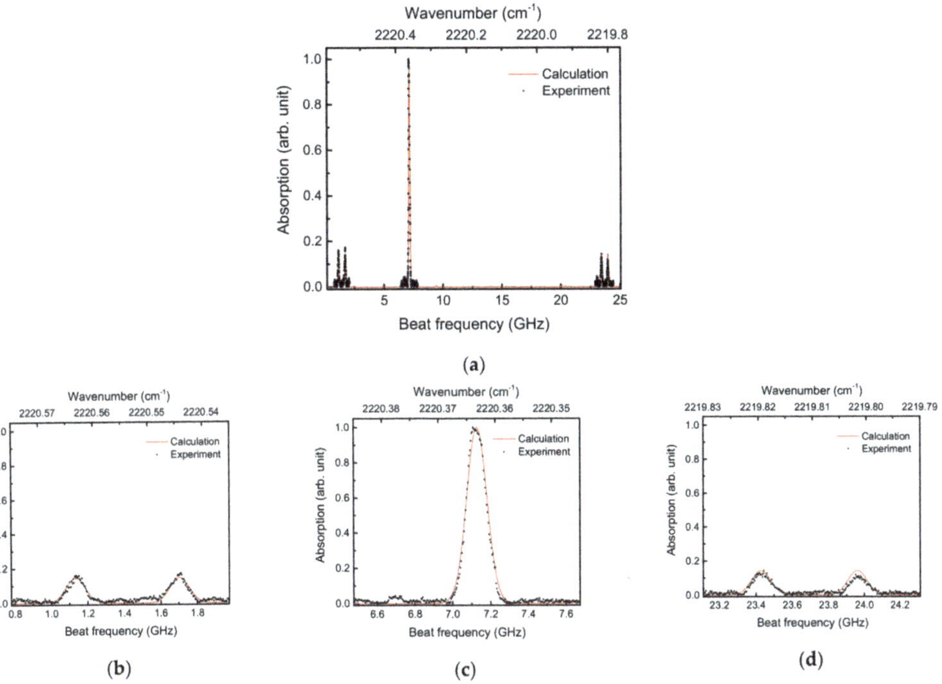

Figure 7. N_2O absorption lines observed with broadband heterodyne spectroscopy. The experimental data are dots, and the red lines are calculated absorptions of N_2O based on SpectraPlot. The experimental and calculated spectra were normalized by maximum values in the determining areas. (**a**) Entire frequency range. (**b**–**d**) Enlarged absorption spectra corresponding to the beat frequency domains of 0.78–1.98 GHz, 6.47–7.67 GHz, and 23.1–24.3 GHz, respectively.

4. Discussion

The operational bandwidth of the QC detector was theoretically predicted to be greater than 150 GHz [11] because of the picosecond electron transit times in the QC structures [25,26]. Here, the QC detector response speed was limited by the parasitic capacitance and circuit inductance. Further improvement is required while maintaining the responsivity. The normal incidence schemes have the potential to overcome this challenge by simultaneously enlarging the acceptance area and reducing the capacitance with a smaller device structure [27–30]. Furthermore, this approach would resolve the polarization dependency of the photoresponse and would make it possible to realize an arrayed QC detector for imaging [31]. However, MIR point-detectors with operational bandwidths of several tens gigahertz are expected to enable innovative technologies. One example would be high-speed spectroscopy [32], based on the strong molecular interaction with MIR light, especially for spectroscopic analysis of instantaneous and non-invertible phenomena, such as explosions, combustion, or chemical reactions. Another example would be free space optical communications that take advantage of the low propagation losses of MIR due to weak absorption in the atmospheric window, and low Rayleigh scattering relative to that in the visible and near-infrared regions [33,34]. QC detectors are highly suitable receivers for any device that requires high-speed communications because they do not consume any power.

Here, broadband heterodyne spectroscopy was demonstrated by using the wide-band QC detector. Although the data acquisition was performed over three separate spectral regions, the entire 1 cm^{-1} tuning range was continuously covered by using a large modulation amplitude for the QC laser injection current and by using optimal resolution. The resolution bandwidth could be optimized for increased resolution and shorter measurement times. Here, they were specified at 5 MHz and 4 min, respectively.

The heterodyne optical system of QC lasers and detectors was a compact spectroscopic module [35,36] for broadband high-resolution gas sensing in cases where absorption lines of several gases are mixed. By combining broadband tunable QC lasers in a CW mode, such as an external cavity configuration [37–40], for signal and local oscillators, the observable spectral range could be significantly extended [41]. Broadband heterodyne detection would also be essential for absolute frequency determinations when MIR optical frequency combs are used [42]. Wide-band QC detectors and their use in broadband heterodyne optical systems have much potential to expand the application of the MIR spectroscopy.

5. Conclusions

High-speed operation with a 3-dB bandwidth over 20 GHz was realized in an uncooled QC detector with a 4.6-μm peak response. This was accomplished by reducing the parasitic capacitance and circuit inductance with a narrow 25 μm × 100 μm ridge-mesa structure, a thick active region of 90 cascade modules, and air-bridge wiring. A noise-equivalent power of 7.7×10^{-11} W/Hz$^{1/2}$ was obtained with a peak responsivity of 5.7 mA/W, and a flat noise level of 4.4×10^{-13} A/Hz$^{1/2}$. By using the high-speed QC detector, a broadband heterodyne absorption spectrum of N$_2$O was obtained over the range of ~0.8 cm^{-1}, with a resolution bandwidth of 5 MHz. The absorption strengths, spectral positions, and spectral widths of the absorption lines were in good agreement with the calculations. In future, a compact high-resolution broadband spectroscopic module could be realized with a heterodyne system incorporating QC lasers and detectors.

Author Contributions: Conceptualization, T.D. and N.A.; methodology, T.D. and N.A.; validation, T.D. and N.A.; formal analysis, T.D.; investigation, T.D.; resources, T.D.; data curation, T.D.; writing—original draft preparation, T.D.; writing—review and editing, N.A.; visualization, T.D. and N.A.; supervision, T.D. and N.A.; project administration, T.D. and N.A. All authors have read and agreed to the published version of the manuscript.

Funding: This research received no external funding.

Institutional Review Board Statement: Not applicable.

Informed Consent Statement: Not applicable.

Data Availability Statement: The data presented in this study are available from the corresponding author upon reasonable request.

Acknowledgments: The authors are grateful to M. Hitaka and A. Ito at Hamamatsu Photonics K.K. for the preparation of the device. The authors also express their gratitude to K. Fujita and M. Yamanishi at Hamamatsu Photonics K.K. for the valuable discussions with respect to the active region design and the physics of high-speed quantum cascade detectors. The authors are also grateful to T. Edamura at Hamamatsu Photonics K.K. for the fruitful discussions and his support throughout this work. We thank Alan Burns, PhD, from Edanz (https://jp.edanz.com/ac, accessed on 29 July 2021) for editing a draft of this manuscript.

Conflicts of Interest: The authors declare no conflict of interest.

References

1. Hofstetter, D.; Beck, M.; Faist, J. Quantum-cascade-laser structures as photodetectors. *Appl. Phys. Lett.* **2002**, *81*, 2683. [CrossRef]
2. Gendron, L.; Carras, M.; Huynh, A.; Ortiz, V.; Koeniguer, C.; Berger, V. Quantum cascade photodetector. *Appl. Phys. Lett.* **2004**, *85*, 2824. [CrossRef]
3. Levine, B.F. Quantum-well infrared photodetectors. *J. Appl. Phys.* **1993**, *74*, R1. [CrossRef]
4. Faist, J.; Capasso, F.; Sivco, D.L.; Sirtori, C.; Hutchinson, A.L.; Cho, A.Y. Quantum Cascade Laser. *Science* **1994**, *264*, 553–556. [CrossRef]
5. Graf, M.; Hoyler, N.; Giovannini, M.; Faist, J.; Hofstetter, D. InP-based quantum cascade detectors in the mid-infrared. *Appl. Phys. Lett.* **2006**, *88*, 241118. [CrossRef]
6. Reininger, P.; Schwarz, B.; Detz, H.; MacFarland, D.; Zederbauer, T.; Andrews, A.M.; Schrenk, W.; Baumgartner, O.; Kosina, H.; Strasser, G. Diagonal-transition quantum cascade detector. *Appl. Phys. Lett.* **2014**, *105*, 091108. [CrossRef]
7. Dougakiuchi, T.; Fujita, K.; Hirohata, T.; Ito, A.; Hitaka, M.; Edamura, T. High photoresponse in room temperature quantum cascade detector based on coupled quantum well design. *Appl. Phys. Lett.* **2016**, *109*, 261107. [CrossRef]
8. Giorgetta, F.R.; Baumann, E.; Hofstetter, D.; Manz, C.; Yang, Q.; Köhler, K.; Graf, M. InGaAs/AlAsSb quantum cascade detectors operating in the near infrared. *Appl. Phys. Lett.* **2007**, *91*, 111115. [CrossRef]
9. Graf, M.; Scalari, G.; Hofstetter, D.; Faist, J.; Beere, H.; Linfield, E.; Ritchie, D.; Davies, G. Terahertz quantum well infrared photodetector. *Appl. Phys. Lett.* **2004**, *84*, 475. [CrossRef]
10. Vardi, A.; Sakr, S.; Mangeney, J.; Kandaswamy, P.K.; Monroy, E.; Tchernycheva, M.; Schacham, S.E.; Julien, F.H.; Bahir, G. Femto-second electron transit time characterization in GaN/AlGaN quantum cascade detector at 1.5 micron. *Appl. Phys. Lett.* **2011**, *99*, 202111. [CrossRef]
11. Dougakiuchi, T.; Ito, A.; Hitaka, M.; Fujita, K.; Yamanishi, M. Ultimate response time in mid-infrared high-speed low-noise quantum cascade detectors. *Appl. Phys. Lett.* **2021**, *118*, 041101. [CrossRef]
12. Hillbrand, J.; Krüger, L.M.; Cin, S.D.; Knötig, H.; Heidrich, J.; Andrews, A.M.; Strasser, G.; Keller, U.; Schwarz, B. High-speed quantum cascade detector characterized with a mid-infrared femtosecond oscillator. *Opt. Express* **2021**, *29*, 5774–5781. [CrossRef]
13. Tittel, F.K.; Richter, D.; Fried, A. Mid-Infrared Laser Applications in Spectroscopy. In *Solid-State Mid-Infrared Laser Sources*, 1st ed.; Sorokina, I.T., Vodopyanov, K.L., Eds.; Springer: Berlin/Heidelberg, Germany, 2003; pp. 458–529.
14. Du, Z.; Zhang, S.; Li, J.; Gao, N.; Tong, K. Mid-Infrared Tunable Laser-Based Broadband Fingerprint Absorption Spectroscopy for Trace Gas Sensing: A Review. *Appl. Sci.* **2019**, *9*, 338. [CrossRef]
15. Waynant, R.W.; Ilev, I.K.; Gannot, I. Mid-infrared laser applications in medicine and biology. *Phil. Trans. R. Soc. Lond. A* **2001**, *359*, 635–644. [CrossRef]
16. Sigrist, M.W. Mid-infrared laser-spectroscopic sensing of chemical species. *J. Adv. Res.* **2015**, *6*, 529–533. [CrossRef] [PubMed]
17. Ycas, G.; Giorgetta, F.R.; Cossel, K.C.; Waxman, E.M.; Baumann, E.; Newbury, N.R.; Coddington, I. Mid-infrared dual-comb spectroscopy of volatile organic compounds across long open-air paths. *Optica* **2019**, *6*, 165–168. [CrossRef]
18. Gong, Y.; Bu, L.; Yang, B.; Mustafa, F. High Repetition Rate Mid-Infrared Differential Absorption Lidar for Atmospheric Pollution Detection. *Sensors* **2020**, *20*, 2211. [CrossRef]
19. Lenth, W. Optical heterodyne spectroscopy with frequency- and amplitude-modulated semiconductor lasers. *Opt. Lett.* **1983**, *8*, 575–577. [CrossRef] [PubMed]
20. Piotrowski, A.; Piotrowski, J. Uncooled infrared detectors in Poland, history and recent progress. In Proceedings of the 26th European Conference on Solid-State Transducers (Eurosensors), Krakow, Poland, 9–12 September 2012; pp. 1506–1512.
21. Dougakiuchi, T.; Edamura, T. High-speed quantum cascade detector with frequency response of over 20 GHz. In Proceedings of the SPIE Future Sensing Technologies, Tokyo, Japan, 12 November 2019.
22. Hofstetter, D.; Graf, M.; Aellen, T.; Faist, J.; Hvozdara, L.; Blaser, S. 23 GHz operation of a room temperature photovoltaic quantum cascade detector at 5.35 μm. *Appl. Phys. Lett.* **2006**, *89*, 061119. [CrossRef]
23. Goldenstein, C.S.; Miller, V.A.; Spearrin, R.M.; Strand, C.L. SpectraPlot.com: Integrated spectroscopic modeling of atomic and molecular gases. *J. Quant. Spectrosc. Radiat. Transf.* **2017**, *200*, 249–257. [CrossRef]

24. SpectraPlot. Available online: https://www.spectraplot.com/ (accessed on 31 May 2021).
25. Choi, H.; Diehl, L.; Wu, Z.K.; Giovannini, M.; Faist, J.; Capasso, F.; Norris, T.B. Gain Recovery Dynamics and Photon-Driven Transport in Quantum Cascade Lasers. *Phys. Rev. Lett.* **2008**, *100*, 167401. [CrossRef] [PubMed]
26. Talukder, M.A. Modeling of gain recovery of quantum cascade lasers. *J. Appl. Phys.* **2011**, *109*, 033104. [CrossRef]
27. Zhai, S.Q.; Liu, J.Q.; Liu, F.Q.; Wang, Z.G. A normal incident quantum cascade detector enhanced by surface plasmons. *Appl. Phys. Lett.* **2012**, *100*, 181104. [CrossRef]
28. Reininger, P.; Schwarz, B.; Harrer, A.; Zederbauer, T.; Detz, H.; Andrews, A.M.; Gansch, R.; Schrenk, W.; Strasser, G. Photonic crystal slab quantum cascade detector. *Appl. Phys. Lett.* **2013**, *103*, 241103. [CrossRef]
29. Wang, F.J.; Zhuo, N.; Liu, S.M.; Ren, F.; Ning, Z.D.; Ye, X.L.; Liu, J.Q.; Zhai, S.Q.; Liu, F.Q.; Wang, Z.G. Temperature independent infrared responsivity of a quantum dot quantum cascade detector. *Appl. Phys. Lett.* **2016**, *108*, 251103. [CrossRef]
30. Palaferri, D.; Todorov, Y.; Bigioli, A.; Mottaghizadeh, A.; Gacemi, D.; Calabrese, A.; Vasanelli, A.; Li, L.; Davies, A.G.; Linfield, E.H.; et al. Room-temperature nine-μm-wavelength photodetectors and GHz-frequency heterodyne receivers. *Nature* **2018**, *556*, 85–88. [CrossRef]
31. Harrer, A.; Schwarz, B.; Schuler, S.; Reininger, P.; Wirthmüller, A.; Detz, H.; MacFarland, D.; Zederbauer, T.; Andrews, A.M.; Rothermund, M.; et al. 4.3 μm quantum cascade detector in pixel configuration. *Opt. Express* **2016**, *24*, 17041. [CrossRef]
32. Kawai, A.; Hashimoto, K.; Dougakiuchi, T.; Badarla, V.R.; Imamura, T.; Edamura, T.; Ideguchi, T. Time-stretch infrared spectroscopy. *Commun. Phys.* **2020**, *3*, 152. [CrossRef]
33. Martini, R.; Whittaker, E.A. Quantum cascade laser-based free space optical communications. *J. Opt. Fiber. Commun. Rep.* **2005**, *2*, 1–14. [CrossRef]
34. Pang, X.; Ozolins, O.; Schatz, R.; Storck, J.; Udalcovs, A.; Navarro, J.R.; Kakkar, A.; Maisons, G.; Carras, M.; Jacobsen, G.; et al. Gigabit free-space multi-level signal transmission with a mid-infrared quantum cascade laser operating at room temperature. *Opt. Lett.* **2017**, *42*, 3646–3649. [CrossRef]
35. Weidmann, D.; Reburn, W.J.; Smith, K.M. Ground-based prototype quantum cascade laser heterodyne radiometer for atmospheric studies. *Rev. Sci. Instrum.* **2007**, *78*, 073107. [CrossRef] [PubMed]
36. Wang, Y.; Soskind, M.G.; Wang, W.; Wysocki, G. High-resolution multi-heterodyne spectroscopy based on Fabry-Perot quantum cascade lasers. *Appl. Phys. Lett.* **2014**, *104*, 031114. [CrossRef]
37. Tan, S.; Zhang, J.C.; Zhuo, N.; Wang, L.J.; Liu, F.Q.; Yao, D.Y.; Liu, J.Q.; Wang, Z.G. Low-threshold, high SMSR tunable external cavity quantum cascade laser around 4.7 μm. *Opt. Quant. Electron.* **2013**, *45*, 1147–1155. [CrossRef]
38. Dougakiuchi, T.; Fujita, K.; Akikusa, N.; Sugiyama, A.; Edamura, T.; Yamanishi, M. Broadband Tuning of External Cavity Dual-Upper-State Quantum-Cascade Lasers in Continuous Wave Operation. *Appl. Phys. Express* **2011**, *4*, 102101. [CrossRef]
39. Mohan, A.; Wittmann, A.; Hugi, A.; Blaser, S.; Giovannini, M.; Faist, J. Room-temperature continuous-wave operation of an external-cavity quantum cascade laser. *Opt. Lett.* **2007**, *32*, 2792–2794. [CrossRef] [PubMed]
40. Dougakiuchi, T.; Fujita, K.; Sugiyama, A.; Ito, A.; Akikusa, N.; Edamura, T. Broadband tuning of continuous wave quantum cascade lasers in long wavelength (>10 μm) range. *Opt. Express* **2014**, *22*, 19930–19935. [CrossRef] [PubMed]
41. Weidmann, D.; Wysocki, G. High-resolution broadband (>100 cm^{-1}) infrared heterodyne spectro-radiometry using an external cavity quantum cascade laser. *Opt. Express* **2009**, *17*, 248–259. [CrossRef] [PubMed]
42. Schliesser, A.; Picqué, N.; Hänsch, T.W. Mid-infrared frequency combs. *Nat. Photonics* **2012**, *6*, 440–449. [CrossRef]

Article

The Short-Term Performances of Two Independent Gas Modulated Refractometers for Pressure Assessments

Clayton Forssén [1,2], **Isak Silander** [1], **Johan Zakrisson** [1], **Ove Axner** [1] **and Martin Zelan** [2,*]

[1] Department of Physics, Umeå University, SE-901 87 Umeå, Sweden; clayton.forssen@umu.se (C.F.); isak.silander@umu.se (I.S.); johan.zakrisson@umu.se (J.Z.); ove.axner@umu.se (O.A.)

[2] Measurement Science and Technology, RISE Research Institutes of Sweden, SE-501 15 Borås, Sweden

* Correspondence: martin.zelan@ri.se

Abstract: Refractometry is a powerful technique for pressure assessments that, due to the recent redefinition of the SI system, also offers a new route to realizing the SI unit of pressure, the Pascal. Gas modulation refractometry (GAMOR) is a methodology that has demonstrated an outstanding ability to mitigate the influences of drifts and fluctuations, leading to long-term precision in the 10^{-7} region. However, its short-term performance, which is of importance for a variety of applications, has not yet been scrutinized. To assess this, we investigated the short-term performance (in terms of precision) of two similar, but independent, dual Fabry–Perot cavity refractometers utilizing the GAMOR methodology. Both systems assessed the same pressure produced by a dead weight piston gauge. That way, their short-term responses were assessed without being compromised by any pressure fluctuations produced by the piston gauge or the gas delivery system. We found that the two refractometer systems have a significantly higher degree of concordance (in the 10^{-8} range at 1 s) than what either of them has with the piston gauge. This shows that the refractometry systems under scrutiny are capable of assessing rapidly varying pressures (with bandwidths up to 2 Hz) with precision in the 10^{-8} range.

Keywords: refractometry; pressure; short-term performance; Fabry–Perot cavity; gas modulation; modulation techniques; metrology

Citation: Forssén, C.; Silander, I.; Zakrisson, J.; Axner, O.; Zelan, M. Short-Term Performances of Two Independent Gas Modulated Refractometers for Pressure Assessments. *Sensors* **2021**, *21*, 6272. https://doi.org/10.3390/s21186272

Academic Editors: Krzysztof M. Abramski and Piotr Jaworski

Received: 28 July 2021
Accepted: 13 September 2021
Published: 18 September 2021

Publisher's Note: MDPI stays neutral with regard to jurisdictional claims in published maps and institutional affiliations.

1. Introduction

Refractometry is a powerful technique for assessing gas pressure. It is based upon measuring, by optical means, the change in refractive index in a measurement compartment as gas is let into it. From the change in refractive index, under the condition that the molar polarizability (and higher order refractive virial coefficients) of the gas is known, the change in gas density can be calculated. From this, provided that the gas temperature is known, the pressure can be assessed by utilizing an equation of state. Moreover, since the Boltzmann constant was given a fixed value (i.e., without uncertainty) in the 2019 revision of the SI system of units [?], refractometry also offers a new and independent route to realizing the SI unit of pressure, i.e., the Pascal [?]. These exciting prospects have spurred a significant increase of interest within the field of refractometry. Work to explore and utilize the potential of optical methods for assessing the molar density and pressure of gas presently takes place at several national metrology institutes and universities [? ? ? ? ? ? ? ?].

Besides being a potential primary method for measuring the Pascal, the technology also has several other highly interesting properties and advantages. As optical measurements do not utilize any mechanical actuators, the highest pressures that can be measured tend to be limited by the gas handling system used. The lowest pressure shifts that can be resolved are in turn limited by the laser locking. In practical terms, the dynamic range can be as high as eight orders of magnitude, typically covering the range from 1 mPa to 100 kPa. As optical measurements are performed by measuring changes in frequency, which can be measured swiftly and continuously with high accuracy, these systems can also be designed to give measurements of pressure with high time resolution. In practice, while the long-term performance is given by the stability of the cavity spacer, the time resolution and short-term performance are given by the acquisition rate and stability of the frequency counter. This combination of an extraordinary large dynamic range and a fast response facilitate accurate assessments of large pressure shifts with short settling times. Such measurements can be used to characterize pressure sensors, resolve the differences between sensor responses and actual pressure changes, study rapidly changing pressures, and investigate processes giving rise to such.

The best performing refractometers are based on Fabry–Perot (FP) cavities, where a laser is used to probe the frequency of a longitudinal cavity mode [? ? ? ? ? ? ? ? ? ? ?]. By measuring the change in frequency between an empty (evacuated) and a gas-filled cavity, the refractivity can be assessed, from which the molar density and the pressure can be calculated. However, since such measurements cannot distinguish changes in refractivity of a gas from drifts in the physical length of the cavity, the most sophisticated FP refractometers utilize a dual FP cavity (DFPC) design in which one cavity acts as the measurement cavity and the other as a reference. This eliminates common-mode drifts in the cavity spacer caused by aging, temperature drifts, mechanical stress, etc. However, since the two cavities in a DFPC can drift dissimilarly, extraordinarily stable conditions are still required to achieve optimal performance. For a 15 cm long cavity, a drift in length of 1 pm gives rise to a shift in the assessed pressure of nitrogen of 2.5 mPa. As a means to remedy this, we developed a measurement methodology denoted gas modulation refractometry (GAMOR) [? ? ? ? ?].

The GAMOR methodology is based on repeated measurements performed on a relatively short timescale (typically using gas filling and evacuating cycles of 100 s) combined with an interpolation procedure in which the empty measurement cavity response is taken as the interpolated value of two such measurements—one taken just before and one directly after the filled measurement cavity measurement. That way, the influences of both long-term drifts and various types of fluctuations can be strongly suppressed [? ? ?]. Furthermore, the influences of leaks and outgassing in the reference cavity can automatically be corrected for.

In order to perform high accuracy refractometry based on the GAMOR principle, care needs to be taken regarding the construction of the refractometers. To enable repeated filling and emptying of gas on relatively short time scales without introducing excessive amounts of PV work, cavities with small volumes have been implemented (<5 cm^3) [? ?]. The cavity spacers in these works were made from Invar which has both higher thermal conductivity and a larger volumetric heat capacity than commonly used glass materials (Zerodur and ULE glass).

This does not only eliminate any possible heat islands in the system; it also facilitates the assessment of gas temperature, which is performed by measurement of the temperature of the cavity spacer by the use of temperature probes placed in drilled holes in, or in direct contact with, the spacer. The use of Invar also eliminates effects of gas permeation which have been reported for ULE glass [? ?]. Furthermore, to allow for fully automatic operation with sturdy laser locking and automated mode jumps, systems based on rugged narrow-banded fiber lasers working in the near IR (NIR) communication region (around 1.55 μm) have been used [?].

This has lead to instrumentation that is capable of providing measurements with precision in the sub-ppm (sub-parts-per-million or sub-10^{-6}) range [? ? ? ?]. By then also using well-calibrated temperature sensors and accurately assessed molecular parameters (molar polarizabilties and virial coefficients), the systems can demonstrate good accuracy. Such a system, denoted the stationary optical pascal (SOP), was recently characterized in terms of its ability to realize the Pascal [?]. It was found that its uncertainty was $[(10 \text{ mPa})^2 + (10 \times 10^{-6}P)^2]^{1/2}$, mainly limited by the uncertainty in the molar polarizability of nitrogen (8 ppm) [?].

To assess the ability to realize a transportable refractometer, a similar system, denoted the transportable optical pascal (TOP), was recently developed and characterized. It was found that its uncertainty was $[(16 \text{ mPa})^2 + (28 \times 10^{-6}P)^2]^{1/2}$, mainly limited by the uncertainty of the temperature probes used for assessment of the temperature (26 ppm) [?].

As was alluded to above, to make viable assessments of large pressure shifts with short settling times, which is needed for a number of applications, it is of importance that the system has a fast response. Although several types of refractometers have been scrutinized over the years [? ? ? ? ? ? ? ? ? ? ? ? ? ? ? ? ?], virtually none of them has yet been assessed with respect to its short-term behavior. Access to two GAMOR-based refractometer systems allows for scrutiny of the short-term behavior of GAMOR-based refractometry in more detail. By comparing two fully independent GAMOR-based refractometer systems (the aforementioned SOP and TOP systems) connected to the same gas system, whose pressure was set by a dead weight piston gauge (DWPG), their short-term performances could be scrutinized in some detail. As the refractometers were completely independent, it could be concluded that deviations that are common to both systems are not inherent to one or the other refractometer, but rather the DWPG and/or the gas handling system. Thereby, we could ascertain the precision of the refractometers without any influence from the DWPG or gas handling system. Indeed, we assessed the short-term performances of two independent gas modulated refractometers regarding their ability to assess pressure. It was found that the refractometers can provide short-term precision on the 1 s time scale of 3×10^{-8}, which is one order of magnitude better than the corresponding stability of the pressure provided by the DWPG. This opens up a number of novel applications for refractometry.

Although the SOP refractometer previously has been well described [? ?], the TOP system has not. This system, including its construction and various components, is therefore described in some detail here. In addition, the theoretical model used for the evaluation of the data gathered is provided.

2. Theory

2.1. Refractivity

As has previously been outlined [?], each DFPC refractometer addresses the empty cavity mode q_{01} or q_{02} with light of frequency ν_{01} or ν_{02}, respectively. The beat frequency between the two lasers, f, which is the measured entity, is given by the difference between the two laser frequencies, defined as $| \nu_1 - \nu_2 |$. Since the lasers have a limited tuning range, automatic mode jumps will take place when the change in pressure becomes large. This implies that f is a non-monotonic (i.e., a wrapped) function. It is therefore convenient to define an unwrapped beat frequency as

$$f_{UW} = \pm f - \left(\frac{\Delta q_1}{q_{01}} \nu_{01} - \frac{\Delta q_2}{q_{02}} \nu_{02} \right), \tag{1}$$

where Δq_1 and Δq_2, counted from q_{01} and q_{02}, are the mode jumps and where the $\pm$ sign refers to the cases when $\nu_1 > \nu_2$ and $\nu_1 < \nu_2$.

The refractivity can then be expressed as a function of the shift of the unwrapped beat frequency when gas is let out of (or into) the measurement cavity, Δf_{UW}. As has been shown recently [?], while denoting the measurement cavity as m, the refractivity can be expressed as a function of the unwrapped beat frequency when GAMOR is used as

$$n - 1 = \frac{| \Delta f_{UW} | / \nu_{0m}}{1 - | \Delta f_{UW} | / \nu_{0m} + \Delta q_m / q_{0m} + \varepsilon_m}, \tag{2}$$

where ε_m is a deformation parameter comprising the refractivity-normalized relative difference in lengths of the two cavities due to pressurization, given by $[(\delta L / L_0)_m - (\delta L / L_0)_r]/(n - 1)$, where $(\delta L / L_0)_m$ and $(\delta L / L_0)_r$ are the relative changes in length of the measurement and reference cavities when the measurement cavity is pressurized [? ? ?].

It is worth noting that ε_m can be assessed with high accuracy by a methodology developed by Zakrisson et al. [?].

In Equation (2) the influences of the mirror dispersion and the finite penetration depth of the mirrors have been neglected. The former since the systems in this work use light in the communication band (around 1.55 µm), for which there are mirrors with a minimum of (linear) dispersion. The latter since the effect is smaller than the uncertainty of the molar polarizability of the gas [? ?].

2.2. Molar Density

For pressures below one atmosphere, the molar density can be calculated by assessing the refractive index and using the extended Lorentz–Lorenz equation

$$\rho = \frac{2}{3A_R}(n-1)[1 + b_{n-1}(n-1)], \tag{3}$$

where A_R and b_{n-1} are the molar dynamic polarizability [? ?]. The latter is given by $-(1 + 4B_R/A_R^2)/6$, where, in turn, B_R is the second refractivity virial coefficient in the Lorentz–Lorenz Equation [? ? ?].

2.3. Pressure

The molar density can then be used to assess the pressure as

$$P = RT\rho[1 + B_\rho(T)\rho], \tag{4}$$

where R is the ideal gas constant, T is the temperature of the gas, and $B_\rho(T)$ is the second density virial coefficient.

For more detailed theoretical descriptions of the Lorentz–Lorenz equation and the equation of state, and for expressions valid for higher pressures, the reader is referred to the literature, e.g., [? ? ? ? ? ? ? ?].

2.4. Molecular Data

In this work, all assessments were performed on nitrogen. Table 1 provides information about the relevant gas constants for nitrogen, A_R, b_{n-1}, and B_ρ.

Table 1. Gas coefficients for N_2 at 302.91 K and 1550.14 nm.

Coefficients	Value (k = 2)	Reference
A_R	$4.396549(34) \times 10^{-6}$ m^3/mol	[? ?]
b_{n-1}	$-0.195(7)$	[? ?]
B_ρ	$-4.00(24) \times 10^{-6}$ m^3/mol	[? ?]

2.5. Set Pressure of the DWPG

In this work a DWPG was used to provide a pressure by loading a known mass on a piston-cylinder ensemble with a known area. The pressure was calculated as

$$P_{DW} = \frac{(m_p + \Sigma_i\, m_i)g \cdot \cos(\theta)}{A_{eff}[1 + \alpha(T_p - T_{ref})]} + P_{hood}, \tag{5}$$

where m_p is the mass of the piston, m_i is the mass of the individual weights, g is the local gravity, θ is the angle between the piston cylinder assembly and the gravity vector, A_{eff} is the effective area of the piston at the temperature T_{ref}, α is the combined temperature expansion of the piston and cylinder, T_p is the measured temperature, and P_{hood} is the hood pressure [?].

3. Experimental Setup

3.1. The Dual Refractometry System Used in This Work

To demonstrate the short-term performance of GAMOR based refractometry, two fully independent Invar-based DFPC refractometers, the aforementioned SOP and TOP, were connected to a DWPG. Figure **??** shows a picture of the experimental setup. While the SOP was firmly placed on an optical table (placed in the rightmost box on the optical table, in the center of the figure), the TOP is designed to fit in a 19-inch transportable rack (the standalone unit to the right). They were both connected to a DWPG (placed in the leftmost box on the optical table).

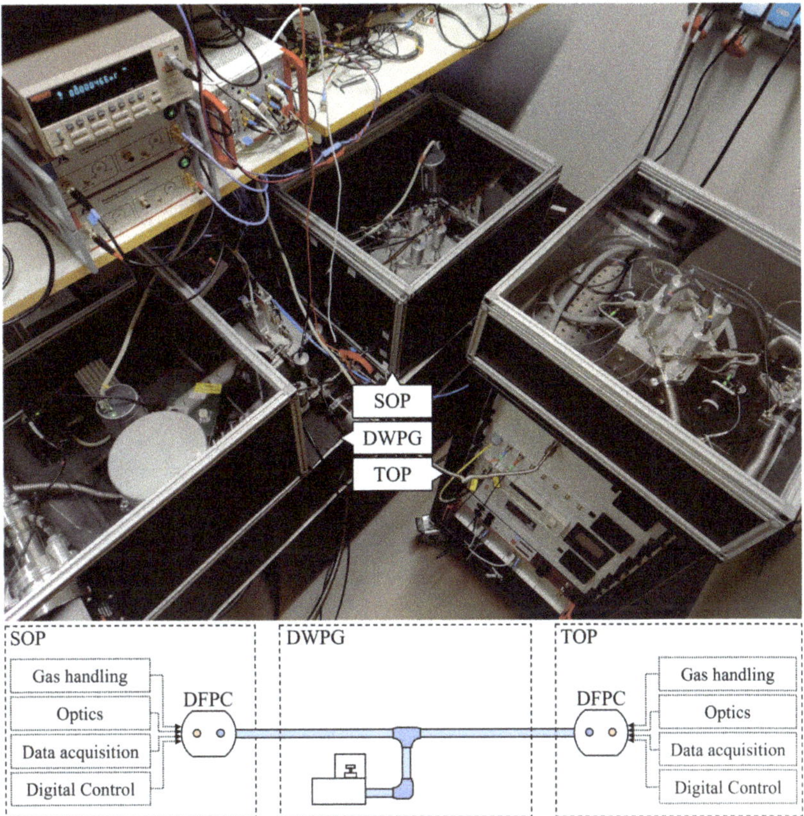

Figure 1. The dual refractometry system scrutinized in this work. It consists of three main components: the SOP (the system in the rightmost box on the optical table); the TOP-refractometer (the standalone system to the right); and the DWPG (the system in the leftmost box on the optical table). In addition to this, it comprises a common gas supply (seen between the SOP and DWPG boxes), a common vacuum system (not in the figure), a computer (for control and data acquisition), and various electronics—for the SOP, partly seen on the shelves, and for the TOP, in the rack. In the bottom part of the figure, a schematic showing the subsystems of the two refractometers and their connection to the DWPG is presented. The blue gas line represents the gas pressure under assessment and the red circles within the DFPCs represent evacuated cavities.

The two refractometry systems are virtually identical in terms of optical and electronic components, including the FP-cavity ensemble. As shown in the schematics above, each system has its own gas handling system; optics (including lasers, electro-optics, passive optics, and locking electronics); and data acquisition and digital control in the form of digital to analogue converter (DAQ) modules and a computer. However, they differ when it comes to the means by which they assess the gas temperature: while the SOP system uses a thermocouple that refers to a miniature fixed point gallium cell to measure the spacer temperature [?], the TOP system assesses it through the use of calibrated Pt-100 sensors. Normally, when working individually, the two refractometry systems use their own designated gas supplies and vacuum systems. In this work, however, they were connected to a common gas supply and vacuum system.

3.2. The TOP Refractometer

While the construction of the SOP, and the role of its various comprised parts, have previously been described in some detail in the literature [? ?], those of the TOP have not. Therefore, we describe those in more detail here. As is shown in Figure **??**, the TOP refractometer system was built within a 19-inch transportable rack. For sturdy transportation and best serviceability, the system has been divided into a number of subsystems. From top to bottom, they comprise a top unit (denoted the cavity unit) and seven subrack enclosures, denoted the modules (A–G), comprising a number of separate mechanical, optical, and electrical entities that play the same role in the system as the corresponding parts do in the SOP system [? ?].

The cavity unit consists of a 60×60 cm breadboard that is firmly attached to the top of the rack. The DFPC Invar spacer sits at the center of this breadboard within an aluminum enclosure (oven) that is temperature controlled by four Peltier elements. The breadboard is, in turn, temperature regulated by a heat mat placed under it. Mounted to the breadboard, surrounding all components in this unit, there is a $60 \times 60 \times 25$ cm aluminum framework with thermally isolated walls (shown in the figure).

Four pneumatic valves, used to control the flow of gas into and out of each cavity, are attached to the top of the oven. As is further described in Section **??**, two of these valves (denoted $V_{T.1}$ and $V_{T.3}$) are connected to the gas supply unit, and two ($V_{T.2}$ and $V_{T.4}$) are connected to a turbo pump. To provide an assessment of the reference pressure, a pressure gauge (denoted $G_{T.2}$ in Section **??**) is mounted on the turbo line (in close proximity to the reference cavity).

The cavity unit also contains customized fiber collimators that mode match the light from the lasers into the cavities; mirrors that direct the light; and detectors (Thorlabs, PDA50B-EC), placed behind each cavity, that detect the light transmitted on resonance.

Module A comprises the gas inlet system, consisting of a mass flow controller and an electronic pressure controller (denoted MFC and EPC in Section **??**, respectively) that provide a continuous flushing of gas and regulation of the pressure; and a pressure gauge, $G_{T.1}$, that provides a rough assessment of the pressure under scrutiny. It also contains a four slot compact DAC (CompactDAQ, National Instruments, cDAQ-9174) that holds an analogue input module (National Instruments, NI-9215) to monitor the feedback voltages sent to the lasers; a temperature input module (National instruments, NI-9216) to measure the Pt-100 readings; a voltage output module (National Instruments, NI-9263) to give feed back to the Peltier drivers; and a digital output module (National Instruments, NI-9474) to control the pilot valves (which also resides in Module A). The front panel is equipped with a VCR port to connect the device to be scrutinized by the TOP (the device under test, DUT).

The rear panel is equipped with 230, 24, and 12 V power supply inputs (in the leftmost part of the figure). Above these, there are two USB connectors to the cDAQ and the MFC/EPCs. At the center there are eight push-in 6 mm pneumatic fittings to provide pressurized air to the seven pilot valves and the gas to the supply unit. Above these, there are three D-sub connectors, which are used to connect the high pressure gauge to the vacuum gauge controller (Oerlikon-Leybold, Graphix Three); the cDAQ with the Peltier driver; and a fill pressure relay with the gas filling valve. In the rightmost part of this panel there are two gas connectors: one VCR that is connected to the valve system inside the cavity unit at the top of the rack, and one Swagelok connector that can be used for rough pumping of the gas system.

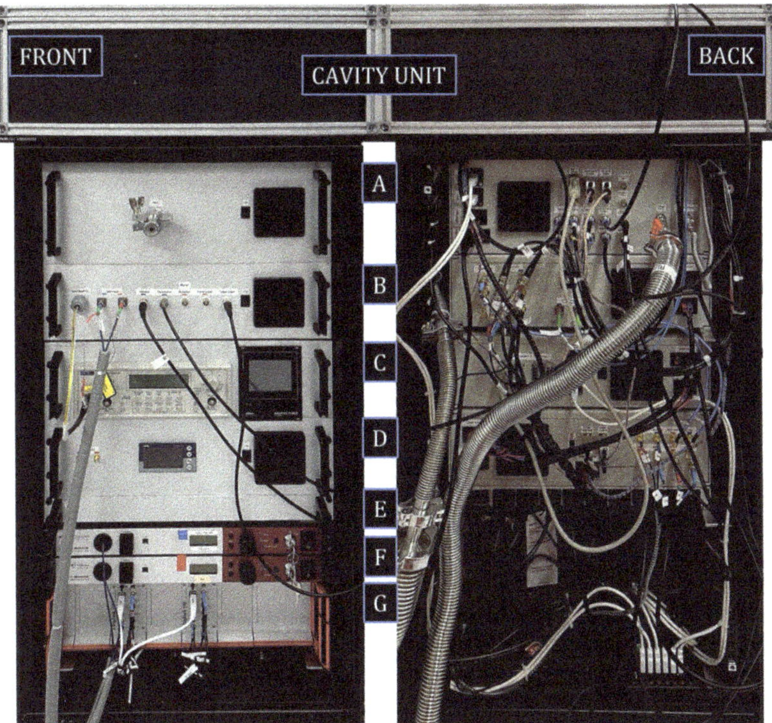

Figure 2. The TOP system seen from the front and rear. All lasers, electronics, and gas connections are placed within a 19-inch rack. On top of the rack, there is a 60 × 60 × 25 cm encapsulated box (denoted the cavity unit) that contains, as its base, an optical breadboard, on which the Invar-based DFPC is placed (in turn, encapsulated in an aluminum enclosure, denoted the "oven"). This unit also comprises four pneumatic valves that control the filling and emptying of gas in the cavity during the GAMOR-cycles (as can be seen in Figure **??**, placed on top of the oven); and collimators, mirrors, and detectors that couple light into the cavities and measure the transmittance. The rack contains thereafter, from the top to the bottom, seven modules, denoted A–G, containing vacuum connectors, a communication hub, fiber-optics, a frequency counter, two fiber lasers, and locking electronics. The rack stands on four wheels that allow the system to be easily moved within the laboratory.

Module B contains most of the optics, passive fiber optical components (e.g., circulators and isolators), and opto-electronics. The leftmost part of the front panel comprises the output from the beat detector and the input for the fibers from the lasers. The light that enters via the fibers is coupled into acousto optic modulators (AOM, AA Opto-Electronic, MT110-IR25-3FIO), after which it is coupled into 90/10 splitters. The light in the 10% outputs of the two splitters is coupled to the beat detector (Thorlabs, PDA8GS) via a 50/50 combiner. The light in the 90% outputs is coupled into electro-optic modulators (EOM, General Photonics, LPM-001-15) for the production of sidebands for the Pound–Drever–Hall locking. The light fields are then coupled into circulators via isolators (to prevent back reflections to the EOM). The forward output of each circulator is coupled via a fiber to the collimator for further passage into the cavity unit, and their rear outputs, which monitor the reflections from the cavities, are connected to reflection detectors (Thorlabs, PDA10CE-EC). The front panel is also equipped with five BNC-ports that are connected to the transmission and reflection detectors of the system. The fifth of these is used as a trigger that enables an oscilloscope to be connected to the other ports for the alignment procedure of the free space optics in the cavity unit.

On the rear panel, nine SMA-connectors are positioned to the left, comprising the control signals for the EOM and AOM; the inputs from the transmission detectors (for the monitoring port on the front panel); and the outputs from the reflection detectors (for the feedback signal to the automatic locking unit). The ninth port is the trigger input from a digital laser locking module (Toptica, DigiLock 110). At the center are the circulator outputs, which are connected to the collimators in the cavity unit, and a USB port (not connected).

Module C holds a frequency counter (Aim-TTi, TF960) that measures the beat frequency detected by the beat detector (positioned in module B) and the vacuum gauge controller that is used to monitor the pressure gauges within the system (the $G_{T.1}$ monitoring the pressure under scrutiny and $G_{T.2}$ recording the reference pressure). The frequency counter and gauge controller are digitally controlled and monitored but can also be reached manually, as their fronts are shown on the front panel of this module. The back panel comprises three D-sub connectors, which are used to connect the two pressure gauges and the fill pressure relay, and two USB-ports, which are the communication interface for the frequency counter and the vacuum gauge controller.

Module D comprises various 12 and 24 V power supplies; the custom made voltage controlled oscillators, which regulate the AOMs in module B; servo circuits for the locking of the lasers to the cavity modes; and the control unit for the heat mat (JUMO, diraTRON 108) that regulates the breadboard under the cavity unit, seen at the center of the front panel. The front panel is also equipped with two SMA ports for monitoring of the VCOs.

In the center of the back panel, the output for the driving current of the Peltier elements and the heat mat can be found. The rightmost part of the back panel comprises the inputs and outputs for the VCO and Laser PZT voltages.

Module E is a power distribution unit, in which the main 230 V input is split into nine 230 V outputs for power distribution to each subsystem.

Module F consists of two Er-doped fiber lasers (EDFL, NTK, Koheras Adjustik E15) that produce the light that is coupled, through fibers, to module C.

Module G, on the very bottom, contains the two digital laser locking modules (Toptica, DigiLock110) used for the automatic locking procedure of the lasers to their respective cavity.

Finally, at the bottom right on the back side of the rack there is a USB-hub that connects various electronics to a laptop accompanying the TOP system. The system is controlled by the laptop that, through the use of custom made LabVIEW software, gathers all data required for analysis.

3.3. The Gas Handling System for the Dual Refractometry System Used in This Work

Figure **??** shows a schematic view of how the two refractometers are connected to the gas system comprising a common gas supply and distribution system.

In the figure, the colored lines represent gas tubes where the red color relates to low pressures while the blue represents high pressures. To fill the system with gas, the gas supply system, consisting of a mass flow controller (MFC, Bronkhorst, FG-201CV) and an electronic pressure controller (EPC Bronkhorst, P-702CV), is connected to a supply (in this work N_2). In the volume to the left of valve $V_{S,5}$, gas constantly circulates to prevent contamination build up.

When the refractometers are to be filled with gas, the valves $V_{S/T.5}$ and $V_{S/T.1}$ are opened. Valve $V_{S,5}$ is opened and closed by a relay controlled by switching the set-point of the vacuum gauge controller. The input for the set-point is the pressure measured by gauge $G_{S,1}$ (Oerlikon-Leybold, CTR 101 N 1000 Torr) in the SOP-refractometer and a set pressure chosen "close" to the nominal set value of the DWPG (as given by Equation (**??**)). This setup means that the gas system will be re-pressurized whenever the pressure drops below the chosen set pressure. After the re-pressurization the DWPG will automatically regulate the pressure to its set-pressure. During the gas filling and stabilization stage, the valves $V_{S/T.2}$ and $V_{S/T.3}$ are closed, and the valves $V_{S/T.4}$ are open, resulting in evacuation (close to vacuum) of the reference cavities in both of the refractometers, represented by the red gas lines. When both measurement cavities are to be evacuated, the valves $V_{S/T.1}$ are closed and the valves $V_{S/T.2}$ are opened, leading to the evacuation of all cavities.

The gas lines are not depicted at an appropriate scale; counted from the common tee (depicted above valve $V_{S.5}$) and the gas molecular turbo pump, the gas lines to the TOP are significantly longer than the corresponding ones to the SOP.

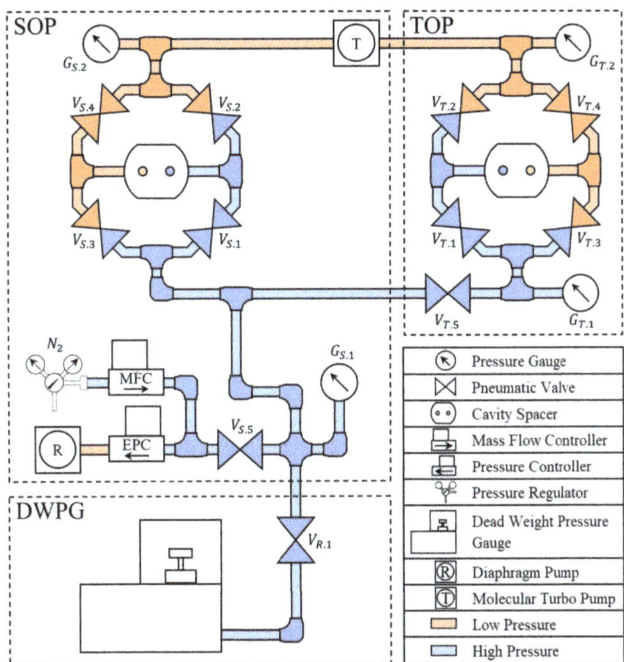

Figure 3. Schematic view of the gas delivery system. The components are described by the legend in the lower right corner. The setup is divided into three sub-systems, viz., the SOP, the TOP, and the DWPG. The SOP, presented in the upper left corner, regulates and controls the gas filling unit, consisting of the MFC and EPC. It also controls the primary fill valve, $V_{S.5}$. The valves $V_{S.1-4}$, which control the filling and evacuation of the cavities, are opened or closed in a given sequence. This subsystem also comprises two pressure gauges: $G_{S.1}$, which monitors the high pressure side; and $G_{S.2}$, which measures the residual pressure on the low pressure side. The TOP, which is displayed in the upper right corner, applies the same logic as the SOP system to its valves ($V_{T.1-4}$) and gauges ($G_{T.1-2}$). This subsystem can additionally be connected to or disconnected from the gas delivery system by use of valve $V_{T.5}$. Finally, the DWPG is presented in the lower left corner. This is connected or disconnected to the rest of the system by the valve $V_{R.1}$. In this presentation, the systems are displayed with the filling/connecting valves, $V_{S.5}$, $V_{S.1}$, $V_{T.5}$, $V_{T.1}$, and $V_{R.1}$ open. This implies that the measurement cavities are filled by the pressure set by the DWPG (represented by the blue gas lines). In addition, the valves $V_{S.4}$ and $V_{T.4}$ are open to allow for an evacuation of the reference cavities (represented by the red gas lines). The measurement cycle is followed by an evacuation state in which the valves $V_{S.1}$ and $V_{T.1}$ are closed and $V_{S.2}$ and $V_{T.2}$ are opened, which evacuates all cavities.

4. Methodology and Results

In order to perform the measurements presented in this work, the GAMOR methodology has been used. Although this methodology has previously been described in the literature [? ? ?], the following two sections, Sections ?? and ??, give a brief overview of its principles. The latter one emphasizes how the methodology can be used to obtain pressure assessments in seconds. Section ?? provides a characterization of the two GAMOR-based

DFPC refractometry systems used, and Section **??** gives an example of an assessment. Finally, Section **??** presents the results of a series of assessments.

4.1. Conventional Realization of the GAMOR Methodology

As has been indicated previously [? ? ?], the GAMOR methodology is based on two cornerstones: viz., (i) frequent referencing of filled measurement cavity beat frequencies to evacuated cavity beat frequencies, and (ii) an assessment of the evacuated measurement cavity beat frequency at the time of the assessment of the filled measurement cavity beat frequency by use of an interpolation between two evacuated measurement cavity beat frequency assessments, one performed before and one after the filled cavity assessments. The principles for the methodology when campaign-persistent drifts take place are schematically illustrated in Figure **??**.

Figure **??**a illustrates the pressure in the measurement cavity, which, according to cornerstone (i), is alternately evacuated and filled with gas (upper red curve), and the reference cavity is held at a constant pressure (lower blue curve). Campaign-persistent drifts will affect the frequencies of both the measurement and the reference lasers (although possibly to dissimilar extent, as shown in Figure **??**b) and thereby both the assessed beat frequency, $f(t)$, and its unwrapped counterpart, $f_{UW}(t)$ (the latter displayed by the upper black curve in Figure **??**c). These curves indicate that the influence of drifts can be reduced by shortening the modulation cycle period; for a given drift rate, the shorter the gas modulation period, the less the assessed beat frequency will be affected by drifts [?].

Furthermore, according to cornerstone (ii), the unwrapped evacuated measurement cavity beat frequency is, for each modulation cycle, not assessed by a single measurement. It is instead estimated by the use of a linear interpolation between two evacuated (unwrapped) measurement cavity beat frequency assessments performed in rapid succession—one taken directly prior to when the measurement cavity is filled with gas (for cycle n, at a time t_n, denoted $f_{UW}^{(0)}(t_n)$), and another directly after it has been evacuated (at a time t_{n+1}, denoted $f_{UW}^{(0)}(t_{n+1})$), both marked by crosses in Figure **??**c. By this, the unwrapped evacuated measurement cavity beat frequency, $\tilde{f}_{UW}^{(0)}(t_n, t, t_{n+1})$, can be estimated at all times t during a modulation cycle. For cycle n, for which $t_n \leq t \leq t_{n+1}$, it is estimated as

$$\tilde{f}_{UW}^{(0)}(t_n, t, t_{n+1}) = f_{UW}^{(0)}(t_n) + \frac{f_{UW}^{(0)}(t_{n+1}) - f_{UW}^{(0)}(t_n)}{t_{n+1} - t_n}(t - t_n). \tag{6}$$

For the case with campaign-persistent drifts, this interpolated value is represented by the green line in Figure **??**c.

By subtracting the estimated (interpolated) unwrapped evacuated measurement cavity beat frequency ($\tilde{f}_{UW}^{(0)}(t_n, t, t_{n+1})$, the green line) from the measured (drift-influenced) unwrapped beat frequency during gas filling ($f_{UW}(t)$, the black curve), both in Figure **??**c, a campaign-persistent, drift-corrected net beat frequency, represented by the black curve in Figure **??**d, can be obtained. The average value of this curve a short time period just before the cavity is evacuated, at a time denoted t_g, represents the Δf_{UW} to be used in the Equation (2) when GAMOR is performed. This shows that it is feasible to interpret GAMOR as "interpolated gas modulated refractometry."

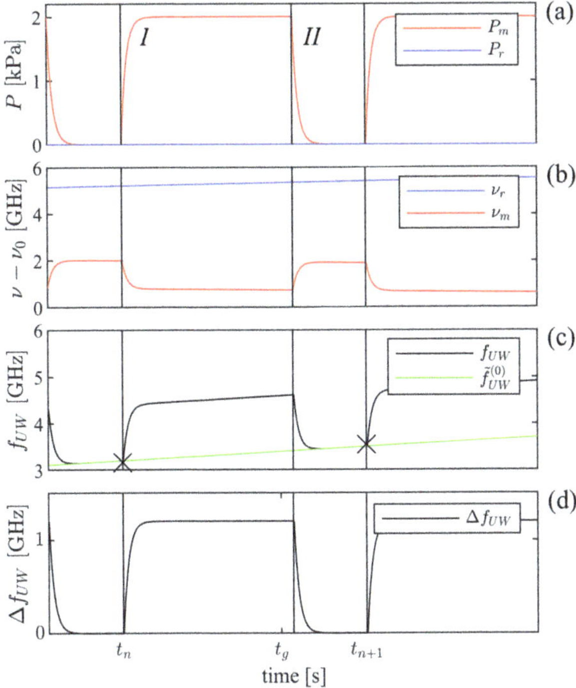

Figure 4. A schematic illustration of the principles of GAMOR implemented on a system exposed to campaign-persistent drifts. Panel (**a**) shows, as functions of time, by the the upper red curve, $P_m(t)$, the pressure in the measurement cavity, and by the lower blue curve, the pressure in the reference cavity, $P_r(t)$. Panel (**b**) depicts the corresponding frequencies of the measurement and reference lasers, $\nu_m(t)$ (the lower red curve) and $\nu_r(t)$ (the upper blue curve), respectively, for display purposes, both offset by a common frequency. Panel (**c**) displays, by the upper black curve, the corresponding unwrapped beat frequency in the presence of gas, $f_{UW}(t)$, and the lower green line depicts the estimated evacuated measurement cavity beat frequency, $\tilde{f}_{UW}^{(0)}(t_n, t, t_{n+1})$. Panel (**d**) displays the drift-corrected shift in unwrapped beat frequency, $\Delta f_{UW}(t)$. While the data that are used in ordinary GAMOR constitute the data points in the last part (10–20%) of section *I* (the time period between t_n and slightly after t_g) in panel (**d**), which are averaged to a single data value, in this work where cycle resolved assessments are performed, a significant part (ca. 80%) of the data in section *I* is used in an unaveraged manner. Note that the drifts have been greatly exaggerated for clarity.

4.2. Use of GAMOR to Assess Short-Term Pressure Fluctuations

GAMOR refractometry has so far been used to assess static pressures through the use of (and averaging over) a series of gas modulation cycles. It has been shown, for example, that, for the case with an Invar-based DFPC system, a minimum deviation could be achieved when averaging was performed over ten modulation cycles (i.e., over 10^3 s) [?]. Such a mode of operation is suitable when static pressures (or slowly varying pressures, those that change slowly over time intervals corresponding to several gas modulation periods) are to be assessed. In such a case, the methodology first calculates a single pressure value for each individual gas modulation cycle (as schematically described in Figure **??**) and then takes the average over n such cycles. For the case when the instrumentation is mainly affected by white noise, this process will improve on the precision (decrease the influence of noise) by a factor of $n^{-1/2}$.

The GAMOR methodology can though also be used for assessing short-term fluctuations of pressure. In this case, the assessment of the pressure, $P(t)$, is continuously carried out from the shift in the incessantly assessed unwrapped beat frequency, $\Delta f_{UW}(t)$, during individual modulation cycles. A calculation of the cycle resolved pressure began by assessing, for each refractometer, from the beat frequency, $f(t)$, and the shifts in cavity mode numbers $\Delta q_1(t)$ and $\Delta q_2(t)$, the unwrapped (i.e., the mode-jump-corrected) beat frequency, $f_{UW}(t)$. The beat frequency was continuously sampled by a frequency counter with a readout rate of 4 Hz. The shift in cavity mode number $\Delta q_i(t)$ was calculated as the nearest integer to $[(V_i(t) - V_{0,i})/V_{FSR,i} + q_{0,i}(P_{cav,i}(t)/P_0)(n_0 - 1)]$, where $V_i(t)$ is the voltage sent to the tuning control of laser i (acting on its piezo stretcher), $V_{0,i}$ is the voltage for an empty cavity locked to mode $q_{0,i}$, $V_{FSR,i}$ is the voltage required to tune the laser FSR, $P_{cav,i}(t)$ is the pressure in the cavity i, and $n_0 - 1$ is the refractivity for the gas addressed that corresponds to the pressure P_0 (here taken as 10^5 Pa). The pressure $P_{cav,i}(t)$ is assessed by the use of pressure gauge $G_{S/T.1}$. Hence, by continuously measuring the voltages sent to the lasers, all information needed to calculate the changes of the cavity mode numbers, $\Delta q_1(t)$ and $\Delta q_2(t)$, is available at all times. Using this information and Equation (1) it is possible to calculate the unwrapped beat frequency, $f_{UW}(t)$, at all times during a modulation cycle.

4.3. System Characterizations

Prior to the measurements, the two refractometers were first individually characterized by assessing the cavity deformations by the use of the methodology presented in [?]. The results are presented in detail in [?]. It was found that the cavity deformation parameters, ε_m, for the SOP and TOP when assessing nitrogen, were 0.001972(1) and 0.001927(1). Since $(n - 1) \propto (1 - \varepsilon_m)$, the measurement uncertainty in the cavity deformations will solely contribute to the total expanded uncertainty in pressure (k = 2) with 1 ppm. Furthermore, using a thorough evaluation, the two refractometers were attributed expanded uncertainties (k = 2) for assessment of pressure of nitrogen, of $((10\text{ mPa})^2 + (10 \times 10^{-6}P)^2)^{1/2}$ for the SOP and $((16\text{ mPa})^2 + (28 \times 10^{-6}P)^2)^{1/2}$ for the TOP [?]. It was found that while the SOP is predominantly limited by the uncertainty in the molar polarizability of nitrogen (8 ppm), the accuracy of the TOP is limited by the uncertainty of the temperature probes used for the temperature assessment (26 ppm). It should be noticed though, that both refractometers had smaller evaluated uncertainties than that of the DWPG, which was assessed to be $((60\text{ mPa})^2 + (41 \times 10^{-6}P)^2)^{1/2}$.

4.4. Cycle Resolved Pressure Assessment

Figure ?? shows cycle resolved raw data from a single cycle from the SOP with a set pressure of the DWPG of 30.7 kPa; Figure ??a displays the measured beat frequency, $f(t)$, Figure ??b shows the changes in cavity mode numbers, $\Delta q_i(t)$, and Figure ??c illustrates the calculated shift of the unwrapped beat frequency $\Delta f_{UW}(t)$. This shows that although mode jumps are seen as steps in the beat frequency in Figure ??a, when the measured shifts in cavity mode numbers displayed in Figure ??b is taken into account, the unwrapped beat frequency illustrated in Figure ??c is a continuous function. The gas modulation had a cycle time of 200 s, distributed over a filled and an evacuated measurement cavity cycle, both lasting 100 s (denoted t_I and t_{II} in Figure ??, respectively).

The filled measurement cavity cycle was initiated at 0 s by the closing of valve $V_{S.2}$ and an opening of valve $V_{S.1}$, which results in a fast increase of the pressure. The MFC was then filling the system (resulting in a constant increase of the pressure) for a time of 20 s (referred to as t_f in Figure ??), until the set pressure was reached.

After the set pressure was reached, the piston in the DWPG was floating, which resulted in a stabilization of the pressure at a constant pressure for 80 s (denoted t_s in Figure ??, given by $t_I - t_f$). The filled measurement cavity assessment, $f_{UW}(n, t)$, was measured during the last 20 s of this period.

Thereafter, valve 1 was closed and valve 2 was opened, which resulted in a fast decrease in pressure (an increase in the unwrapped beat frequency). Both cavities were then evacuated for 100 s. The empty measurement cavity assessment, $f_{UW}(n = 1, t)$, was measured during the last 20 s of this period. After this, the cycle was repeated.

These signals were then converted into pressure by Equations (2)–(4) and (**??**). Figure **??** shows the cycle resolved pressure calculated by these means from the $f_{UW}(t)$ data displayed in Figure **??**. Note that the data in Figure **??** include several mode jumps during the filling (t_f) and the emptying stages that produced short "sparks" in the unwrapped pressure. Since the evaluation procedure did not use data points during the these stages (i.e., when the mode hops take place), they did not affect the final assessments.

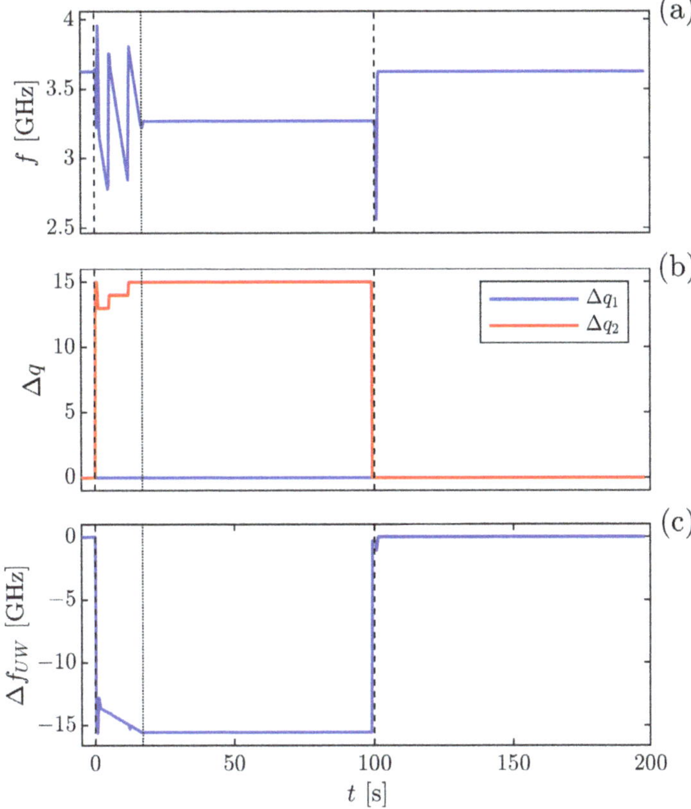

Figure 5. The time evolution of (**a**) the raw beat frequency [$f(t)$]; (**b**) the evaluated shift in mode number ($\Delta q_1(t)$ and $\Delta q_2(t)$); and (**c**) the corresponding unwrapped beat frequency ($f_{UW}(t)$) from the SOP over a 200 s long modulation cycle. For descriptions of the various time intervals of the modulation cycle, see the caption of Figure **??**.

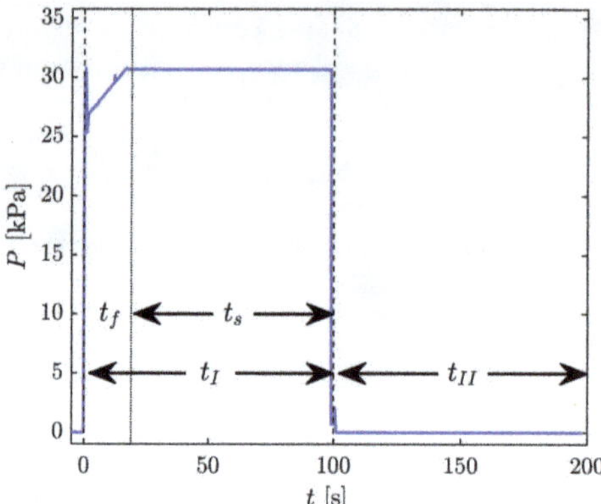

Figure 6. The time evolution of the pressure assessed in the SOP during the 200 s long gas modulation cycle displayed in Figure **??**, $P(t)$. t_I represents the time of filling and t_{II} the time of evacuation during the modulation cycle, here set to 100 and 100 s, respectively. t_f is the time at which the MFC was re-filling the DWPG, and t_s is the time during which the DWPG was stabilizing the pressure (i.e., when the piston was floating).

4.5. Evaluation of the Degree of Short-Term Concordance between the Two Refractometers

To properly evaluate the degree of short-term concordance between assessments made by the two refractometers, they were jointly connected to the DWPG while extraordinarily long gas modulation cycles (300 s) were used (each comprising filling and evacuation periods of 250 and 50 s, respectively). A series of 20 such gas modulation cycles (which thus took 100 min) were performed. To reduce the load on the turbo pump, which was significantly affected by the repeated out pumping of gas at high pressure, this evaluation was performed at a set pressure of 16 kPa.

Figure **??** shows two typical consecutive modulation cycles. Figure **??**a–c encompass the same information, although Figure **??**b,c are zooms of the data with 10^2 and 10^4 times magnification of (a), respectively.

The set pressure of the DWPG, estimated by use of Equation (**??**), is marked with the black (almost fully horizontal) curve. The pressure readings from the pressure gauges in the SOP and TOP systems, $G_{S.1}$ and $G_{T.1}$, respectively, are represented by the green and purple curves, respectively. The SOP and TOP refractometry signals are represented by the blue and red curves, respectively. In this comparison, the gauges and the refractometers have, for clarity, been adjusted by an offset (for SOP and TOP by 0.11 and 0.12 Pa, respectively) to overlap the set pressure of the DWPG. It is worth noting that said adjustments are well within their uncertainty budgets and the uncertainty of the DWPG (from the uncertainty presented in Section **??**, at 16 kPa the SOP, TOP, and DWPG uncertainties are 0.16, 0.45, and 0.66 Pa, respectively). It also does not affect their short-term performance.

It is noteworthy that, in Figure **??**a,b, the refractometer signals are not visible. This is due to the fact that they fluctuated less than the thickness of the DWPG curve and are hence hidden behind it. In Figure **??**b, the pressure gauge readings (the green and purple curves) show bit noise; i.e., they fluctuate between two bits, an amount of 4 Pa (corresponding to 250 ppm of 16 kPa).

In Figure **??**c, the gauges are off-scale, but the fluctuations of the refractometer signals are clearly visible. It is also worth noting that although there are significant fluctuations in both refractometer signals, there is a large degree of concordance between them. There is a slight tendency (predominantly seen during the first 50 s of the cycles) that the response of the TOP drifts with respect to that of the SOP. This is attributed to the fact that the gas lines to the TOP, because of practical reasons, had to be significantly longer than those to the SOP. This implies that the evacuation of the TOP during the evacuation cycle was not as efficient as that of the SOP. Hence, when the gas filling cycle commenced, there was a slightly higher residual pressure in the reference cavity of the TOP than that of the SOP. During the gas filling cycle, in which the data in Figure **??**c were taken, and during which the reference cavities were constantly evacuated, the reference cavity of the TOP was further pumped down, resulting in an artificial drift of the TOP signal during the first 50 s of the gas filling cycle. It is important to note that this neither affects the level of correlation, nor does it imply that the GAMOR refractometers drift under normal working conditions; this drift takes place only because of the fact that the TOP system in this work, because of practical reasons, had to be connected to the gas system with unusually long gas lines.

To emphasize the degree of concordance of the two refractometer signals, 70 s of the pressures assessed by the refractometers depicted in the first cycle of Figure **??** are plotted at an enlarged scale in Figure **??**a.

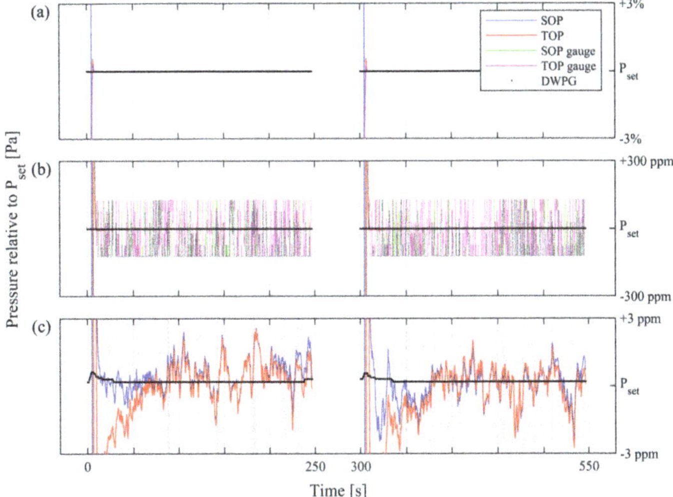

Figure 7. The 250 s parts of two consecutive (300 s long) gas modulation cycles when the measurement cavities contain gas for a DWPG set pressure of 16 kPa. The three panels display the same data centered around the set pressure but with dissimilar y-scales: in panel (**a**) with a scale of ±3%; panel (**b**) with a scale of ±300 ppm; and panel (**c**) with a scale of ±3 ppm. Black curves: the DWPG; green curves: the SOP gauge; purple curves: the TOP gauge; blue curves: the SOP; and red curves: the TOP.

The degree of concordance between the two refractometer signals in Figure **??**a is striking and impressive; they jointly and concurrently illustrate a common fluctuating pressure. The fact that they detected in unison the same fluctuation indicates that the signals originated from the DWPG and gas delivery system rather than from the refractometers.

To assess the degree of correlation between the two refractometer signals, they are plotted against one another in Figure **??**b. The data show that while the pressure produced by the DWPG and the gas delivery system fluctuated more than 5 ppm, the two refractometer signals differed significantly less than 0.5 ppm. A correlation analysis of the data provides a remarkable correlation coefficient of the two refractometry signals of 0.995.

To complement the correlation analysis above, the measurement data were also exposed to an Allan variance evaluation. Figure **??** displays, in terms of Allan deviations, the pressure assessments made by the SOP (blue curves), the TOP (the red curves), and their difference (yellow curves). Figure **??**a displays all 20 assessments separately, and Figure **??**b shows their average.

This figure shows a large degree of concordance also between the Allan plots of the 20 individual assessments for both the SOP and the TOP (as well their difference).

This indicates that both refractometry systems were stable over the total period of the measurements. The standard deviation of the individual readings of each refractometer (the left-most points in Figure **??**b) was 0.1 ppm, and their difference only had a standard deviation of 0.04 ppm.

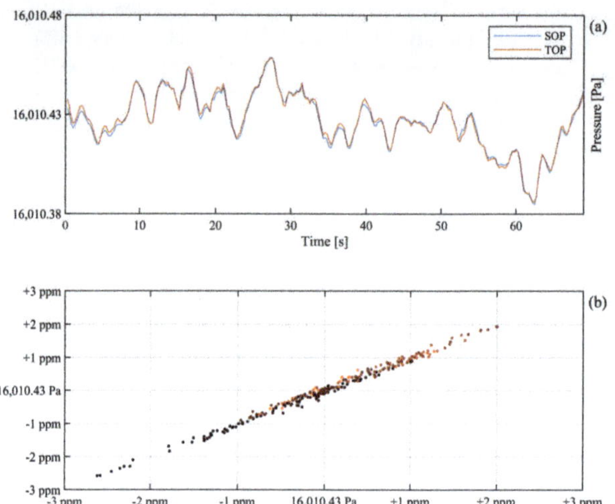

Figure 8. (**a**) An enlargement of 70 s of the refractometry data shown in the first cycle of Figure **??**. (**b**) A correlation plot of the same data. The *x* and the *y*-axes represent the pressures assessed by the SOP and the TOP, respectively. In the latter, time is represented by the color, where the first data points are marked with orange and the last ones are in black.

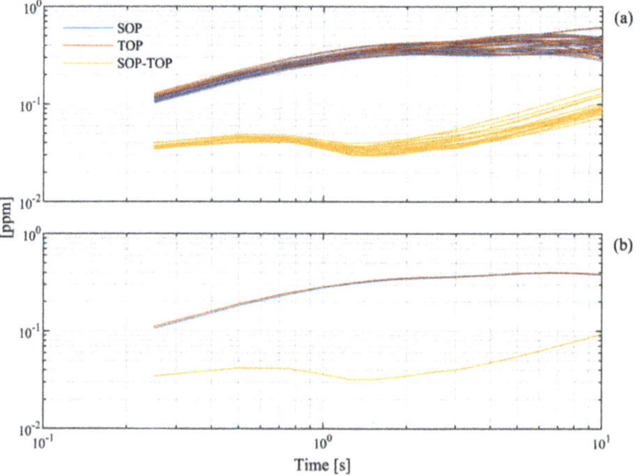

Figure 9. (**a**) The Allan deviation (as a function of averaging time) for the SOT refractometer (blue curves), the TOP refractometer(red curves), and their difference (yellow curves) assessed over 20 cycles. (**b**) The average of each set of data in panel (**a**).

5. Conclusions and Discussion

This paper provided scrutiny of the short-term performance capabilities of refractometry instrumentation based on the GAMOR methodology. We did so by comparing two independent refractometry systems coupled to a common dead weight piston gauge (DWPG). In contrast to conventional GAMOR-based refractometry, in which static pressures are assessed by the use of (and averaging over) a series of gas modulation cycles, the short-term assessments are performed *within* individual modulation cycles. More precisely, the short-term response is scrutinized through the use of a methodology in which the pressure, $P(t)$, is continuously assessed by the shift in the incessantly measured unwrapped beat frequency, $\Delta f_{UW}(t)$, during individual modulation cycles. By this, GAMOR based instrumentation can assess fluctuations of pressures on time scales below the gas modulation time period. In addition, the methodology allows for an investigation of the ability of GAMOR to assess short-term fluctuations of pressures without any influence from the pressure producing and gas delivery system.

Figures **??** and **??** display the typical cycle resolved response from one of the GAMOR instrumentations. The data show that, despite discrete mode jumps, the unwrapped beat frequency provides a continuous signal. The data also show that, in agreement with findings to be mediated by Rubin et al. [?], the response settled, within the resolution of the figure, within a fraction of the gas modulation period (the data do not show any visible drifts over the 80 s long time period denoted t_s). This vouches for the possibility to assess rapid changes in pressure during this time period.

Figure **??** shows that the simultaneous assessments performed by the two refractometry systems had a high degree of correlation; the deviations in their assessments were not only significantly smaller than those provided by the pressure gauges (which are limited by bit noise on a level orders of magnitude above that of the refractometers), they were also markedly smaller than those of the pressure assessed by either of them. A similarly excellent correlation between the pressure assessments performed by the two refractometers is shown in Figure **??**. This indicates that their precision is significantly better than the stability of the pressure they assess, which implies that the deviations of the assessed pressure are attributable to fluctuations in the pressure in the DWPG and the gas delivery system, rather than to the performance of the refractometers.

These fluctuations can potentially have several causes, e.g., the ambient pressure, the gas temperature, vibrations, or fluctuations in the pressure produced by the DWPG. However, since Figure **??** shows that they take place over seconds, most of these potential causes are improbable. Instead, we presently attribute the most likely cause of the pressure fluctuations to the pressure produced by the DWPG.

Since Figure **??**b indicates that the deviation of the difference assessment for time scales up to a few seconds was 0.04 ppm, it can be concluded that the short-term deviation of the pressure assessment by the use of a single refractometer was 0.03 ppm. Although pressure assessments of 4303 Pa have been demonstrated with an Allan deviation of 0.08 ppm assessed over 10^4 s (corresponding to a standard deviation) [?], this shows that the refractometer systems have better precision than what so far has been demonstrated.

Since data were collected at 4 Hz (given by the finite updating time of the frequency counter) and there was no averaging process in the data acquisition, the bandwidth of the assessments shown in Figure **??** was 2 Hz (given by the Nyqvist theorem). Fundamentally though, this was limited by the cavity linewidth which, in this work, was in the order of 10's of kHz.

With the extraordinary temporal response of the refractometer, this type of instrumentation can not only be used to measure rapid pressure changes and fluctuations, to investigate processes giving rise to such, and resolve the difference between sensor responses and actual pressure changes, it can also be used for characterization of the dynamic responses of pressure gauges (such as Pirani gauges).

Another application is that if the pressure can be kept constant, e.g., within a system regulated by a DWPG or another type of pressure regulator, it can serve as an instrument to characterize the temporal responses of temperature sensors. Finally, under stable conditions, one can isolate acoustic effects in the infrasound region, and hence be used in relation to the dB-scale.

Author Contributions: Conceptualization, C.F. and I.S.; methodology, C.F. and M.Z.; software, C.F., I.S., and J.Z.; validation, C.F., I.S., O.A., and M.Z.; formal analysis, C.F.; investigation, C.F.; resources, O.A. and M.Z.; data curation, C.F.; writing—original draft preparation, C.F. and M.Z.; writing—review and editing, C.F., I.S., J.Z., O.A., and M.Z.; visualization, C.F. and I.S.; supervision, O.A. and M.Z.; project administration, O.A. and M.Z.; funding acquisition, O.A and M.Z. All authors have read and agreed to the published version of the manuscript.

Funding: This project (QuantumPascal, 18SIB04) has received funding from the EMPIR programme co-financed by the Participating States and from the European Union's Horizon 2020 research and innovation programme; Vetenskapsrådet (VR) (621-2015-04374 and 621-2020-05105); the UmeåUniversity Industrial doctoral school (IDS-18); the Vinnova Metrology Programme (2017-05013, 2018-04570, and 2019-05029); the Kempe Foundations (1823.U12).

Institutional Review Board Statement: Not applicable.

Informed Consent Statement: Not applicable.

Data Availability Statement: Data is contained within the article.

Conflicts of Interest: The authors declare no conflict of interest.

References

. Stock, M.; Davis, R.; de Mirandés, E.; Milton, M.J.T. The revision of the SI—The result of three decades of progress in metrology. *Metrologia* **2019**, *56*, 022001; Corrigendum in **2019**, *56*, 49502. [CrossRef]
. Jousten, K.; Hendricks, J.; Barker, D.; Douglas, K.; Eckel, S.; Egan, P.F.; Fedchak, J.; Flügge, J.; Gaiser, C.; Olson, D.; et al. Perspectives for a new realization of the pascal by optical methods. *Metrologia* **2017**, *54*, 146–161. [CrossRef]
. Andersson, M.; Eliasson, L.; Pendrill, L.R. Compressible Fabry-Perot refractometer. *Appl. Opt.* **1987**, *26*, 4835–4840. [CrossRef]
. Mari, D.; Bergoglio, M.; Pisani, M.; Zucco, M. Dynamic vacuum measurement by an optical interferometric technique. *Meas. Sci. Technol.* **2014**, *25*, 125303. [CrossRef]
. Egan, P.F.; Stone, J.A.; Hendricks, J.H.; Ricker, J.E.; Scace, G.E.; Strouse, G.F. Performance of a dual Fabry–Perot cavity refractometer. *Opt. Lett.* **2015**, *40*, 3945–3948. [CrossRef]
. Silander, I.; Hausmaninger, T.; Zelan, M.; Axner, O. Gas modulation refractometry for high-precision assessment of pressure under non-temperature-stabilized conditions. *J. Vac. Sci. Technol. A* **2018**, *36*, 03E105. [CrossRef]
. Silvestri, Z.; Bentouati, D.; Otal, P.; Wallerand, J.P. Towards an improved helium-based refractometer for pressure measurements. *Acta IMEKO* **2020**, *9*, 303–309. [CrossRef]
. Takei, Y.; Arai, K.; Yoshida, H.; Bitou, Y.; Telada, S.; Kobata, T. Development of an optical pressure measurement system using an external cavity diode laser with a wide tunable frequency range. *Measurement* **2020**, *151*, 107090. [CrossRef]
. Yang, Y.; Rubin, T.; Sun J. Characterization of a vacuum pressure standard based on optical refractometry using nitrogen developed at NIM. *Vacuum* **2021**, 110598, in press. [CrossRef]
. Fang, H.; Picard, A.; Juncar, P. A heterodyne refractometer for air index of refraction and air density measurements. *Rev. Sci. Instrum.* **2002**, *73*, 1934–1938. [CrossRef]
. Pendrill, L.R. Refractometry and gas density. *Metrologia* **2004**, *41*, 40–51. [CrossRef]
. Stone, J.A.; Stejskal, A. Using helium as a standard of refractive index: Correcting errors in a gas refractometer. *Metrologia* **2004**, *41*, 189–197. [CrossRef]
. Egan, P.F.; Stone, J.A. Absolute refractometry of dry gas to ±3 parts in 10^9. *Appl. Opt.* **2011**, *50*, 3076–3086. [CrossRef]
. Silander, I.; Zelan, M.; Axner, O.; Arrhén, F.; Pendrill, L.; Foltynowicz, A. Optical measurement of the gas number density in a Fabry–Perot cavity. *Meas. Sci. Technol.* **2013**, *24*, 105207. [CrossRef]
. Egan, P.F.; Stone, J.A.; Ricker, J.E.; Hendricks, J.H. Comparison measurements of low-pressure between a laser refractometer and ultrasonic manometer. *Rev. Sci. Instrum.* **2016**, *87*, 053113. [CrossRef]
. Silander, I.; Hausmaninger, T.; Forssén, C.; Zelan, M.; Axner, O. Gas equilibration gas modulation refractometry for assessment of pressure with sub-ppm precision. *J. Vac. Sci. Technol. B* **2019**, *37*, 042901. [CrossRef]
. Silander, I.; Forssén, C.; Zakrisson, J.; Zelan, M.; Axner, O. Invar-based refractometer for pressure assessments. *Opt. Lett.* **2020**, *45*, 2652–2655. [CrossRef]
. Zelan, M.; Silander, I.; Forssén, C.; Zakrisson, J.; Axner, O. Recent advances in Fabry-Perot-based refractometry utilizing gas modulation for assessment of pressure. *Acta IMEKO* **2020**, *9*, 299–304. [CrossRef]
. Silander, I.; Forssén, C.; Zakrisson, J.; Zelan, M.; Axner, O. Optical realization of the Pascal—Characterization of two gas modulated refractometers. *J. Vac. Sci. Technol. B* **2021**, *39*, 044201. [CrossRef]

. Axner, O.; Silander, I.; Forssén, C.; Zakrisson, J.; Zelan, M. Ability of gas modulation to reduce the pickup of fluctuations in refractometry. *J. Opt. Soc. Am. B* **2020**, *37*, 1956–1965. [CrossRef]

. Axner, O.; Forssén, C.; Silander, I.; Zakrisson, J.; Zelan, M. Ability of gas modulation to reduce the pickup of drifts in refractometry. *J. Opt. Soc. Am. B* **2021**, *38*, 2419–2436. [CrossRef]

. Rubin, T.; Silander, I; Bernien, M; Forssén, C.; Zakrisson, J.; Hao, M.; Kussicke, A; Asbahr, P.; Zelan, M.; Axner, O. Thermodynamic effects in a gas modulated Invar-based dual Fabry-Pérot cavity refractometer. **2021**, in manuscript.

. Avdiaj, S.; Yang, Y.; Jousten, K.; Rubin, T. Note: Diffusion constant and solubility of helium in ULE glass at 23 °C. *J. Chem. Phys.* **2018**, *148*, 116101. [CrossRef] [PubMed]

. Forssén, C.; Silander, S.D.; Jönsson, G.; Bjerling, M.; Hausmaninger, T.; Axner, O.; Zelan, M. A transportable refractometer for assessment of pressure in the kPa range with ppm level precision. *Acta IMEKO* **2020**, *9*, 287–292. [CrossRef]

. Silander, I.; Forssén, C.; Zakrisson, J.; Zelan, M.; Axner, O. An Invar-based Fabry-Perot cavity refractometer with a gallium fixed-point cell for assessment of pressure. *Acta IMEKO* **2020**, *9*, 293–298. [CrossRef]

. Zakrisson, J.; Silander, I.; Forssén, C.; Zelan, M.; Axner, O. Procedure for robust assessment of cavity deformation in Fabry–Pérot based refractometers. *J. Vac. Sci. Technol. B* **2020**, *38*, 054202. [CrossRef]

. Axner, O.; Silander, I.; Hausmaninger, T.; Zelan, M. Drift-free Fabry-Perot-cavity-based optical refractometry—Accurate expressions for assessments of gas refractivity and density. *arXiv* **2017**, arXiv:1704.01187v2.

. Brovelli, L.R.; Keller, U. Simple analytical expressions for the reflectivity and the penetration depth of a Bragg mirror between arbitrary media. *Opt. Commun.* **1995**, *116*, 343–350. [CrossRef]

. Buckingham, A.D.; Graham, C. The Density Dependence of the Refractivity of Gases. *Proc. R. Soc. A* **1974**, *337*, 275–291. [CrossRef]

. Achtermann, H.J.; Magnus, G.; Bose, T.K. Refractivity virial coefficients of gaseous CH_4, C_2H_4, C_2H_6, CO_2, SF_6, H_2, N_2, He, and Ar. *J. Chem. Phys.* **1991**, *94*, 5669–5684. [CrossRef]

31. Jaeschke, M.; Hinze, H.M.; Achtermann, H.J.; Magnus, G. PVT data from burnett and refractive index measurements for the nitrogen–hydrogen system from 270 to 353 K and pressures to 30 MPa. *Fluid Phase Equilibria* **1991**, *62*, 115–139. [CrossRef]

. Egan, P.F.; Stone, J.A.; Scherschligt, J.K.; Harvey, A.H. Measured relationship between thermodynamic pressure and refractivity for six candidate gases in laser barometry. *J. Vac. Sci. Technol. A* **2019**, *37*, 031603. [CrossRef]

. Sutton, C.M. The pressure balance as an absolute pressure standard. *Metrologia* **1994**, *30*, 591–594. [CrossRef]

. Axner, O.; Silander, I.; Forssén, C.; Zakrisson, J.; Zelan, M. Assessment of gas molar density by gas modulation refractometry: A review of its basic operating principles and extraordinary performance. *Spectrochim. Acta B* **2021**, *179*, 106121. [CrossRef]

Article

Direct Comb Vernier Spectroscopy for Fractional Isotopic Ratio Determinations

Mario Siciliani de Cumis [1,2,3,*,†], **Roberto Eramo** [2,3,†], **Jie Jiang** [4], **Martin E. Fermann** [4] and **Pablo Cancio Pastor** [2,3,†]

[1] Agenzia Spaziale Italiana, Contrada Terlecchia SNC, 75100 Matera, Italy
[2] Istituto Nazionale di Ottica, INO-CNR, Via N. Carrara 1, 50019 Sesto Fiorentino, Italy; roberto.eramo@ino.cnr.it (R.E.); pablo.canciopastor@ino.cnr.it (P.C.P.)
[3] Dipartimento di Fisica, Universitá degli Studi di Firenze, Via G. Sansone 1, 50019 Sesto Fiorentino, Italy
[4] IMRA America, Inc., 1044 Woodridge Avenue, Ann Arbor, MI 48105, USA; jjiang@imra.com (J.J.); mfermann@imra.com (M.E.F.)
* Correspondence: mario.sicilianidecumis@asi.it; Tel.: +39-0835 377553
† These authors contributed equally to this work.

Abstract: Accurate isotopic composition analysis of the greenhouse-gasses emitted in the atmosphere is an important step to mitigate global climate warnings. Optical frequency comb–based spectroscopic techniques have shown ideal performance to accomplish the simultaneous monitoring of the different isotope substituted species of such gases. The capabilities of one such technique, namely, direct comb Vernier spectroscopy, to determine the fractional isotopic ratio composition are discussed. This technique combines interferometric filtering of the comb source in a Fabry–Perot that contains the sample gas, with a high resolution dispersion spectrometer to resolve the spectral content of each interacting frequency inside of the Fabry–Perot. Following this methodology, simultaneous spectra of ro-vibrational transitions of $^{12}C^{16}O_2$ and $^{13}C^{16}O_2$ molecules are recorded and analyzed with an accurate fitting procedure. Fractional isotopic ratio $^{13}C/^{12}C$ at 3% of precision is measured for a sample of CO_2 gas, showing the potentialities of the technique for all isotopic-related applications of this important pollutant.

Keywords: isotopic ratio; frequency comb; Vernier spectroscopy

Citation: Siciliani de Cumis, M.; Eramo, R.; Jiang, J.; Fermann, M.E.; Cancio Pastor, P. Direct Comb Vernier Spectroscopy for Fractional Isotopic Ratio Determinations. *Sensors* **2021**, *21*, 5883. https://doi.org/10.3390/s21175883

Academic Editor: Krzysztof M. Abramski

Received: 30 June 2021
Accepted: 25 August 2021
Published: 31 August 2021

Publisher's Note: MDPI stays neutral with regard to jurisdictional claims in published maps and institutional affiliations.

1. Introduction

Measuring the isotope ratio of chemical substances (carbon, water, chlorine and so on) has a large variety of applications in environmental sciences [1–5]. Providing a monitor of the emission in the atmosphere of greenhouse gases has potential application in global climate warnings, as well as in general monitoring [6–9], water cycle studies [10–12], and in general for establishing formation mechanisms and strategies [13–17]. Additionally, biomedical applications benefit from accurate fractional isotopic ratio measurements, particularly in human breath analysis [18,19], where it is possible to detect biomarkers related to specific diseases or metabolic processes, or even in pharmacological research [20]. Finally, isotopic ratio measurements are employed in space research, in star dynamics and planet studies [21]; the particular measurement of the $^{13}C/^{12}C$ ratio, which we consider in this work, has additional interests in carbon capture and storage monitoring [22] and in volcanic gas processes studies [23]. According to the type of application, the desirable target for accuracy and precision in such measurements could be different. For example, in very demanding biomedical applications (i.e., breath test for disease diagnosis or metabolic status monitoring), the accuracy and precision level for the $\mathcal{R}_{^{13}C/^{12}C}$ isotopic ratio could be lower than 0.5% [24,25]

Broadband spectroscopic techniques that use optical frequency combs (OFC) as an interaction laser source have become very popular for multiplexed detection of molecular species. Among all molecular parameters measurable with such techniques, the accurate determination of the isotopic composition of a gas sample is surely one of the most challenging applications.

OFC-based spectroscopic techniques with dual comb [26], Fourier transform [27] and spatially dispersive [28–31] detection schemes, sometimes combined with Fabry–Perot enhanced spectroscopy [29–32], are widely used for trace gas detection [19,30,31,33–37]. However, the most accurate results regarding fractional isotopic ratio measurements were obtained by direct frequency comb dispersive spectroscopy (DFCDS) [25,38]. Bailey and coworkers [38] used a MIR-FC-based cross-dispersed spectrometer to measure a fractional isotopic abundance of nitrous oxide (N_2O) with a precision of 6.7×10^{-6} in 1 s of acquisition time. Such measurements can be used to determine sources, skins and mechanisms of formation of this potent greenhouse gas and ozone-depleting agent, helping to improve current mitigation strategies. Similarly, accurate optical number density of ^{12}C and ^{13}C single substituted isotopic variants of $C^{16}O_2$ gas with a precision of, respectively, 0.03% and 1.24%, were measured by using DFCDS in the near-IR [25]. These kinds of measurements open the way for environmental monitoring and biomedical sciences applications of this OFC-based technology. DFCDS exploits the broadband coverage and spectroscopic resolution and irradiance of the laser source to allow simultaneous detection of almost all isotopic components of the targeted gas with short acquisition times and with a compact technology. The accuracy of the resulting fractional isotopic ratio measurements are comparable to single-frequency-laser-based spectrometers [7,8,18,39–42] and to the mass spectrometry performance currently used for these kinds of measurements. Such accuracy is ensured by a not-trivial calibration of the spectral instrumental response of the DFCDS apparatus in order to identify the contribution of each interacting frequency of the comb with the sample gas. In this article, we report the capabilities of a slightly modified DFCDS approach, called direct comb Vernier spectroscopy (DCVS) [29,35], to perform fractional ratio isotopic measurements in $^{12}C^{16}O_2$ and $^{13}C^{16}O_2$ components of a CO_2 gas sample around $5005\ \mathrm{cm}^{-1}$.

Our DCVS combines efficient comb filtering by using interferometric Fabry–Perot (FP) techniques in an adequate Vernier configuration with the spectral resolution of the FP-transmitted comb by using a high-resolution dispersive spectrometer. In addition, our spectrometer well isolates each of the sample-interacting teeth of the OFC from the others, and their spectral contribution can be extracted by using a simpler instrumental response approach, opening the way to very accurate lineshape studies. As a drawback, the detected OFC portions are quite limited compared to DFCDS, reducing the possibilities to reach a larger number of isotopologues of the gas in a single acquisition. Nevertheless, the reported fractional $^{13}C/^{12}C$ at 3% of precision is a proof of the principle of the capabilities of this technique for these kinds of measurements.

2. Materials and Methods

The DCVS apparatus used for the present measurements is described in detail elsewhere [29,35]. Basically, an OFC is spectrally filtered by means of a Fabry–Perot (FP) cavity acting as an interaction cell containing the absorbing gas. The Vernier ratio (V) between the FP and the OFC (i.e., the ratio between the FP's free spectral range Δ_{FSR} and the OFC's repetition rate, f_r) is established to be enough to resolve the FP-transmitted OFC modes with a high-resolution dispersion spectrometer (SOPRA, resolution 2 GHz @ 2 μm). Indeed, after FP-filtering, spectral fractions of the OFC of about $5\ \mathrm{cm}^{-1}$ are dispersed at the output of the SOPRA instrument and detected with an InGaAs linear array detector. Different from the experimental setup described in [29,35], the OFC is a thulium-doped fiber-based mode-locked laser from IMRA America, Inc. It has a spectral coverage of about 40 nm to around 1970 nm with a repetition rate of about 400 MHz, with a maximum averaged power of about 4 W after the final amplification stage. The OFC is a self-referenced system, employing the first amplification stage (about 1 W of avg. power) to generate OFC emission around 1 μm through to the non-linear optical processes in optical fibers. The carrier-offset-frequency parameter of the OFC, f_0, is consequently detected by beating the teeth frequencies of the duplicated fraction of the fundamental comb emission with those of the 1 μm harmonic, and controlled by means of the phase-lock-loop (PLL) against the stable RF clock. The repetition frequency, f_r, is controlled by detecting the beat of the comb modes with a fast InSb photodetector and by mixing it with RF synthesized frequency (f_{RF}) to obtain the RF note at 150 MHz locked with a second PLL circuit. The reference clock for both f_r and f_0 locks is a 10 MHz quartz-Rb-GPS system with a relative frequency stability of 6×10^{-13} in 1 s and an accuracy of 10^{-12}, at worst. Hence, the frequency of the OFC modes is directly traceable to the primary frequency standard with these precision and accuracy figures when locked.

The FP is a linear cavity with silver-coated mirrors. The finesse is about $F = 200$ @ 2 μm with a transmission coefficient around 0.2%. The cavity length is variable and coarsely controllable by means of step-motors mounted in the kinematic mount of one of the two cavity mirrors. It's value is established by the chosen Vernier ratio V and by the requirement to be long enough to obtain an efficient enhancement of the absorption path. For the present measurements, where a simultaneous detection of the comb modes resonant with transitions of different isotopologues of the targeted molecule is needed, V should be fixed to a value which gives a Δ_{FSR} as close as possible to a multiple of the isotope shift between the probed molecular transitions. Transitions of the ^{12}C and ^{13}C isotopes of the carbon dioxide molecule around 5005 cm^{-1} are used to test the spectroscopic performances of the technique. In particular, the $(20^01–00^00)$ R(18) of ^{13}C^{16}O$_2$ @ 5004.84 cm^{-1} and the $(20^01–00^00)$ P(45) of ^{12}C^{16}O$_2$ @ 5005.27 cm^{-1} ro-vibrational transitions are selected in order to match the condition of simultaneous recording of both absorptions in the detected range of 5 cm^{-1} of the cavity-transmitted and dispersed fraction of the interacting OFC for our spectrometer . The frequency shift between these two transitions is around $\Delta \nu^{IS} = 12{,}863.55$ MHz. A Δ_{FSR} of the order of half of $\Delta \nu^{IS}$ allows one to obtain an enhanced absorption path of about 2 m (i.e., $L_{FP} = 2.4$ cm), while keeping the FP mode separation more than three times larger than the SOPRA spectral resolution. In addition, a non-integer V ratio is chosen to obtain rarefied resonance between OFC and FP while keeping the condition that the two FP transmissions are in resonance with the two selected CO$_2$ transitions. Indeed, choosing $V = 15.5$ (i.e., 31 comb modes each 2 FP modes) tailors this condition, further helping in better FP-filtering of the OFC and a better resolved image of the transmitted modes by the SOPRA diffraction spectrometer. In Figure 1 is shown a scheme of the measurements: in one case the comb/FP configuration shown is resonant with the center of ^{13}C^{16}O$_2$ transition. In such a case as shown in Figure 1a, both transitions are simultaneously probed by different comb teeth, while in Figure 1b, only the ^{12}C^{16}O$_2$ transition shows an heavily saturated absorption.

The DCVS is performed under the condition of perfect resonance between the FP and the transmitted OFC modes for each value of their optical frequency. Consequently, the FP length is actively locked to set it on resonance with the maximum transmission of the Vernier resonant modes. To this aim, one of the FP mirrors is mounted in a PZT to obtain fine control of the cavity length by detecting a small part of the FP-transmitted light in a InGaAS detector. A 3 kHz modulation is applied to the PZT, and the first derivative of the FP output detected light is obtained by means of locking detection and used as an error signal to control the cavity length to the maximum transmission condition.

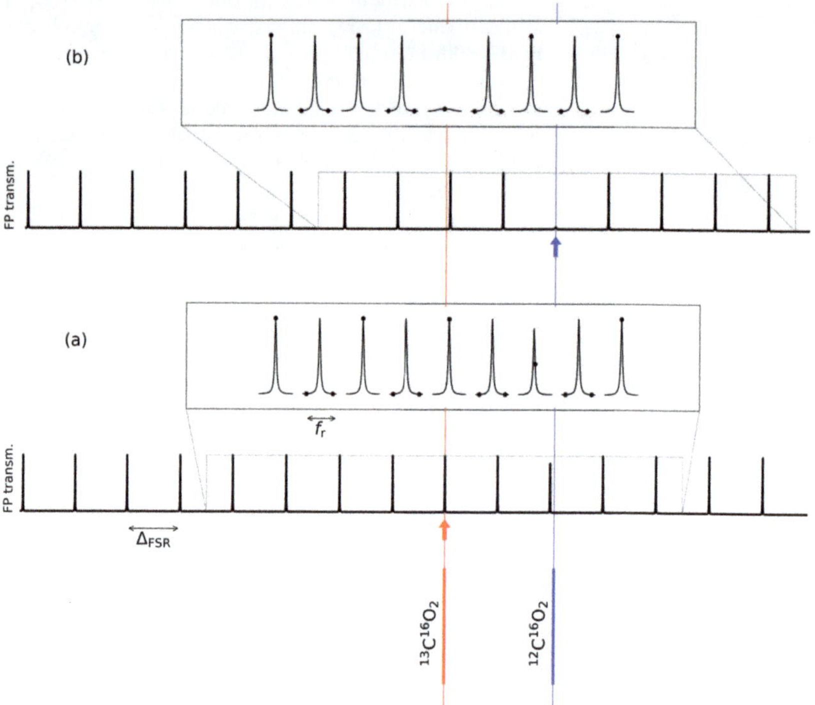

Figure 1. The two roto-vibrational transitions are probed around the two comb/FP configurations shown in figure, where the FP transmission as a function of frequency is shown in two cases. In case (**a**), one comb tooth and one FP mode are resonant with the $^{13}C^{16}O_2$ transition, while in case (**b**), they are resonant with the $^{12}C^{16}O_2$ transition. Far from resonances, the FP mode is an Airy function; around a resonance frequency, the mode is modified by the resonant refractive index of the FP medium. Here, the simulation is made by taking Voigt profiles for the two investigated transitions (Hitran parameters), and with f_r = 400 MHz, Δ_{FSR} = 6.2 GHz (i.e., $2\Delta_{FSR}/f_r = 31$), and a FP finesse $F = 200$. In each inset immediately above the FP transmission, each FP mode is zoomed (broken axis plot). We observe that one FP mode every two transmits a comb tooth, and that between adjacent zoomed FP modes, there are 15 comb modes not transmitted and not shown in the inset. We observe that in the (**a**) case, both transitions are simultaneously probed by different comb teeth, the $^{13}C^{16}O_2$ transition giving an effect not visible in the graph. The (**b**) case shows a heavily saturated $^{12}C^{16}O_2$ transition.

The majority of the FP-transmitted light is sent to the SOPRA input slit, and the diffracted light at the output slit, for a given position of the diffraction grating, is detected by a liquid N_2-cooled InGaAs array detector (PyloN-IR:1024, Princeton Instr.). The arrangement of the optical components before and after SOPRA is devised to obtain the largest spectral fraction in a single array's image, while keeping the maximum SOPRA resolution. In Figure 2d, an example of such an image is shown for f_r = 398.99 MHz. For each image, 13 transmitted modes are detected, which is a spectral portion of the OFC of about 5 cm^{-1} taking into account the 31 comb modes for each transmitted interval. The vertically integrated intensity profile of the dispersed image (Figure 2e) is used to calculate the transmitted contribution of each detected mode. As thoroughly described in [29], a knowledge of the FP-SOPRA system resolution function, i.e., its response to a monochromatic input, is necessary in order to obtain successively the OFC modes transmissions. In this paper, in order to work with an analytical resolution function, we adopt a kind of continuous wavelets [43] approximation variant of the approach followed in [29]: we fit a single mode diode laser response by a superposition of Gaussian functions, treating their centers and widths as free parameters. In practice, a superposition made of seven of such wavelets is sufficient to reproduce the experimental diode laser response within the measurement noise. Once the resolution is determined in this way, the data analysis proceeds as in [29]: the resolution function, which is now an analytical function, is replicated on the set of $N = 13$ transmitted OFC modes, giving a total fitting function for the integrated image with free parameters given by the position of one of the peaks, the peak's separation, and by the N peaks intensities, which we write as $\mathcal{I}_M(f_r, P, T)$, where the arguments f_r, and the pressure P and temperature T of the gas sample identify the experimental conditions, and where the subscript M identifies each one of the N recorded FP modes per image. Figure 2e shows an example of the fit.

The absorption spectrum of the tested molecular transitions is determined by recording the set of array images for a scan of the OFC frequencies around each transition frequency. The synthesized scan of f_r is accomplished with a change of the f_{RF} frequency in the f_r-lock chain. Custom software was used to obtain an automated acquisition of such spectral images as a function of f_r. In Figure 2b, the behaviors of the detected images as a function of the Δf_r for the FP-transmitted orders involved in the determination of the spectra of the $^{12}CO_2$ and $^{13}CO_2$ transitions are shown. For the given grating position, we label the relevant FP orders as the $M = 0$ order, resonant with the $(20^01\text{-}00^00)$ R(18) of $^{13}C^{16}O_2$ transition, the $M = +2$ order, resonant with the $(20^01\text{-}00^00)$ P(45) of $^{12}C^{16}O_2$ transition, and the $M = -2$ and $M = +4$ orders, used to calculate the transmittance spectra not-resonant with the CO_2 transitions. In addition, the scan behaviors for the $M = -12$ order, partially resonant with the $(20^01\text{–}00^00)$ R(14) transition of $^{13}C^{16}O_2$, as well as those of the not CO_2 resonant $M = -14$ and $M = -10$ modes, are shown. Figure 2c shows details of the scan of these modes as a function of the transmitted comb mode frequency, while the corresponding integrated intensities $\mathcal{I}_M$ are shown in Figure 2a.

The ratio between the integrated intensity for the molecule absorbed modes (M = 0 and +2) with the averaged intensity of the not-absorbed modes (M = −2 and +4) is used to calculate the transmittance spectrum of the $^{13}CO_2$ and $^{12}CO_2$ transitions, respectively, as shown in Figure 3 and in the inside graph of Figure 2a:

$$\mathcal{T}_M^{(s)}(\Delta\nu, P, T) \;=\; 2\frac{\mathcal{I}_M(f_r, P, T)}{\mathcal{I}_{-2}(f_r, P, T) + \mathcal{I}_{+4}(f_r, P, T)} \; (M = 0, 2) \tag{1}$$

On the right hand-side of Equation (1) we leave as arguments the parameters f_r, P and T that set the experimental conditions of the acquisition, while for the task of the successive analysis procedure where optical frequencies are required to be compared with other results, the optical detuning $\Delta\nu$ is used instead of f_r. $\Delta\nu$ is calculated as the optical detuning of the M mode with respect to the frequency of the $M = 0$ mode at the center of the f_r scan:

$$\Delta\nu \;=\; N_M f_r - N_0 f_{r_c} \; (M = 0, 2) \tag{2}$$

with the f_{r_c} repetition rate frequency at the center of the scan and with $N_2 = N_0 + 31$. N_0 is calculated from the integer ratio between the reported frequency [44] of the $^{13}C^{16}O_2$ transition and f_{r_c}.

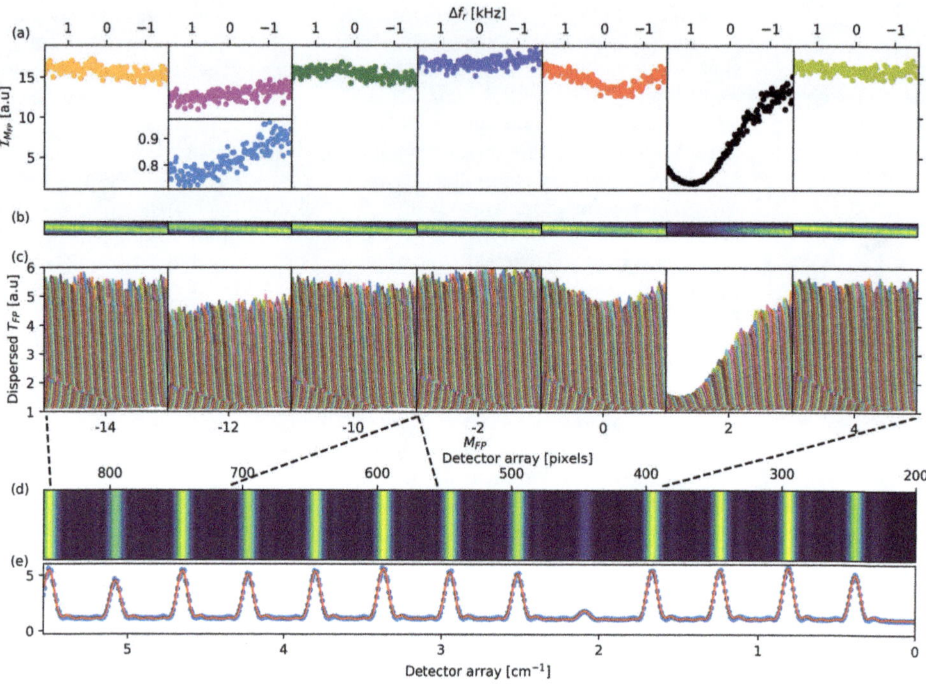

Figure 2. Spectra of the OFC modes filtered by the FP as a function of the OFC's $\Delta f_r = f_r - f_{r_c}$, with f_{r_c} repetition rate frequency at the center of the scan. The spectra are the result of the analysis of the recorded images of the SOPRA-dispersed fraction of the FP-filtered OFC vs f_r. Three of these modes are resonant with the $(20^01\text{-}00^00)$ R(14) of $^{13}C^{16}O_2$ @ 5002.21 cm^{-1}, $(20^01\text{-}00^00)$ R(18) of $^{13}C^{16}O_2$ @ 5004.84 cm^{-1}, and $(20^01\text{-}00^00)$ P(45) of $^{12}C^{16}O_2$ @ 5005.27 cm^{-1} ro-vibrational transitions; the other modes are not absorbed by the CO_2 molecule. The modes are identified by the FP order scale of panel (**c**): the -12, 0 and $+2$ orders are the modes absorbed by $^{13}C^{16}O_2$ and $^{12}C^{16}O_2$ transitions, respectively; the -14, -10, -2 and $+4$ orders are the not-absorbed ones. (**a**) Integrated transmission spectrum. (**b**) Spectral behavior of the array images for each mode. (**c**) Spectral behavior of the SOPRA-resolution wavelet for each mode. (**d**) Detector array image of the portion of the FP-transmitted OFC modes for a given f_r. (**e**) Vertically integrated intensity of the transmitted modes of image in panel (**d**) and the fit to a function resulting form the addition of 13 SOPRA-resolution functions; the intensity of the fitted resolution provides the transmission spectra values for each f_r of panel (**a**). The inside graph of panel (**a**) shows the transmittance spectrum of the R(14) transition of the $^{13}C^{16}O_2$ from the $M = -12$ mode normalized by the average intensity of the -14 and -10 modes.

Due to the limited continuous scan of the f_r in the OFC lock condition, the high frequency wings of the P(45) transition of $^{12}C^{16}O_2$ and of the R(14) transition of $^{13}C^{16}O_2$ are not recorded. Such a limitation should be overcome by using a better combination of the frequency parameters of OFC and FP. Uncertainties of the measured spectral parameters for such transitions are expected to be affected by this issue, as discussed in the following. The situation is critical for the R(14) transition of $^{13}C^{16}O_2$ because it is recorded by less than one half, as shown in the inside graph of Figure 2a. Consequently, it is not considered in the present isotopic ratio measurements.

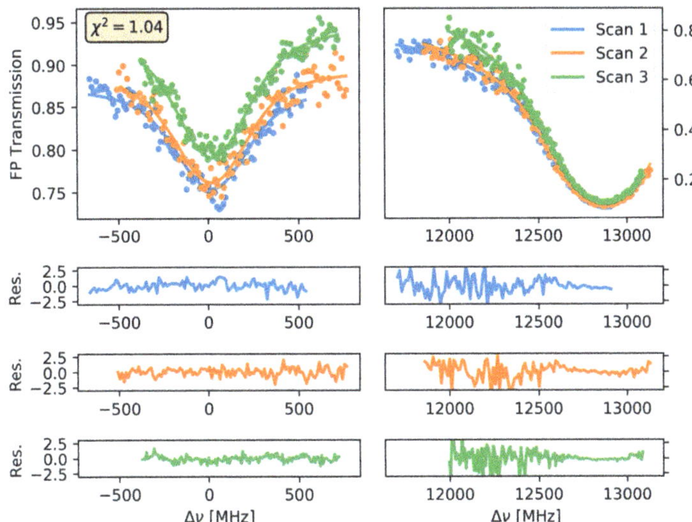

Figure 3. FP transmittance of the absorbed comb modes resonant with the $(20^01\text{-}00^00)$ R(18) of $^{13}C^{16}O_2$ (left side graphs) and $(20^01\text{-}00^00)$ P(45) of $^{12}C^{16}O_2$ (right side graphs) transitions and the fit to the $\mathcal{T}(\Delta\nu, P, T)$ (Equation (4)) of all recorded spectra by using the global fit procedure. P = 55.0 mbar and T = 296 K for the CO_2 gas sample for all acquisitions. The reduced χ^2 of the global fit is also shown.

3. Results

Three DCVS spectra of the $(20^01\text{–}00^00)$ P(45) of $^{12}C^{16}O_2$ and $(20^01\text{–}00^00)$ R(18) of $^{13}C^{16}O_2$ transitions are recorded with CO_2 gas at a pressure of P = 55 mbar and at room temperature (T = 296 K). The total transmittance of the two absorbed modes for each recording is given by the following:

$$\mathcal{T}^{(s)}(\Delta\nu, P, T) = \mathcal{T}^{(s)}_{M=0}(\Delta\nu, P, T) + \mathcal{T}^{(s)}_{M=+2}(\Delta\nu, P, T) \tag{3}$$

with $\mathcal{T}^{(s)}_M$ calculated as described above (Equation (1)).

In order to determine isotope dependent relevant spectroscopic parameters between transitions, namely, frequency shift $\Delta\nu^{IS} = \nu^o(^{12}C) - \nu^o(^{13}C)$ and natural isotopic concentration ratio $\mathcal{R}_{^{13}C/^{12}C} = a^{^{13}C}/a^{^{12}C}$ with a^{IS} isotopic abundance of the IS isotope in the gas sample, the $\mathcal{T}^{(s)}$ data are fitted to a function that describes the absorption modified FP Airy transmission at maximal optical resonance [29,35,45]:

$$\mathcal{T}(\Delta\nu, P, T) = K \sum_{M=0,+2} \mathcal{T}_M(\Delta\nu, P, T) + (A + B\Delta\nu) \tag{4}$$

where a linear spectral background, with A and B as the frequency independent and slope parameters, respectively, and a scale factor K are considered to take into account possible not-compensated instrumental effects, due to the transmittance normalization. In Equation (4), $\mathcal{T}_M$ is given by the following:

$$\mathcal{T}_M(\Delta\nu, P, T) = \frac{\mathcal{T}_{\max}(\alpha_M)}{1 + F(\alpha_M)\sin^2\left(\frac{\pi}{\Delta_{FSR}}(\Delta\nu\Delta n_M + \Delta_{lock})\right)}$$

$$\mathcal{T}_{\max}(\alpha_M) = \frac{T_m^2 e^{-\alpha_M L}}{(1 - R_m e^{-\alpha_M L})^2} \tag{5}$$

$$F(\alpha_M) = 4R_m \frac{e^{-\alpha_M L}}{(1 - R_m e^{-\alpha_M L})^2}$$

where T_m and R_m are the transmission and reflectivity coefficients of the FP mirrors, and $L = c/2\Delta_{FSR}$ is the cavity length. The argument of $\sin^2$ in the Airy function is the FP round trip phase shift, which is written as the sum of the empty cavity contribute, which is $\pi\Delta_{lock}/\Delta_{FSR}$ for the locked cavity, and $\pi\Delta\nu\Delta n_M/\Delta_{FSR}$, representing the contribution due to the gas dispersion for the M mode. Δn_M and α_M are the dispersion and absorption coefficients, respectively, induced on the M-mode by resonant absorption transitions of the gas sample, which are calculated as a function of the laser detuning $\Delta\nu$ and the thermodynamic conditions, P and T, from the real and imaginary part of FP's refraction index variation: $(n-1)_M = \Delta n_M + i\alpha_M/2k_M$, with k_M being the wave vector module of the absorbed M comb mode. Because $\Delta\nu^{IS}$ between the transitions is very large when compared to their linewidth in our thermodynamic conditions, any spectral interference effects between the two CO_2 transitions can be safely neglected. Consequently, $(n-1)_M$ contributions for the $M = 0$ and $M = 2$ modes can be considered to be only induced separately by the $^{13}CO_2$ and $^{12}CO_2$ transitions, respectively, simplifying the analysis. If we label as $t = a$ and $t = b$ these $^{13}CO_2$ and $^{12}CO_2$ transitions, and considering a Voigt profile for the CO_2 absorptions, we have the following:

$$\Delta n_M + i\frac{\alpha_M}{2k} = -\frac{cN\,a^{IS}\,S_t}{2\pi^{3/2}k_M\Delta\nu_t^D} \int_{-\infty}^{\infty} du\, \frac{e^{-u^2}}{\frac{\Delta\nu - \Delta\nu_t^o}{\Delta\nu_t^D} - u + i\frac{\Gamma_t/2}{\Delta\nu_t^D}} \tag{6}$$

with the correspondence $M = 0,2 \Leftrightarrow t = a,b$. In Equation (6), c is the light speed, N is the numeric density of the gas at the given thermodynamic conditions, a^{IS} is the abundance of the isotopologue IS in the gas mixture, S_t is the linestrength of the transition t per molecule $[S_t = S_{HI}/a^{IS}$, with S_{HI} the linestrength value of the HITRAN database [44]. In this way, the isotopic abundances and their ratios can be measured independently of the transition's levels], $\Delta\nu_t^D$ is the FWHM Doppler linewidth of the transition t of the isotopologue IS at T, $\Delta\nu_t^o$ is the transition frequency detuning, $\Gamma_t = \frac{\delta\Gamma_t}{\delta P}P$ is the FWHM collisional linewidth contribution at P and T and u represents the Doppler shift of each molecular class velocity.

The least-squared fit procedure of $\mathcal{T}$ (Equation (4)) to $\mathcal{T}^{(s)}$ determines the best values of the molecular parameters, $\Delta\nu_t^o$, S_t, a^{IS}, $\Delta\nu_t^D$ and Γ_t for both transitions as well as of the instrumental-related parameters Δ_{FSR}, F, Δ_{lock}, K, A, and B. As in previous measurements [29,35,45], two different fit strategies are performed. In the first approach, the final values of the relevant parameters are calculated from the weighted average of those values resulting from the fits of each individual scan at pressure P. In the other approach, all acquired spectra are considered in a single global fit, where some fit parameters (i.e., molecular-related parameters, Δ_{FSR}, and F) are considered shared between all the scans, while Δ_{lock}, K, A, and B are local parameters, considered to be different for each scan. In addition, parameters that can be evaluated independently, such as Δ_{FSR}, S_t, and $\Delta\nu_t^D$ for both transitions are kept fixed during the fit procedure. The result of this global fit is graphically shown in Figure 3. A summary of the ν_t^o, a^{IS} and $\delta\Gamma_t/\delta P$ parameters for both transitions and the determined values $\Delta\nu^{IS}$ and $\mathcal{R}_{^{13}C/^{12}C}$ from them are reported in Table 1. Their differences against the values reported in the HITRAN database [44] are also tabulated.

Table 1. $\Delta \nu^{IS}$ and $\mathcal{R}_{^{13}C/^{12}C}$ determinations from the measured spectral parameters of the $(20^01\text{-}00^00)$ R(18) of $^{13}C^{16}O_2$ and $(20^01\text{-}00^00)$ P(45) of $^{12}C^{16}O_2$ transitions from DCVS measurements @ 5005 cm^{-1} and comparison with HITRAN database values [44]. Absolute frequencies are calculated by $\nu_f^o = N_0 f_{r_c} + f_o + \Delta \nu_f^o$ with N_0 order number of the OFC tooth transmitted by the M=0 FP mode and resonant with the $^{13}C^{16}O_2$ transition and f_{r_c} and f_o repetition rate frequency at the center of the scan and offset frequency of the OFC, respectively. (Errors reported in parentheses). $\Delta_{(DCVS-HI)}$ are the differences between HITRAN database values and the present measured values for each tabulated parameter. The HITRAN database values are $\nu_o = 150041403.1\ (2)$ MHz, $a^{13}C = 0.01106\ (1)$, $\delta\Gamma/\delta P = 5.92\ (5)$ MHz/mbar for the $(20^01\text{-}00^00)$ R(18) transition of $^{13}C^{16}O_2$ and $\nu_o = 150{,}054{,}266.5\ (2)$ MHz, $a^{12}C = 0.9842\ (9)$, $\delta\Gamma/\delta P = 4.56\ (4)$ MHz/mbar for the $(20^01\text{-}00^00)$ P(45) transition of $^{12}C^{16}O_2$. The frequency shift between them is $\Delta \nu^{IS} = 12{,}863.5\ (3)$ and the natural isotopic concentration ratio $\mathcal{R}_{^{13}C/^{12}C} = 0.01123\ (2)$. Errors of this differences take into account the error of the HITRAN values, which are added in quadrature to the measured ones.

Parameter	Global Fit	$\Delta_{(DCVS-HI)}$	Weighted-Average of Individual Fits	$\Delta_{(DCVS-HI)}$
$\Delta\nu^{IS}$ [MHz]	12,868.8 (6.4)	5.3 (6.4)	12,860.8 (6.0)	−2.7 (6.0)
$\mathcal{R}_{^{13}C/^{12}C}$	0.0116(4)	0.0004 (4)	0.0116 (4)	0.0004 (4)
Transition a	$(20^01\text{--}00^00)$ R(18)	$^{13}C^{16}O_2$		
ν^0 [MHz]	150,041,400.3 (6.0)	−2.6 (6.0)	150,041,404.3 (5.6)	1.4 (5.6)
$a^{13}C$	0.0109 (3)	−0.0002(3)	0.0107 (2)	−0.0004(2)
$\delta\Gamma/\delta P$ [MHz/mbar]	5.3 (3)	−0.6 (3)	4.5 (3)	−1.4 (3)
Transition b	$(20^01\text{--}00^00)$ P(45)	$^{12}C^{16}O_2$		
ν^0 [MHz]	150,054,269.2 (2.0)	2.7 (2.0)	150,054,265.2 (2.1)	−1.3 (2.1)
$a^{12}C$	0.94 (2)	−0.04(2)	0.92 (3)	−0.06 (3)
$\delta\Gamma/\delta P$ [MHz/mbar]	4.0 (1)	−0.6 (1)	3.8 (2)	−0.8 (2)

4. Discussion

Abundances of $^{12}C^{16}O_2$ and $^{13}C^{16}O_2$ are measured with a precision of 2.1% and 2.7%, respectively, in agreement with the HITRAN values. Consequently, a fractional isotope ratio $\mathcal{R}_{^{13}C/^{12}C} = 0.0116\ (4)$ is determined, as expected for a CO_2 gas sample with the current natural isotopic content [44]. The present measurements can be considered a proof of principle of the DCVS applied to fractional isotopic ratio determinations, even if the experiment was performed at only one pressure, which was selected for giving the best precision performance for both isotopes in a single OFC scan. The final uncertainty is due to several issues: S/N ratio of each transmittance spectrum, limited scan of the full spectral profile of the transitions, and saturated absorption effects. The last two issues are particularly present in the $^{12}C^{16}O_2$ transition, which shows a quasi total saturation of the absorption, and the high frequency side is unrecorded. The final precision of $a^{12}C$ determination is strongly limited by these experimental issues, which must be avoided to obtain the determinations for this isotope that are comparable to those achieved in other DFCDS [25]. Measurements at lower pressures could avoid saturation effects for this transition, paying for a decrease in the S/N ratio of the $^{13}C^{16}O_2$ transition by considering that the gas sample and absorption path are shared for both transitions. We are confident that measurements with gas sample pressure between 10 and 55 mbar could lead to a result that is compatible with the present uncertainties of the a^{13C}. In addition, detection of the complete transition profiles would allow a significant improvement of the spectral parameters determination, including a^{12C}, even at the gas pressure of the experiment. A more accurate choice of the frequency parameters of the OFC and FP could allow to center both transitions in the present maximum continuous scan of the OFC frequency for a single acquisition. Alternatively, consecutive OFC-shifted scans could be combined to increase the scan range. Through global fits, spectra recorded at different gas pressures, and involving other transitions of the sample gas, could be considered all together, increasing the final precision. Finally, combining detection and scan schemes as those described in [30,31] with the high Vernier filtering of our DCVS spectrometer, would allow faster broadband acquisitions.

Besides the fractional isotopic ratio, other spectral parameters of the targeted transitions are determined in our measurements. We find the transition frequencies to be in agreement with the HITRAN values, considering one standard deviation uncertainty. Their absolute value is reported with a precision of 1×10^{-8} and 4×10^{-8} for the $^{12}C^{16}O_2$ and $^{13}C^{16}O_2$ transitions, respectively. Instead, a small disagreement with the HITRAN values of the collisional broadening coefficients should be noted. Nevertheless, we believe that for our single pressure measurement, such discrepancies could be expected for a parameter that is just measuring a pressure-induced effect.

The present results show the capabilities of DCVS for precise measurements of the fractional isotopic ratios in a sample gas, with potential applicability in the detection of rare isotopologues [46], where absorption background from the others must be avoided. In principle, our technique could be applied to different kinds of OFC (ICL and QCL combs) in order to realize a compact setup toward the mid-infrared region.

Author Contributions: Conceptualization and methodology, R.E., M.S.d.C. and P.C.P.; comb laser development, J.J. and M.E.F., experimental data recording, M.S.d.C. and P.C.P.; formal analysis and software, R.E.; data analysis, P.C.P.; writing—original draft preparation, P.C.P.; writing—review and editing, M.S.d.C. and R.E. All authors have read and agreed to the published version of the manuscript.

Funding: This research was funded by the project Extreme Light Infrastructure-Italy (ELI-Italy) and by the QOMBS project (FET Flagship on Quantum Technologies grant No. 820419).

Institutional Review Board Statement: Not applicable.

Informed Consent Statement: Not applicable.

Data Availability Statement: The data supporting this paper are available from the corresponding authors upon reasonable request.

Conflicts of Interest: The authors declare no conflict of interest.

References

1. Kerstel, E.; Gianfrani, L. Advances in laser-based isotope ratio measurements: selected applications. *Appl. Phys. B* **2008**, *92*, 439–449. [CrossRef]
2. Torn, M.S.; Biraud, S.C.; Still, C.J.; Riley, W.J.; Berry, J.A. Seasonal and interannual variability in 13C composition of ecosystem carbon fluxes in the U.S. Southern great plains. *Tellus B Chem. Phys. Meteorol.* **2011**, *63*, 181–195. [CrossRef]
3. Bauska, T.K.; Baggenstos, D.; Brook, E.J.; Mix, A.C.; Marcott, S.A.; Petrenko, V.V.; Schaefer, H.; Severinghaus, J.P.; Lee, J.E. Carbon isotopes characterize rapid changes in atmospheric carbon dioxide during the last deglaciation. *Proc. Natl. Acad. Sci. USA* **2016**, *113*, 3465–3470. [CrossRef]
4. Yakir, D.; da Sternberg, L.S.L. The use of stable isotopes to study ecosystem gas exchange. *Oecologia* **2000**, *123*, 297–311. [CrossRef]
5. Yoshida, N.; Toyoda, S. Constraining the atmospheric N_2O budget from intramolecular site preference in N_2O isotopomers. *Nature* **2000**, *405*, 330–334. [CrossRef]
6. Assonov, S.S., Brenninkmeijer, C.A.M.T.; Schuck, J.; Taylor, P. Analysis of ^{13}C and ^{18}O isotope data of CO_2 in CARIBIC aircraft samples as tracers of upper troposphere/ lower stratosphere mixing and the global carbon cycle. *Atmos. Chem. Phys.* **2010**, *10*, 8575-8599. [CrossRef]
7. Wolf, B.; Merbold, L.; Decock, C.; Tuzson, B.; Harris, E.; Six, J.; Emmenegger, L.; Mohn, J. First on-line isotopic characterization of N2O above intensively managed grassland. *Biogeosciences* **2015**, *12*, 2517–2531. [CrossRef]
8. Harris, E.; Henne, S.; Hüglin, C.; Zellweger, C.; Tuzson, B.; Ibraim, E.; Emmenegger, L.; Mohn, J.J. The isotopic composition of atmospheric nitrous oxide observed at the high-altitude research station Jungfraujoch, Switzerland. *Geophys. Res. Atmos.* **2017**, *122*, 1850–1870. [CrossRef]
9. Delli Santi, M.G.; Bartalini, S.; Cancio, P.; Galli, I.; Giusfredi, G.; Haraldsson, C.; Mazzotti, D.; Pesonen, A.; De Natale, P. Biogenic Fraction Determination in Fuel Blends by Laser-Based $^{14}CO_2$ Detection. *Adv. Photonics Res.* **2021**, *2*, 2000069-7. [CrossRef]
10. Casado, M.; Landais, A.; Masson-Delmotte, V.; Genthon, C.; Kerstel, E.; Kassi, S.; Arnaud, L.; Picard, G.; Prie, F.; Cattani, O.; et al. Continuous measurements of isotopic composition of water vapour on the East Antarctic Plateau. *Atmos. Chem. Phys.* **2016**, *13*, 8521–8538. [CrossRef]

11. Wu, T.; Chen, W.; Fertein, E.; Masselin, P.; Gao, X.; Zhang, W.; Wang, Y.; Koeth, J.; Brückner, D.; He, X. Measurement of the D/H, ^{18}O/ ^{16}O, and 1 ^{17}O/ ^{16}O Isotope Ratios in Water by Laser Absorption Spectroscopy at 2.73 µm. *Sensors* **2014**, *14*, 9027–9045. [CrossRef]

12. Wei, Z.; Lee, X.; Aemisegger, F. A global database of water vapor isotopes measured with high temporal resolution infrared laser spectroscopy. *Sci. Data* **2019**, *6*, 180302. [CrossRef]

13. Galfond, B.; Riemer, D.; Swart, P. Analysis of signal-tonoise ratio of *delta*^{13}C-CO2 measurements at carbon capture, utilization and storage injection sites. *Int. J. Greenh. Gas Control* **2015**, *42*, 307–318. [CrossRef]

14. Krevor, S.; Perrin, J.C.; Esposito, A.; Rella, C.; Benson, S. Rapid detection and characterization of surface CO2 leakage through the real-time measurement of 13C signatures in CO2 flux from the ground. *Int. J. Greenh. Gas Control* **2010**, *4*, 811–815. [CrossRef]

15. Ciais, P.; Chris, S.; Govindasamy, B.; Bopp, L.; Brovkin, V.; Canadell, J. Chhabra, A.; Defries, R.; Galloway, J.; Heimann, M.; Carbon and other biogeochemical cycles. In *Climate Change 2013: The Physical Science Basis. Contribution of Working Group I to the Fifth Assessment Report of the Intergovernmental Panel on Climate Change*; Cambridge University Press: Cambridge, UK; New York, NY, USA, 2013; pp. 465–470.

16. Denk, T.R.A.; Mohn, J.; Decock, C.; Lewicka-Szczebak, D.; Harris, E.; Butterbach-Bahl, K.; Kiese, R.; Wolf, B. The nitrogen cycle: A review of isotope effects and isotope modeling approaches. *Soil Biol. Biochem.* **2017**, *105*, 121–137. [CrossRef]

17. Ostrom, N.E.; Ostrom, P.H. Mining the isotopic complexity of nitrous oxide: a review of challenges and opportunities. *Biogeosciences* **2017**, *132*, 359–372. [CrossRef]

18. Crosson, E.R.; Ricci, K.N.; Richman, B.A.; Chilese, F.C.; Owano, T.G.; Provencal, R.A.; Todd, M.W.; Glasser, J.; Kachanov, A.A.; Paldus, B.A.; et al. Stable isotope ratios using cavity ring-down spectroscopy: Determination of 13C/12C for carbon dioxide in human breath. *Anal. Chem.* **2002**, *74*, 2003–2007. [CrossRef]

19. Thorpe, M.J.; Balslev-Clausen, D.; Kirchner, M.S.; Ye, J. Cavity-enhanced optical frequency comb spectroscopy: Application to human breath analysis. *Opt. Express* **2008**, *1*, 2387–2397. [CrossRef]

20. Yang, T.H.; Heinzle, E.; Wittmann, C. Theoretical aspects of ^{13}C metabolic flux analysis with sole quantification of carbon dioxide labeling. *Comput. Biol. Chem.* **2005**, *29*, 121–133. [CrossRef]

21. Schild, H.; Boyle, S.J.; Schmid, H.M. Infrared spectroscopy of symbiotic stars: Carbon abundances and 12C/13C isotopic ratios. *Mon. Not. R. Astron. Soc.* **1992**, *258*, 95–102. [CrossRef]

22. van Geldern, R.; Nowak, M.E.; Zimmer, M.; Szizybalski, A.; Myrttinen, A.; Barth, J.A.C.; Jost, H.-J. Field-Based Stable Isotope Analysis of Carbon Dioxide by Mid-Infrared Laser Spectroscopy for Carbon Capture and Storage Monitoring. *Anal. Chem.* **2014**, *86*, 12191–12198. [CrossRef]

23. Gagliardi, G.; Castrillo, A.; Iannone, R.; Kerstel, E.T.; Gianfrani, L. High-precision determination of the ^{13}CO2/ ^{12}CO2 isotope ratio using a portable 2.008-µm diode-laser spectrometer. *Appl. Phys.* **2003**, *77*, 119–124. [CrossRef]

24. Zhou, T.; Wu, T.; Wu, Q.; Ye, C.; Hu, R.; Chen, W.; He, X. Real-time measurement of CO2 isotopologue ratios in exhaled breath by a hollow waveguide based mid-infrared gas sensor. *Opt. Express* **2020**, *28*, 10970–10980. [CrossRef] [PubMed]

25. Scholten S.K.; Perrella C.; Anstie J.D.; White R.T.; Luiten A.N. Accurate Optical Number Density Measurement of ^{12}CO$_2$ and ^{13}CO$_2$ with Direct Frequency Comb Spectroscopy. *Phys. Rev. Appl.* **2019**, *12*, 034045. [CrossRef]

26. Coddington, I.; Newbury, N.; Swann, W. Dual-comb spectroscopy. *Optica* **2016**, *3*, 414–426. [CrossRef] [PubMed]

27. Mandon, J.; Guelachvili, G.; Picqué, N. Fourier transform spectroscopy with a laser frequency comb. *Nat. Photon.* **2009**, *3*, 99–102. [CrossRef]

28. Diddams, S.A.; Hollberg, L.; Mbele, V. Molecular fingerprinting with the resolved modes of a femtosecond laser frequency comb. *Nature* **2007**, *445*, 627–630. [CrossRef]

29. Siciliani de Cumis, M.; Eramo, R.; Coluccelli, N.; Cassinerio, M.; Galzerano, G.; Laporta, P.; De Natale, P.; Cancio Pastor, P. Tracing part-per-billion line shifts with direct-frequency-comb Vernier spectroscopy. *Phys. Rev. A* **2015**, *91*, 012505. [CrossRef]

30. Rutkowskia, L.; Morvilleb, J. Continuous Vernier filtering of an optical frequency comb for broadband cavity-enhanced molecular spectroscopy. *J. Quant. Spec. Rad. Trans.* **2017**, *187*, 204–215. [CrossRef]

31. Khodabakhsh, A.; Rutkowski, L.; Morville, J.; Foltynowicz, A. Mid-infrared continuous-filtering Vernier spectroscopy using a doubly resonant optical parametric oscillator. *Appl. Phys. B* **2017**, *123*, 210. [CrossRef]

32. Adler, F.; Thorpe, M.J.; Cossel, K.C.; Ye, J. Cavity enhanced direct frequency comb spectroscopy: Technology and applications. *Annu. Rev. Anal. Chem.* **2010**, *3*, 175–205. [CrossRef]

33. Rieker, G.B.; Giorgetta, F.R.; Swann, W.C.; Kofler, J.; Zolot, A.M.; Sinclair, L.C.; Baumann, E.; Cromer, C.; Petron, G.; Sweeney, C.; et al. Frequency-comb-based remote sensing of greenhouse gases over kilometer air paths. *Optica* **2014**, *1*, 290–298. [CrossRef]

34. Cossel, K.C.; Waxman, E.M.; Finneran, I.A.; Blake, G.A.; Ye, J.; Newbury, N.R. Gas-phase broadband spectroscopy using active sources: progress, status, and applications. *J. Opt. Soc. Am. B* **2017**, *34*, 104–129. [CrossRef] [PubMed]

35. Siciliani de Cumis, M.; Eramo, R.; Coluccelli, N.; Galzerano, G.; Laporta, P.; Cancio Pastor, P. Multiplexed direct-frequency-comb Vernier spectroscopy of carbon dioxide $2\nu_1 + \nu_3$ ro-vibrational combination band. *J. Chem. Phys.* **2018**, *148*, 114303. [CrossRef]

36. Picqué, N.; Hänsch, T.W. Frequency comb spectroscopy. *Nat. Photonics* **2019**, *13*, 146–157. [CrossRef]

37. Vodopyanov, K.L. Isotopologues Detection and Quantitative Analysis by Mid-Infrared Dual-Comb Laser Spectroscopy. In *Encyclopedia of Analytical Chemistry*; John Wiley and Sons, Ltd.: Hoboken, NJ, USA, 2020.

38. Bailey, D.M.; Zhao, G.; Fleisher, A.J. Precision Spectroscopy of Nitrous Oxide Isotopocules with a Cross-Dispersed Spectrometer and a Mid-Infrared Frequency Comb. *Anal. Chem.* **2020**, *92*, 13759–13766 [CrossRef] [PubMed]

39. Kantnerová, K.; Tuzson, B.; Emmenegger, L.; Bernasconi, S. M.; Mohn, J. Quantifying Isotopic Signatures of N_2O Using Quantum Cascade Laser Absorption Spectroscopy. *Chimia* **2019**, *73*, 232–238. [CrossRef]

40. Harris, S.J.; Liisberg, J.; Xia, L.; Wei, J.; Zeyer, K.; Yu, L.; Barthel, M.; Wolf, B.; Kelly, B.F.J.; Cendón, D. I.; et al. N_2O isotopocule measurements using laser spectroscopy: Analyzer characterization and intercomparison. *Atmos. Meas. Tech.* **2020**, *13*, 2797–2831. [CrossRef]

41. Zare, R.N.; Kuramoto, D.S.; Haase, C.; Tan, S.M.; Crosson, E.R.; Saad, N.M.R. High-precision optical measurements of $^{13}C/^{12}C$ isotope ratios in organic compounds. *Proc. Natl. Acad. Sci. USA* **2009**, *106*, 10928–10932. [CrossRef]

42. Dickinson, D.; Bodé, S.; Boeckx, P. Measuring ^{13}C-enriched CO_2 in air with a cavity ring-down spectroscopy gas analyser: Evaluation and calibration. *Rapid Commun. Mass Spectrom.* **2017**, *31*, 1892–1902 [CrossRef]

43. Available online: https://en.wikipedia.org/wiki/Wavelet (accessed on 11 August 2021).

44. Gordon, I.E.; Rothman, L.S.; Hill, C.; Kochanov, R.V.; Tan, Y.; Bernath, P.F.; Birk, M.; Boudon, V.; Campargue, A.; Chance, K.V.; et al. The HITRAN2016 Molecular Spectroscopic Database. *J. Quan. Spec. Rad. Trans.* **2017**, *203*, 3–69. [CrossRef]

45. Eramo, R.; Cancio Pastor, P.; Siciliani de Cumis, M. Accurate fit of pressure-broadened molecular line shapes in direct-frequency-comb Vernier spectroscopy. **2021**, to be submitted.

46. Galli, I.; Bartalini, S.; Ballerini, R.; Barucci, M.; Cancio, P.; De Pas, M.; Giusfredi, G.; Mazzotti, D.; Akikusa, N.; De Natale, P. Spectroscopic detection of radiocarbon dioxide at parts-per-quadrillion sensitivity. *Optica* **2017**, *3*, 385–388. [CrossRef]

Communication

Development of a Laser Gas Analyzer for Fast CO_2 and H_2O Flux Measurements Utilizing Derivative Absorption Spectroscopy at a 100 Hz Data Rate

Mingxing Li [1,2], Ruifeng Kan [1,*], Yabai He [1,*], Jianguo Liu [1], Zhenyu Xu [1], Bing Chen [1], Lu Yao [1], Jun Ruan [1], Huihui Xia [1], Hao Deng [1], Xueli Fan [1], Bangyi Tao [3] and Xueling Cheng [4]

[1] AnHui Institute of Optics and Fine Mechanics, HeFei Institutes of Physical Sciences, Chinese Academy of Sciences, Hefei 230031, China; mxli@aiofm.ac.cn (M.L.); jgliu@aiofm.ac.cn (J.L.); zyxu@aiofm.ac.cn (Z.X.); bchen@aiofm.ac.cn (B.C.); lyao@aiofm.ac.cn (L.Y.); ruanjun@aiofm.ac.cn (J.R.); hhxia@aiofm.ac.cn (H.X.); hdeng@aiofm.ac.cn (H.D.); xlfan@aiofm.ac.cn (X.F.)

[2] University of Science and Technology of China, Hefei 230026, China

[3] State Key Laboratory of Satellite Ocean Environment Dynamics, Second Institute of Oceanography, Ministry of Natural Resources, Hangzhou 310012, China; taobangyi@sio.org.cn

[4] State Key Laboratory of Atmospheric Boundary Layer Physics and Atmospheric Chemistry, Institute of Atmospheric Physics, Chinese Academy of Sciences, Beijing 100029, China; chengxl@mail.iap.ac.cn

* Correspondence: kanruifeng@aiofm.ac.cn (R.K.); yabaihe@hotmail.com (Y.H.)

Citation: Li, M.; Kan, R.; He, Y.; Liu, J.; Xu, Z.; Chen, B.; Yao, L.; Ruan, J.; Xia, H.; Deng, H.; et al. Development of a Laser Gas Analyzer for Fast CO_2 and H_2O Flux Measurements Utilizing Derivative Absorption Spectroscopy at a 100 Hz Data Rate. *Sensors* **2021**, *21*, 3392. https://doi.org/10.3390/s21103392

Academic Editors: Krzysztof M. Abramski and Piotr Jaworski

Received: 27 March 2021
Accepted: 10 May 2021
Published: 13 May 2021

Abstract: We report the development of a laser gas analyzer that measures gas concentrations at a data rate of 100 Hz. This fast data rate helps eddy covariance calculations for gas fluxes in turbulent high wind speed environments. The laser gas analyzer is based on derivative laser absorption spectroscopy and set for measurements of water vapor (H_2O, at wavelength ~1392 nm) and carbon dioxide (CO_2, at ~2004 nm). This instrument, in combination with an ultrasonic anemometer, has been tested experimentally in both marine and terrestrial environments. First, we compared the accuracy of results between the laser gas analyzer and a high-quality commercial instrument with a max data rate of 20 Hz. We then analyzed and compared the correlation of H_2O flux results at data rates of 100 Hz and 20 Hz in both high and low wind speeds to verify the contribution of high frequency components. The measurement results show that the contribution of 100 Hz data rate to flux calculations is about 11% compared to that measured with 20 Hz data rate, in an environment with wind speed of ~10 m/s. Therefore, it shows that the laser gas analyzer with high detection frequency is more suitable for measurements in high wind speed environments.

Keywords: laser gas analyzer; flux measurement; eddy covariance method; derivative absorption spectroscopy; gas sensors

1. Introduction

The exchange of energy and mass between the ocean and atmosphere has significant impacts on the global environment, climate, and ecological balance. Flux measurements of heat, water, carbon dioxide, and methane, as well as other trace gases have been widely used to estimate the exchange of energy and mass [1–5]. With decades of technological development, the eddy covariance method has become a preferred method for direct flux estimations in turbulent motions without parametric assumptions, and is widely used in ecological flux observations [6].

Generally, the physical principle for the eddy covariance method is to measure the quantity of molecules moving upward or downward over time, and the speed in which they travel. Mathematically it can be represented as a covariance between measurements of vertical velocity of the upward or downward movements, and the concentration of the entity of interest [7]. The basic equipment for a flux measurement system mainly includes a three-dimensional ultrasonic anemometer and a gas analyzer. In the last decade, substantial progress has been made in the development of spectroscopic trace gas sensing technologies. This includes non-dispersive infrared spectroscopy (NDIR), tunable diode laser absorption spectroscopy (TDLAS), quantum cascade laser absorption spectroscopy (QCL-TDLAS), cavity ring-down spectroscopy (CRDS), and photoacoustic spectroscopy (PAS). Spectroscopic methods have advantages of high selectivity, high sensitivity, long-term stability, and have been applied for eddy covariance measurements. For example, Fortuniak et al. measured the greenhouse gases (CO_2, CH_4, H_2O) at the wetlands of Biebrza National Park in Poland by using a sonic anemometer and gas analyzers (LI-COR LI-7500-H_2O/CO_2 and LI-7700-CH_4) operating with 10 Hz frequency [8]. Kormann et al. developed a novel tunable diode laser absorption spectrometer for trace gas flux measurements based on micrometeorological techniques where the spectrometer was used to measure CH_4 and N_2O fluxes from rice paddies and tropical ecosystems [9]. Christian et al. tested a performance of a quantum cascade laser (QCL)-based N_2O flux measurements against gas chromatography (GC) [10]. Crosson developed an analyzer based on cavity ring-down spectroscopy to measure the concentrations of CO_2, H_2O and CH_4 [11]. He et al. developed a unique open-path CRDS technique for atmospheric sensing [12]; and Gong et al. recently developed a high-sensitivity resonant photoacoustic sensor for remote CH_4 gas detection at ppb-levels [13,14].

Turbulent changes happen very quickly, and the respective changes are very small in concentration, density, or temperature. It is therefore necessary to use an instrument with high precision and fast data rate of measurements, especially in high wind environments. Nevertheless, data rates of flux measurements reported in literature are typically around 20 Hz or slower. The 20 Hz frequency detection may cause data loss and inaccuracy for analyzing the gas exchange and flux. For trace gases measurements, tunable laser absorption spectroscopy was developed decades ago as an ideal analysis and measurement technology, which has the advantages of high resolution, high selectivity, and high sensitivity [15,16]. It is widely used in the fields of greenhouse gas detection, toxic and hazardous gas detection in chemical parks, respiratory diagnosis, aero-engine combustion flow field diagnosis, deep-sea dissolved gas, and isotope detection [17–24].

In this work, we have developed a simple and compact laser gas analyzer with a data rate of 100 Hz, based on laser absorption spectroscopy and derivative absorption spectroscopy. The analyzer is designed by using two diode DFB lasers operating at wavelengths of ~2004 nm for CO_2 and ~1392 nm for H_2O measurements. Meanwhile, we have designed a multi-pass cell with a 20 m optical path length for CO_2 absorption measurements and a single-path cell of 15 cm optical path length for H_2O absorption measurements, as well as a miniaturized TDLAS electronics system. By developing a fast data processing of derivative absorption spectroscopy, we were able to achieve gas concentration measurements at a 100 Hz data rate. The system was tested in high and low wind speed environments by field measurements on an offshore platform in the Yellow Sea near the Yan-tai city in Shandong province and on the Jue-hua Island near Huludao city in Liaoning province. We compare the accuracy of results between our laser gas analyzer and a commercial instrument LI-COR LI-7500. Finally, we analyzed and compared the impact of data rates between 100 Hz and 20 Hz in high and low wind speeds to verify the contribution of high frequency detection.

2. Materials and Methods

2.1. Transmission-Intensity-Normalized Second-Derivative Spectroscopy

A direct tunable diode laser absorption spectroscopy (dTDLAS) is a reliable means for trace gas detections as it is relatively simple in construction, easy to handle, and reliable to use [25]. The technology is based on an attenuation of laser radiation due to absorption as descript by the Lambert–Beer's law, which can be written as:

$$I(\nu) = I_0(\nu) \cdot \exp[-\varepsilon(\nu) \cdot L \cdot C], \tag{1}$$

where $I_0(\nu)$ is the incident intensity of the laser radiation of frequency ν. After passing through an absorbing medium, where optical path length is L and gas concentration is C, the transmitted intensity $I(\nu)$ is detected. The concentration-normalized absorption coefficient $\varepsilon(\nu)$ can be described by Equation (2):

$$\varepsilon(\nu) = S(T) \cdot P \cdot \phi(P,T,\nu),\tag{2}$$

where P is the total pressure, $S(T)$ is the temperature-dependent line strength, $\phi(P,T,\nu)$ is the line shape function which is pressure and temperature dependent [26,27].

However, the detection limit of dTDLAS is affected by noise contributions in the measurement signal. The data analysis during concentration inversion also involves numerical division, logarithmic calculations, and possible nonlinear least-squares fitting. This type of calculation-intensive analysis is a challenge for the simple microcontrollers typically used in such measurement instruments, and slows the data acquisition rate. To improve on this, a derivative spectroscopy technique [28,29] can be applied. By processing spectral signal with second-order differential, the derivative spectral signal is obtained, and correlated with gas concentration. The transmission-intensity-normalized first and second derivatives of measurement signals can be written by Equations (3) and (4), respectively.

$$\frac{dI}{d\nu}/I = \frac{dI_0}{d\nu}/I_0 - L \cdot C \cdot \frac{d\varepsilon}{d\nu},\tag{3}$$

$$\frac{d^2I}{d\nu^2}/I = \frac{d^2I_0}{d\nu^2}/I_0 + \left(L \cdot C \cdot \frac{d\varepsilon}{d\nu}\right)^2 - 2 \cdot L \cdot C \cdot \frac{d\varepsilon}{d\nu} \times \frac{dI_0}{d\nu}/I_0 - L \cdot C \cdot \frac{d^2\varepsilon}{d\nu^2},\tag{4}$$

When a linearly ramp is used as drive current to a diode laser, the first term of Equation (4) is zero in an ideal case when changes of laser intensity are proportional to changes in its drive current. The residual deviation from zero is not dependent on the gas absorption and can be treated as an offset background. The values of the second and third term are zero at the center frequency of an absorption line, where the curvature slope (i.e., first derivative) is zero. The fourth term (second derivative) reaches a maximum value at the line center. Therefore, the transmission-intensity-normalized second derivative spectra have a linear relationship with the concentration of the absorbing medium.

2.2. Method of Flux Measurements

The eddy covariance (EC) method has been widely used for direct measurements of surface atmosphere exchange. It uses the covariance between vertical velocity w_i in wind speed and fast variations of C_i in trace concentration. The EC flux F can be calculated from a recorded time series of N measurements as:

$$F = \overline{C' \cdot w'} = \frac{1}{N}\sum_{i=1}^{N}[(C_i - \overline{C}) \cdot (w_i - \overline{w})] = \frac{1}{N}\sum_{i=1}^{N}C'_i \cdot w'_i,\tag{5}$$

where C' and w' are the instantaneous deviations from their corresponding mean values $\overline{C}$ and $\overline{w}$, respectively. For this work, an ultrasonic anemometer (GILL-HS100) was used to determine vertical wind speeds. The instantaneous gas concentrations of the species of interest (i.e., H_2O vapor and CO_2 gas) were measured with our newly developed laser gas analyzer at a fast data rate of 100 Hz. As eddies occur on a wide range of timescales, it is necessary to use sufficiently long averaging time for calculating mean values. For this study, a time interval of about 5 min was chosen for calculating the average value when operating at a data rate of 100 Hz (resulting in a total of 30,000 data points). Alternatively, the effective data sampling can be slowed to a lower rate (e.g., 20 Hz), or the averaging time base can be extended.

2.3. Selection of Spectral Absorption Lines

To achieve reliable measurements of trace gas concentrations, a spectral simulation is performed to determine whether the selected lines have sufficient strengths for measurements and are well isolated from the absorptions by other gas species without any serious interference. H_2O and CO_2 have several strong absorption bands in the infrared spectral range between 1.0 μm and 2.5 μm, as shown in Figure 1 [30]. For example, the line strengths of CO_2 near 2.0 μm wavelength region are much stronger than in 1.6 μm region. So in this study, we used a ~2.0 μm diode laser for more sensitive detections. Figure 1c,d show the simulation of spectral absorption around 4991 cm^{-1} and 7181 cm^{-1}, based on the HITRAN2016 database [31] for 2% H_2O and 400 ppmv CO_2 in air under nominal conditions (P = 1 atm, T = 296 K, path length L = 15 cm for H_2O or 20 m for CO_2, respectively). The results indicate that the target lines for H_2O and CO_2 detection are appropriate with minimum spectral inference. So the diode lasers of 1.392 μm wavelength (NEL, NLK1E5GAAA) and 2.004 μm wavelength (NEL, KELD1G5BAAA) are used in this work.

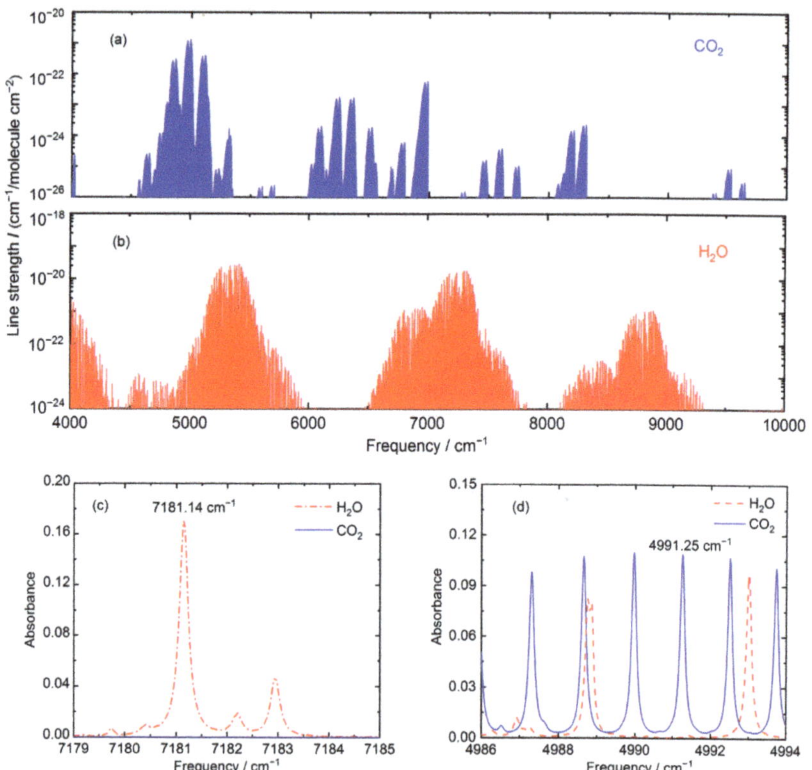

Figure 1. (**a**,**b**) Spectral line strengths; (**c**,**d**) simulation of spectral absorption with 2% H_2O and 400 ppmv CO_2 in air at a temperature of 296 K and pressure of 1 atm, in the near infrared wavelength regions (**c**) 1.392 μm, path length 15 cm, and (**d**) 2.004 μm, path length 20 m, respectively.

2.4. Experimental Setup

A schematic of the laser gas analyzer and experimental setup developed for flux measurement is shown in Figure 2, comprised of three units: a laser gas analyzer, an ultrasonic anemometer for three-dimensional wind speeds, power supply, and data acquisition. Among them, our developed laser gas analyzer consists of a miniaturized TDLAS measuring system and two gas absorption cells. The two diode lasers are driven by current controllers and temperature controllers with precision setting voltages, generated by digital-to-analog converters (DAC, Analog Devices AD5682, 14 bits) in combination with a microcontroller (ST, STM32H743VIT6). The wavelengths of diode lasers are ramped at a rate of 2 kHz via their operation currents. The fiber-coupled laser output is collimated and focused to either a single-pass absorption cell (15 cm optical path length) for H_2O measurements, or a multi-pass absorption cell (Herriott style, 50 passes, total 20 m optical path length, 51-mm dimeter mirrors with 99.99% highly-reflective dielectric coatings around 2.0 µm) for CO_2 measurements. The laser radiation is detected by a wavelength-extended (up-to 2.6 µm) InGaAs photodiode (GPD Optoelectronics), and then amplified (Analog Devices AD8065) and recorded by an analog-to-digital converter (ADC, 16 bits, on the STM32H743 microcontroller). The final results of gas concentrations and wind speeds are sent to a laptop computer by an Ethernet port and saved to a memory card. A GPS receiver is used to provide time information for synchronizing the data between the ultrasonic anemometer and the laser gas analyzer.

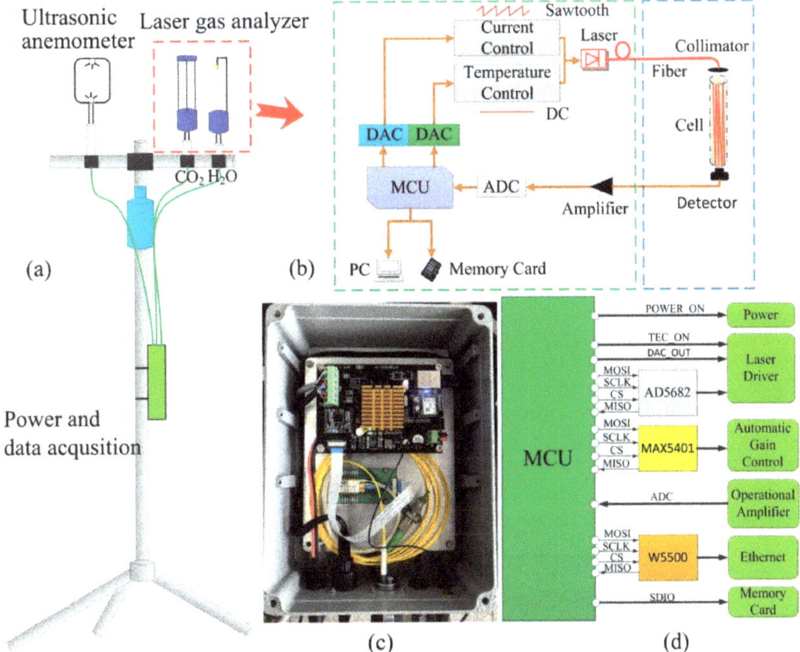

Figure 2. Schematic of the laser gas analyzer and experimental setup for flux measurements. (**a**) the three major components of the system; (**b**–**d**) details of the miniaturized TDLAS system.

3. Results and Discussion

3.1. Signal Processing

To quickly process the measurement data for this project, we applied the Savitzky–Golay filter method [32] for signal smoothing and differentiation. This digital filtering technique fits successive sub-sets of adjacent data points with a low-degree polynomial by the method of linear least squares. As a result of this moving-window data smoothing process, it increased the precision of the data without distorting the signal tendency. Another important aspect of this Savitzky–Golay filtering technique is that it also obtains derivative information of the signal profile based on the fitted polynomials. For spectroscopy applications, the Savitzky–Golay filtering technique can help to reduce signal noise and identify structure components in complex spectra [33,34]. It enables us to achieve fast numerical data analysis of recorded measurement transmission absorption spectra for determining gas concentrations in gas sensing applications.

Figure 3a shows one example of a CO_2 measurement spectrum and the associated 1st and 2nd derivatives obtained via a Savitzky–Golay filtering. The transmission spectrum was acquired for samples of ~500 ppmv CO_2 by the ADC with 450 data points, and averaged 16 times in successive laser current scans. Simulations of the absorbance and 2nd differentiation spectrum at 1 ppmv CO_2 concentration are displayed in Figure 3b. The 16 bits ADC for recording the photodetector signal has sufficient resolution (i.e., 1/65,535) to cover concentrations from ~2000 ppmv down to sub ppmv. Our noise-limited detection sensitivity corresponds to an absorbance level of ~1.5×10^{-4}.

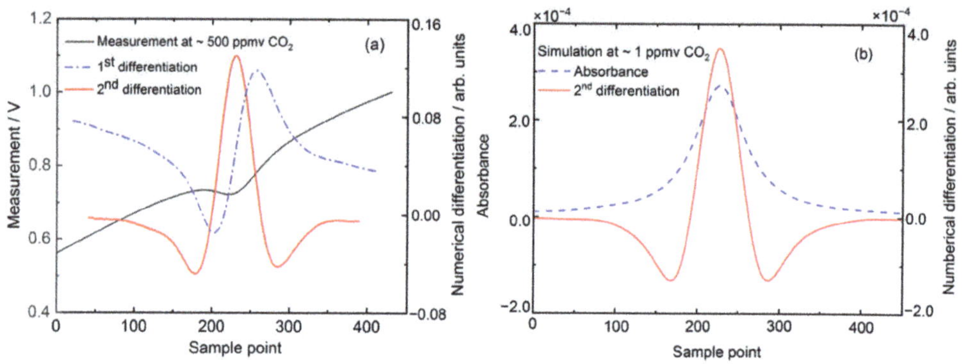

Figure 3. (**a**) A measurement example of tunable diode laser transmission spectrum of CO_2 absorption, and the subsequent numerical analysis of its transmission-intensity-normalized 1st and 2nd differentiations by Savitzky–Golay filtering; (**b**) simulations of absorbance and 2nd differentiation at 1 ppmv CO_2 concentration.

3.2. Gas Analyzer Performance

To evaluate the performance of numerical analysis of derivative absorption spectroscopy, we prepared a series of reference gas mixtures of CO_2 and H_2O with different concentrations and made measurements at atmospheric pressure. The subsequent second derivative absorption signals were calculated by using the Savitzy–Golay filter method, as shown in Figure 4a for CO_2 and Figure 4c for H_2O. As expected, a linear correlation between the signal peak magnitude of the second derivative spectra and the CO_2 and H_2O concentrations were confirmed. The results presented in Figure 4b,d show a good linear dependence (adj. R^2 = 0.995 for CO_2 and adj. R^2 = 0.999 for H_2O), and demonstrate that the algorithm is valid for trace gas measurements. The slope of the fitted straight-line also serves as a conversion coefficient between the experimental measurements and the resulting gas concentrations.

The detection limit of the developed laser gas analyzer is evaluated by using Allan variance plots [35]. Figure 5 shows the results of Allan deviations for CO_2 and H_2O concentration measurements with a sample containing 500 ppmv CO_2 and 2% H_2O, plotted in a log–log scale. The measurement noise at 100 Hz data rate (i.e., 0.01 s averaging time) is about 0.40 ppmv for CO_2, and 8.17 ppmv for H_2O, respectively. As the averaging time increases, the minimum reaches about 0.026 ppmv for CO_2 at 6 s integration time, and 3.12 ppmv for H_2O at 0.13 s integration time. Table 1 summarizes the performance comparison between the laser gas analyzer we developed, and the commercial instrument LI-7500-CO_2/H_2O based on non-dispersive infrared technology [36]. The TDLAS gas analyzer we developed performs slightly better for H_2O than the LI-7500 instrument, but was slightly worse for CO_2. The major advancement of our device is its fast maximum measurement data rate of 100 Hz, corresponding to a time resolution of 10 ms, which would enable observation of fast turbulent motions for eddy covariance.

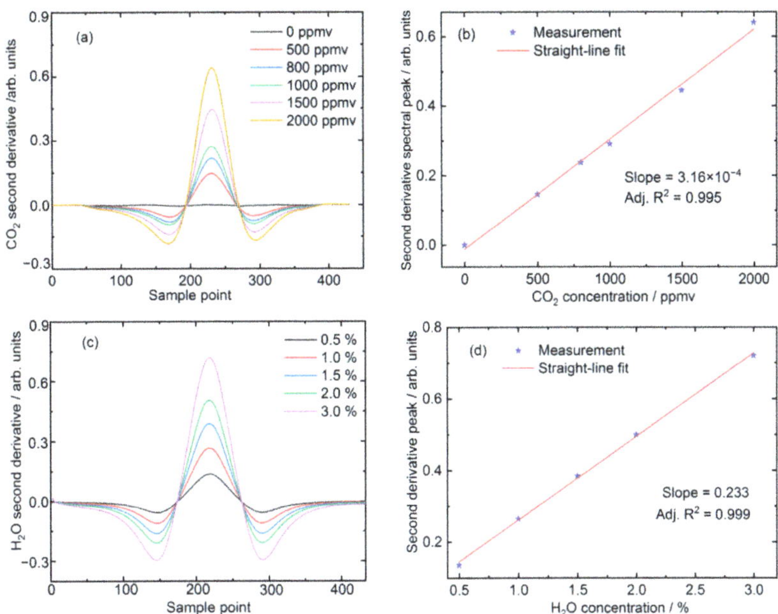

Figure 4. Numerically processed second derivative spectra of measurements at different gas concentrations of (**a**) CO_2 and (**c**) H_2O. The corresponding straight-line fits of the spectral peak magnitudes to the gas concentrations are displayed in (**b**,**d**).

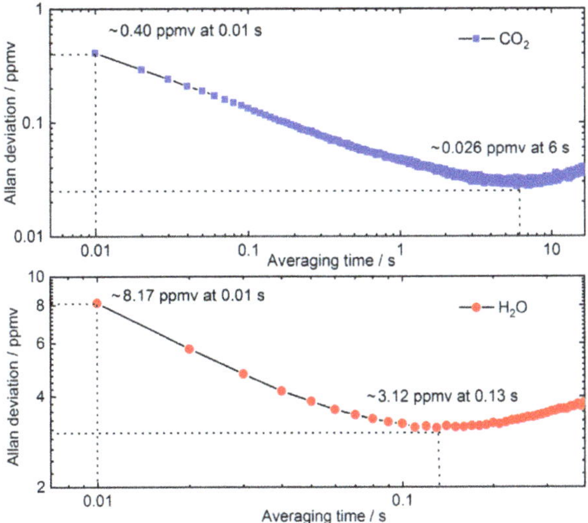

Figure 5. Allan deviation analyses of CO_2 and H_2O detection limits as a function of averaging time.

Table 1. Performance comparison between our laser gas analyzer and a commercial instrument.

	Our TDLAS Analyzer	**LI-7500 Analyzer**
Method	Laser absorption spectroscopy	Non-dispersive infrared spectroscopy
Detection limit, H_2O	3.25 ppmv at 10 Hz 8.17 ppmv at 100 Hz	4.70 ppmv at 10 Hz 6.70 ppmv at 20 Hz
Detection limit, CO_2	0.13 ppmv at 10 Hz 0.40 ppmv at 100 Hz	0.11 ppmv at 10 Hz 0.16 ppmv at 20 Hz
Maximum data rate	100 Hz	20 Hz

3.3. Experimental Field Measurements

In order to test the performance of the developed gas analyzer for H_2O and CO_2 fluxes measurements under low and high wind speed environments, we conducted the field measurements at two distinctive sites, as depicted in Figure 6. Figure 6a shows the installation of our integrated instruments and an ultrasonic anemometer on an offshore platform in the Yellow Sea of Yan-tai city in Shandong province (Site-A) with high-wind-speed marine environment. Figure 6b shows the installation on the Jue-hua Island of Huludao city in Liaoning province (Site-B) with low-wind-speed terrestrial environment. The commercial LI-7500-CO_2/H_2O instrument is also installed nearby to calibrate and compare the accuracy of measurements.

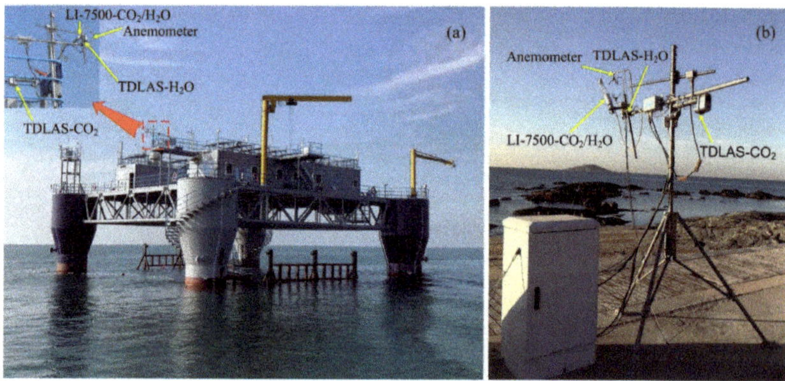

Figure 6. Photographs of the installation for field measurements at two sites of different environmental conditions. (**a**) Site-A of marine environment. (**b**) Site-B of terrestrial environment.

The results of comparison measurements for an hour in the marine environment (Figure 6a) with wind speed of about 13 m/s are displayed in Figure 7a,b, which show CO_2 and H_2O concentrations determined by our laser gas analyzer and the commercial LI-7500 analyzer. Our H_2O sensing unit was installed close to the LI-7500, while the CO_2 sensing unit was slightly further away. The H_2O concentration measurements show good agreement between the two instruments. The difference of CO_2 concentration trend between our laser gas analyzer and the LI-7500 analyzer was partly due to its installation location, which is slightly further away, therefore had different wind conditions. Another contributing factor might be the mounting and housing of the multi-pass sensing unit, which we will investigate in future studies. Furthermore, we measured CO_2 absorption in the wavelength region of its ~2.0 μm absorption band, while the LI-7500 analyzer operated around the ~4.2 μm stronger absorption band, resulting in a higher sensitivity with shorter optical path length. The detected concentration of CO_2 for both analyzers in the marine environment ranges from 380 ppmv to 410 ppmv, and the detected concentration of H_2O ranges for 22 mmol/mol to 25 mmol/mol.

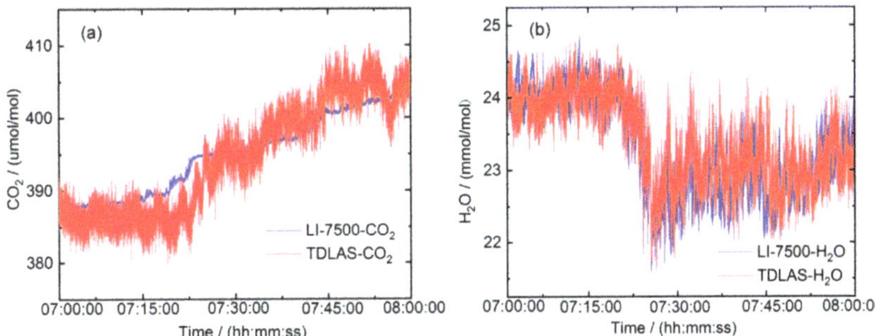

Figure 7. Comparison measurements between the laser gas analyzer and an LI-7500-CO_2/H_2O for an hour. (**a**) CO_2 and (**b**) H_2O measurements. The data rate of the TDLAS was at 100 Hz (with only one-fifth of the data points plotted in the graphs), whereas the LI-7500 was at its max 20 Hz.

For the investigation of the frequency characteristics of the measurements and the time response of the instruments, we applied the frequency power density method to analyze the recorded time series of the concentration data [37]. Figure 8a,b show the comparison of power density spectra for CO_2 and H_2O concentrations between the laser gas analyzer and LI-7500 analyzer via a fast Fourier transform method. The slope of spectra (ploted in log–log scales) is approximate to $-5/3$ of the Kolmogorov theory in the inertial subrange [38], indicating that the laser gas analyzer is capable of measuring turbulence fluxes of CO_2 and H_2O via EC method.

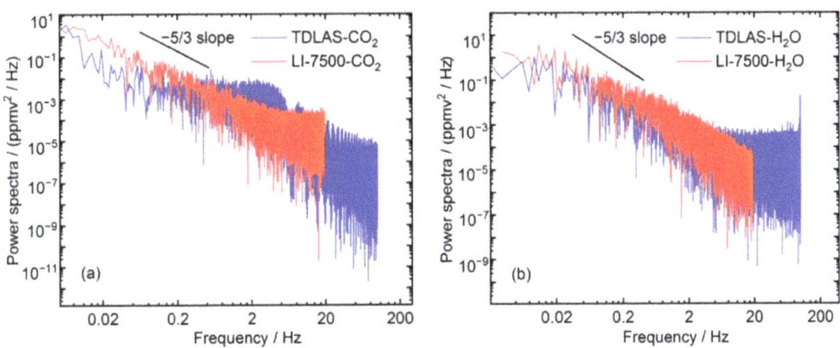

Figure 8. Power density analysis of concentration measurements between laser gas analyzer and LI-7500 analyzer for (**a**) CO_2 and (**b**) H_2O concentration measurements.

3.4. Comparison of Flux Measurements under Low and High Wind Speeds

Measurements of wind speed form an integral part in the determination of flux of gas emissions and movements. Three-dimensional wind speeds are measured via an ultrasonic anemometer. However, the mounting of the anemometer may not be exactly vertical or horizontal to the ground. A coordinate rotation is needed to transform the measurement values of the anemometer to the three velocity components u, v, w in respect to the ground, where w is the vertical wind velocity and is expected to have a mean value of zero. The components u and v are the two wind velocities in the horizontal plane. The horizontal component v is also aligned during the coordinate rotation to have a mean value of zero. Therefore, the component u represents the speed along horizontal wind direction. The results of wind speeds after rotating and H_2O concentrations are plotted in Figure 9. The maximum horizontal wind speed in the marine environment of Site-A is about 13 m/s, whereas the maximum vertical wind speed is about 4 m/s in the terrestrial environment of Site-B.

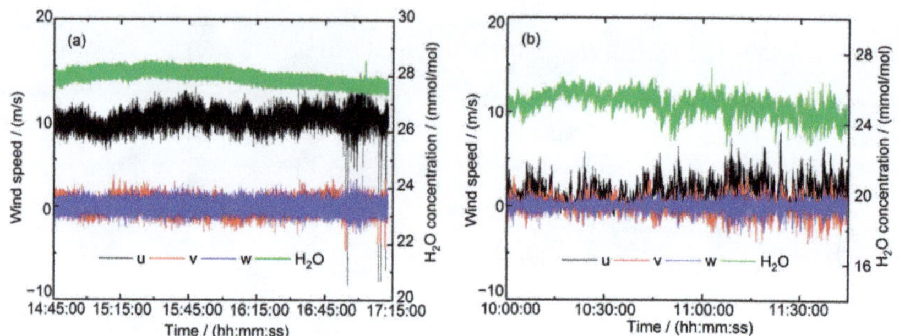

Figure 9. Measurement data of H_2O concentrations and three-dimensional wind velocities (u, v for the two horizontal components, w for vertical). (**a**) Data of Site-A. (**b**) Data of Site-B.

Fast measurements can capture more details of rapid small-scale turbulence in air movements, especially in a high wind speed environment. Subsequently, this is expected to be an advantage for flux determination. We numerically analyzed the impact of data sampling rate by block averaging every 5 data samples of the original 100 Hz data set to obtain a 20 Hz data set. This 20 Hz data set loses frequency contributions above 20 Hz. Both data sets are computed for flux using a 5 min time base (see Equation (5)). The results of H_2O fluxes are shown in Figure 10, for both high and low wind speeds.

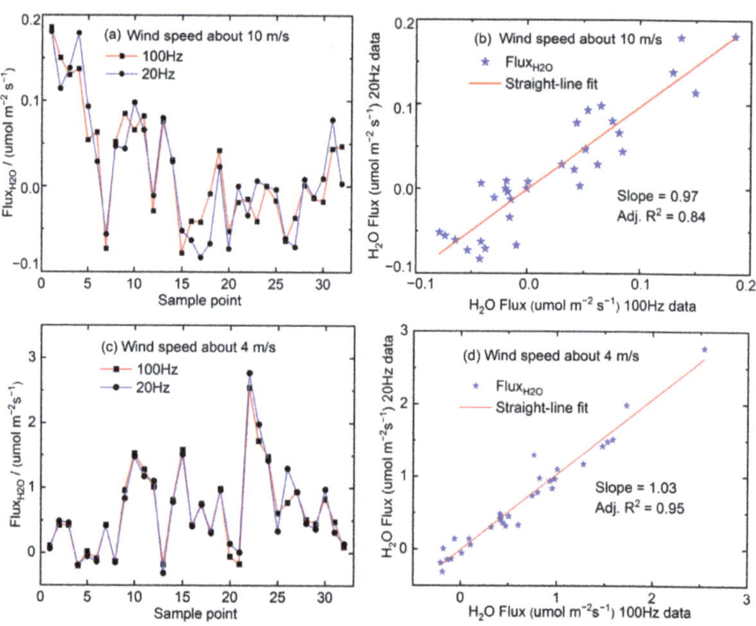

Figure 10. Comparison of H_2O fluxes between 20 Hz and 100 Hz. (**a**) The comparison of H_2O fluxes between 20 Hz and 100 Hz in 10 m/s wind speed environment. (**b**) The correlation analysis of H_2O fluxes between 20 Hz and 100 Hz in 10 m/s wind speed environment. (**c**) The comparison of H_2O fluxes between 20 Hz and 100 Hz in 4 m/s wind speed environment. (**d**) The correlation analysis of H_2O fluxes between 20 Hz and 100 Hz in 4 m/s wind speed environment.

As described in Figure 10a,b, the H_2O fluxes computed for data rates of 20 Hz and 100 Hz can differ by up to 16% (adjustable $R^2 = 0.84$) in the 10 m/s wind speed environment. As a comparison, the difference of H_2O fluxes is about 5% (adjustable $R^2 = 0.95$) in the

4 m/s wind speed environment, shown in Figure 10c,d. The difference suggests that the contribution of 100 Hz flux measurements increases by about 11%, as wind speed changes from 4 m/s to 10 m/s. The data points spread out further away from the straight-line in Figure 10b than in Figure 10d.

4. Conclusions

In this paper, a laser gas analyzer based on a second derivative laser absorption spectroscopy method has been developed to achieve a 100 Hz fast data rate, which is faster than that of a well-established 20 Hz commercial instrument. In combination with an ultrasonic anemometer, we applied the new laser gas analyzer for measurements of gas fluxes by using the eddy covariance method. Two DFB lasers operating at ~2.004 µm for CO_2 and ~1.392 µm for H_2O were used as optical sources. We built a multi-pass absorption cell of 20 m optical path length for CO_2 measurements, and a single-pass absorption cell of 15 cm optical path length for H_2O detection. A miniaturized TDLAS electronics system was designed for the operation and analysis of the gas concentration measurements. The gas analyzer achieves a detection limit of 8.17 ppmv for H_2O and 0.40 ppmv for CO_2 at 0.01 s integration time. Meanwhile, we made field measurements by installing the integrated instruments in two different environments to verify the influence of different wind speeds on flux measurements against a commercial instrument LI-7500. In general, the 100-Hz gas analyzer we developed has a wide prospect for flux measurement applications, especially when rapid turbulence is involved.

Author Contributions: Conceptualization, M.L. and B.C.; Methodology, M.L., L.Y. and R.K.; Project management, J.L.; Software, M.L., H.X. and H.D.; Validation, Z.X. and R.K.; Formal Analysis, J.R. and X.F.; Investigation, B.T. and X.C.; Resources, J.L., B.T. and X.C.; Writing-Initial Draft Preparation, M.L.; Writing-Revising, Y.H. and M.L. All authors have read and agreed to the published version of the manuscript.

Funding: This research was funded by the National Program on Key Research and Development Project of China under Grant 2018YFC0213103, the National Natural Science Foundation of China under Grant 41901081.

Institutional Review Board Statement: Not applicable.

Informed Consent Statement: Not applicable.

Data Availability Statement: All data will be made available on request to the correspondent author's email with appropriate justification.

Conflicts of Interest: The authors declare no conflict of interest.

References

1. Large, W.G.; Pond, S. Sensible and latent heat flux measurements over the ocean. *J. Phys. Oceanogr.* **1982**, *12*, 464–482. [CrossRef]
2. Fisher, J.B.; Malhi, Y.; Bonal, D.; Da Rocha, H.R.; De Araujo, A.C.; Gamo, M.; Von Randow, C. The land–atmosphere water flux in the tropics. *Glob. Chang. Biol.* **2009**, *15*, 2694–2714. [CrossRef]
3. Baldocchi, D.; Falge, E.; Gu, L.; Olson, R.; Hollinger, D.; Running, S. FLUXNET: A new tool to study the temporal and spatial variability of ecosystem-scale carbon dioxide, water vapor, and energy flux densities. *Bull. Am. Meteorol. Soc.* **2001**, *82*, 2415–2434. [CrossRef]
4. Denmead, O.T. Approaches to measuring fluxes of methane and nitrous oxide between landscapes and the atmosphere. *Plant Soil* **2008**, *309*, 5–24. [CrossRef]
5. DiGangi, J.P.; Boyle, E.S.; Karl, T.; Harley, P.; Turnipseed, A.; Kim, S.; Cantrell, C.; Maudlin, R.L., III; Zheng, W.; Flocke, F.; et al. First direct measurements of formaldehyde flux via eddy covariance: Implications for missing in-canopy formaldehyde sources. *Atmos. Chem. Phys.* **2011**, *11*, 10565–10578.
6. Aubinet, M.; Vesala, T.; Papale, D. *Eddy Covariance: A Practical Guide to Measurement and Data Analysis*; Springer Science & Business Media: London, UK, 2012; pp. 292–305.
7. Burba, G. *Eddy Covariance Method for Scientific, Industrial, Agricultural and Regulatory Applications: A Field Book on Measuring ecosystem Gas Exchange and Areal Emission Rates*; LI-COR Biosciences: Lincoln, NE, USA, 2013.

8. Fortuniak, K.; Pawlak, W. Preliminary Results of Net Ecosystem Exchange of Greenhouse Gases (CO_2, CH_4, H_2O) at Wetland of Biebrza National Park, Poland. In *West Siberian Peatlands and Carbon Cycle: Past and Present*; Tomsk University Press: Tomsk, Russia, 2014; pp. 141–143.

9. Kormann, R.; Fischer, H.; Wienhold, F.G. Compact multi-laser TDLAS for trace gas flux measurements based on a micrometeorological technique. In *Application of Tunable Diode and Other Infrared Sources for Atmospheric Studies and Industrial Processing Monitoring II*; International Society for Optics and Photonics: Denver, CO, USA, 1999; Volume 3758, pp. 162–169.

10. Bruemmer, C.; Lyshede, B.; Lempio, D.; Delorme, J.P.; Rueffer, J.J.; Fuss, R.; Moffat, A.M.; Hurkuck, M.; Ibrom, A.; Ambus, P.; et al. Gas chromatography vs. quantum cascade laser-based N_2O flux measurements using a novel chamber design. *Biogeosciences* **2017**, *14*, 1365–1381. [CrossRef]

11. Crosson, E.R. A cavity ring-down analyzer for measuring atmospheric levels of methane, carbon dioxide, and water vapor. *Appl. Phys. B* **2008**, *92*, 403–408. [CrossRef]

12. He, Y.; Jin, C.J.; Kan, R.F.; Liu, J.G.; Liu, W.Q.; Hill, J.; Jamie, I.M.; Orr, B.J. Remote open-path cavity-ringdown spectroscopic sensing of trace gases in air, based on distributed passive sensors linked by km-long optical fibers. *Opt. Express* **2014**, *22*, 13170–13189. [CrossRef] [PubMed]

13. Gong, Z.F.; Wu, G.J.; Jiang, X.; Li, H.; Gao, T.L.; Guo, M.; Ma, F.X.; Chen, K.; Mei, L.; Peng, W.; et al. All-optical high-sensitivity resonant photoacoustic sensor for remote CH_4 gas detection. *Opt. Express* **2021**, *29*, 13600–13609. [CrossRef]

14. Gong, Z.F.; Gao, T.L.; Mei, L.; Chen, K.; Chen, Y.W.; Zhang, B.; Peng, W.; Yu, Q.X. Ppb-level detection of methane based on an optimized T-type photoacoustic cell and a NIR diode laser. *Photoacoustics* **2021**, *21*, 100216. [CrossRef]

15. Reid, J.; Shewchun, J.; Garside, B.K.; Ballik, E.A. High sensitivity pollution detection employing tunable diode lasers. *Appl. Opt.* **1978**, *17*, 300–307. [CrossRef]

16. Hinkley, E.D. High-resolution infrared spectroscopy with a tunable diode laser. *Appl. Phys. Lett.* **1970**, *16*, 351–354. [CrossRef]

17. Yang, C.G.; Mei, L.; Wang, X.P.; Deng, H.; Hu, M.; Xu, Z.Y.; Chen, B.; He, Y.; Kan, R.F. Simultaneous measurement of gas absorption and path length by employing the first harmonic phase angle method in wavelength modulation spectroscopy. *Opt. Express* **2020**, *28*, 3289–3297. [CrossRef] [PubMed]

18. Wei, M.; Kan, R.F.; Chen, B.; Xu, Z.Y.; Yang, C.G.; Chen, X.; Xia, H.H.; Hu, M.; He, Y.; Liu, J.G.; et al. Calibration-free wavelength modulation spectroscopy for gas concentration measurements using a quantum cascade laser. *Appl. Phys. B* **2017**, *123*, 149. [CrossRef]

19. Frish, M.B.; White, M.A.; Allen, M.G. Handheld laser-based sensor for remote detection of toxic and hazardous gases. *Int. Soc. Opt. Photonics* **2001**, *4199*, 19–28.

20. Ghorbani, R.; Schmidt, F.M. ICL-based TDLAS sensor for real-time breath gas analysis of carbon monoxide isotopes. *Opt. Express* **2017**, *25*, 12743–12752. [CrossRef]

21. Nie, W.; Xu, Z.Y.; Kan, R.F.; Ruan, J.; Yao, L.; Wang, B.; He, Y. Development of a Dew/Frost Point Temperature Sensor Based on Tunable Diode Laser Absorption Spectroscopy and Its Application in a Cryogenic Wind Tunnel. *Sensors* **2018**, *18*, 2704. [CrossRef]

22. Li, X.; Fan, X.T.; He, Y.; Chen, B.; Yao, L.; Hu, M.; Kan, R.F. Development of a compact tunable diode laser absorption spectroscopy based system for continuous measurements of dissolved carbon dioxide in seawater. *Rev. Sci. Instrum.* **2019**, *90*, 065110. [CrossRef] [PubMed]

23. Li, X.; Yuan, F.; Hu, M.; Chen, B.; He, Y.; Yang, C.G.; Shi, L.F.; Kan, R.F. Compact open-path sensor for fast measurements of CO_2 and H_2O using scanned-wavelength modulation spectroscopy with 1f-phase method. *Sensors* **2020**, *20*, 1910. [CrossRef]

24. Kühnreich, B.; Wagner, S.; Habig, J.C.; Mohler, O.; Saathoff, H.; Ebert, V. Time-multiplexed open-path TDLAS spectrometer for dynamic, sampling-free, interstitial $H_2^{18}O$ and $H_2^{16}O$ vapor detection in ice clouds. *Appl. Phys. B* **2015**, *119*, 177–187. [CrossRef]

25. Seidel, A.; Wagner, S.; Dreizler, A.; Ebert, V. Robust, spatially scanning, open-path TDLAS hygrometer using retro-reflective foils for fast tomographic 2-D water vapor concentration field measurements. *Atmos. Meas. Tech.* **2015**, *8*, 2061–2068. [CrossRef]

26. Webber, M.E.; Claps, R.; Englich, F.V.; Tittel, F.K.; Jeffries, J.B.; Hanson, R.K. Measurements of NH_3 and CO_2 with distributed-feedback diode lasers near 2.0 μm in bioreactor vent gases. *Appl. Opt.* **2001**, *40*, 4395–4403. [CrossRef]

27. Herbert, F. Spectrum line profiles: A generalized Voigt function including collisional narrowing. *J. Quant. Spectrosc. Radiat. Transf.* **1974**, *14*, 943–951. [CrossRef]

28. Deng, H.; Sun, J.; Li, P.; Liu, Y.; Yu, B.L.; Li, J.S. Sensitive detection of acetylene by second derivative spectra with tunable diode laser absorption spectroscopy. *Opt. Appl.* **2016**, *46*, 353–363.

29. France, J. *High Temperature and Pressure Measurements from TDLAS through the Application of 2nd Derivative Fitting and the Aggregate Boltzmann Method*; University of Michigan: Ann Arbor, MI, USA, 2019.

30. Rothman, L.S.; Rinsland, C.P.; Goldman, A.; Massie, S.T.; Edwards, D.P.; Flaud, J.M. The HITRAN molecular spectroscopic database and HAWKS (HITRAN atmospheric workstation): 1996 edition. *J. Quant. Spectrosc. Radiat. Transf.* **1998**, *60*, 665–710. [CrossRef]

31. Gordon, I.E.; Rothman, L.S.; Hill, C.; Kochanov, R.V.; Tan, Y. The HITRAN2016 molecular spectroscopic database. *J. Quant. Spectrosc. Radiat. Transf.* **2017**, *203*, 3–69. [CrossRef]

32. Press, W.H.; Teukolsky, S.A.; Vetterling, W.T.; Flannery, B.P. *Numerical Recipes in Fortran 77: The Art of Scientific Computing*, 2nd ed.; Cambridge University Press: Cambridge, UK, 1992; pp. 644–649. ISBN 0-521-43064-X.

33. Li, J.S.; Deng, H.; Li, P.F.; Yu, B.L. Real-time infrared gas detection based on an adaptive Savitzky–Golay algorithm. *Appl. Phys. B* **2015**, *120*, 207–216. [CrossRef]

34. Luo, J.; Ying, K.; Bai, J. Savitzky–Golay smoothing and differentiation filter for even number data. *Signal Process.* **2005**, *85*, 1429–1434. [CrossRef]
35. Werle, P.O.; Mücke, R.; Slemr, F. The limits of signal averaging in atmospheric trace-gas monitoring by tunable diode-laser absorption spectroscopy (TDLAS). *Appl. Phys. B* **1993**, *57*, 131–139. [CrossRef]
36. LI-COR, Inc. *LI-7500RS Open-Path CO2/H2O Gas Analyzer Instruction Manual*; LI-COR Biosciences: Lincoln, NE, USA, 2019.
37. Velasco, E.; Pressley, S.; Allwine, E.; Westberg, H.; Lamb, B. Measurements of CO_2 fluxes from the Mexico City urban landscape. *Atmos. Environ.* **2005**, *39*, 7433–7446. [CrossRef]
38. Kolmogorov, A.N. The local structure of turbulence in incompressible viscous fluid for very large Reynolds numbers. *Proc. R. Soc. Lond. Ser. A Math. Phys. Sci.* **1991**, *434*, 9–13.

sensors

MDPI

Article

Fast Quantification of Air Pollutants by Mid-Infrared Hyperspectral Imaging and Principal Component Analysis

Juan Meléndez and Guillermo Guarnizo *

LIR–Infrared Laboratory, Department of Physics, Universidad Carlos III de Madrid, 28911 Leganés, Spain; juan.melendez@uc3m.es
* Correspondence: guillermoandres.guarnizo@uc3m.es

Abstract: An imaging Fourier-transform spectrometer in the mid-infrared (1850–6667 cm^{-1}) has been used to acquire transmittance spectra at a resolution of 1 cm^{-1} of three atmospheric pollutants with known column densities (Q): methane (258 ppm·m), nitrous oxide (107.5 ppm·m) and propane (215 ppm·m). Values of Q and T have been retrieved by fitting them with theoretical spectra generated with parameters from the HITRAN database, based on a radiometric model that takes into account gas absorption and emission, and the instrument lineshape function. A principal component analysis (PCA) of experimental data has found that two principal components are enough to reconstruct gas spectra with high fidelity. PCA-processed spectra have better signal-to-noise ratio without loss of spatial resolution, improving the uniformity of retrieval. PCA has been used also to speed up retrieval, by pre-calculating simulated spectra for a range of expected Q and T values, applying PCA to them and then comparing the principal components of experimental spectra with those of the simulated ones to find the gas Q and T values. A reduction in calculation time by a factor larger than one thousand is achieved with improved accuracy. Retrieval can be further simplified by obtaining T and Q as quadratic functions of the two first principal components.

Keywords: infrared imaging; multispectral and hyperspectral imaging; air pollution monitoring; remote sensing and sensors; spectroscopy; fourier transform; image processing

Citation: Meléndez, J.; Guarnizo, G. Fast Quantification of Air Pollutants by Mid-Infrared Hyperspectral Imaging and Principal Component Analysis. *Sensors* **2021**, *21*, 2092. https://doi.org/10.3390/s21062092

Academic Editor: Krzysztof M. Abramski

Received: 29 January 2021
Accepted: 8 March 2021
Published: 17 March 2021

Publisher's Note: MDPI stays neutral with regard to jurisdictional claims in published maps and institutional affiliations.

1. Introduction

Public concern about the adverse health effects of air pollution has increased considerably in recent years. This growing concern is being progressively translated into more restrictive legislation [1]: new emission limit values (ELVs) are set for previously unregulated pollutants, and more stringent levels are established for those already regulated. There is thus an increasing need to develop reliable methods for the measurement of atmospheric gases at immission levels. An example of this trend is the IMPRESS 2 project, funded by the research program EMPIR (European Metrology Programme for Innovation and Research) of the European Association of National Metrology Institutes (EURAMET), with the aim of improving measurement of pollutant gases at several levels: to develop new reference measurement methods for gases not yet regulated, to improve hyperspectral techniques, to determine uncertainty and traceability of mass emission measurements, etc. [2].

Ideally, a measurement method for air pollutants should be both versatile and accurate. Since all pollutant gases show characteristic absorption–emission bands in the infrared (IR) spectral region, IR optical techniques are such a versatile method with the additional advantage of providing remote and non-intrusive measurements. There are many techniques for IR optical gas sensing (see [3] for a comprehensive review) but high resolution spectroscopy is the most wide ranging in its applications, being able to detect several gases at the same time, and has the potential for high accuracy, since the dependence of line intensities on temperature and concentration is very well known.

Due to these features, Fourier transform spectrometry has been used for a long time to measure emissions from smokestack effluents and other industrial sources [4–6], but in

135

recent years, imaging spectrometers have conferred additional power to this technique [7]: it has become possible to map column densities Q (concentration·path product) of pollutants and plume temperatures T [8] over a large area, or to track gas flows and estimate effluent mass flow rates [9]. Cooler sources, such as automobile exhaust emissions, have also been measured in absorption mode [10], as well as ambient-temperature greenhouse emissions [11].

These studies apply techniques originally developed for non-imaging absorption spectroscopy to each pixel of the acquired datacube. It is possible, however, to take advantage of the large amount of data provided by imaging instruments to improve the sensitivity and signal to noise ratio. The objective of this paper is to study the absorption spectroscopy of pollutant gases in the atmosphere in the context of hyperspectral imaging, taking advantage of those possibilities. In particular, the well-known statistical technique of principal component analysis (PCA) is applied to gas spectra in the datacube, first to filter out noise and then to fasten retrieval of T and Q values. A simple radiative model applicable to field measurements is defined, although in this work it has been used only for laboratory measurements with a gas cell in order to evaluate its accuracy for the determination of gas concentrations.

Three gases have been studied: methane (CH_4), nitrous oxide (N_2O) and propane (C_3H_8). The first two are greenhouse gases and the third is a hydrocarbon that frequently appears jointly with methane and whose spectral features are in the same spectral region. For each of them, a mixture of known concentration has been prepared, and measured with a hyperspectral imager that operates in the mid-infrared band. Values of T and Q have been retrieved by fitting experimental spectra with simulated ones, and have been compared with the nominal values to assess the accuracy of the method. It has been demonstrated that processing with PCA increases signal to noise ratio which, in turn, improves the accuracy of retrieval, without losing spatial resolution or increasing acquisition time.

The basics of our approach are described in Section 2. After briefly explaining the radiative model in Section 2.1, the retrieval procedure is outlined in Section 2.2 and detailed in Section 2.3. The experimental setup and the measurements performed are described in Section 3. Principal component analysis is exposed and applied to noise filtering of spectra in Section 4; then it is applied, in Section 5, to reduce the dimensionality of spectra, thus making possible a faster retrieval of column density Q and temperature T. Retrieval is further simplified in Section 5.3 by defining polynomial functions that provide Q and T directly as functions of the principal components of the spectra. Finally, conclusions are summarized in Section 6.

2. Radiative Model and Retrieval Method

Nearly all gas molecules have characteristic absorption/emission spectra in the infrared (IR) spectral region, due to transitions between ro-vibrational levels. For a specific line at wavenumber ν with absorptivity a, gas transmittance is given by the Lambert–Beer law:

$$\tau_g(\nu, C_g, T_g) = e^{-a(\nu, T_g)C_g L_g} \equiv e^{-a(\nu, T_g)Q_g} \tag{1}$$

where L_g is the gas optical path, C_g is the concentration, $Q_g = C_g L_g$ is the column density, and the dependence of a on wavenumber and temperature has been shown explicitly. If there is more than one absorbing species, $\tau(\nu)$ is just a product of terms, as in Equation (1), one for each species; if the concentration is not homogeneous, the product aCL is replaced by an integral. Since absorptivities are well-known parameters that can be extracted from spectroscopic databases such as HITRAN [12], a transmittance measurement over a spectral range provides, in principle, an accurate way to identify gases in a sample and to determine their concentrations.

This is the basis of IR absorption spectroscopy, a classical method of analytical chemistry. In its most straightforward laboratory implementation, a gas cell in a spectrophotometer is filled with the sample to be measured, and then with a reference gas without absorption lines in the spectral region of interest (typically N_2). Transmittance is obtained as the ratio of the two spectra.

However, the full potential of absorption spectroscopy is displayed in remote measurements. In a typical field measurement with an imaging spectrometer, a gas cloud is observed against a background, and the instrument provides a measurement of the spectral radiance incoming to each pixel. In order to relate this radiance with the gas parameters, a radiative model of the measurement configuration is needed (Figure 1).

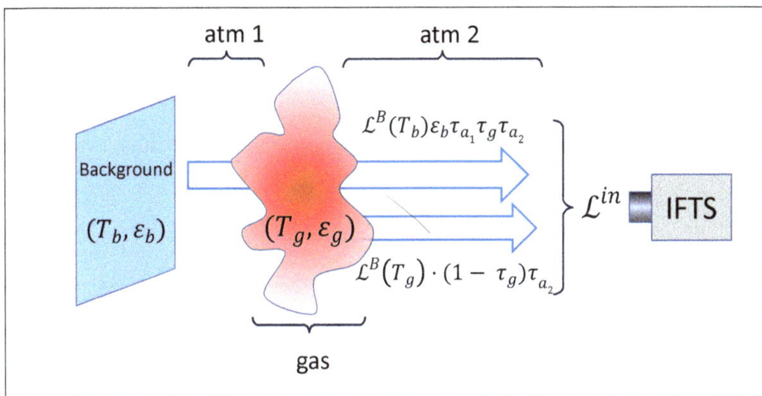

Figure 1. Schematics of the radiative model.

2.1. Radiative Model

The following simplifying assumptions will be made:

1. The gas is in local thermal equilibrium, so that Boltzmann distribution holds and absorptance α equals emittance ε (Kirchhoff's Law).
2. The effects of absorption and scattering by particulate matter are negligible.
3. For each pixel, the gas is modeled by a single temperature, and a single value of concentration for each species (these values are considered as line-of-sight averages); therefore, the gas cloud can be characterized by a single transmittance τ_g and emittance $\varepsilon_g = \alpha_g = 1 - \tau_g$ at each pixel.
4. The background emissivity ε_b is large, so that the reflection of ambient radiation in the background is negligible.
5. The emission of the atmosphere is negligible (i.e., near transparent spectral region, and/or ambient temperature T_a much lower than those of gas cloud and background).

With these approximations, the radiance measured by the radiometer can be expressed as:

$$\mathcal{L}_m = \mathcal{L}^B(T_b)\cdot\varepsilon_b\cdot\tau_{a_1}\tau_g\tau_{a_2} + \mathcal{L}^B(T_g)\cdot(1 - \tau_g)\tau_{a_2} \tag{2}$$

where τ_g, τ_{a_1} and τ_{a_2} are, respectively, the transmittances of the gas cloud and the first and second atmospheric paths (atm 1 and atm 2 in Figure 1), $\mathcal{L}^B$ stands for Planck's blackbody radiance, and T_b and T_g are, respectively, the temperatures of background and gas cloud.

To obtain a transmittance measurement, a reference spectrum must be measured without gas:

$$\mathcal{L}_r = \mathcal{L}^B(T_b)\cdot\varepsilon_b\cdot\tau_{a_1}\tau_{g_0}\tau_{a_2} \tag{3}$$

where τ_{g_0} stands for the transmittance of the region of atmosphere that was previously occupied by gas cloud; it will be assumed that $\tau_{g_0} \approx 1$.

A nominal transmittance is obtained as the ratio:

$$\tau_{nom} \equiv \frac{\mathcal{L}_m}{\mathcal{L}_r} = \tau_g + \frac{\mathcal{L}^B(T_g)}{\mathcal{L}^B(T_b)} \cdot (1-\tau_g) \cdot \frac{1}{\varepsilon_b \tau_{a_1}} \equiv \tau_g + \tau' \tag{4}$$

The positive term τ' is negligible if $\varepsilon_b \mathcal{L}^B(T_b) >> \mathcal{L}^B(T_g)$, i.e., when the background is much hotter than the gas; otherwise, the equation can be solved for τ_g if T_g, T_b and ε_b are known (it will be generally assumed that in the spectral region considered, $\tau_{a_1} \approx 1$).

2.2. Temperature and Column Density Retrieval

Our aim is to obtain the values of gas concentration C_g from experimental measurements of $\mathcal{L}_r(\nu)$ and $\mathcal{L}_m(\nu)$ but, since only the product CL appears in the equations (cf. Equation (1)), the result can only be the column density $Q_g \equiv C_g L_g$ rather than the concentration C_g. The amount of gas will be measured, as usual by spectroscopic remote sensing methods, in units of ppm·m (parts per million per meter).

Since absorptivity $a(\nu, T_g)$ is a known parameter, the most straightforward method to recover Q_g for each gas is to solve Equation (4) for τ_g and then use Lambert–Beer law (1) to obtain Q_g. However, in many practical cases the gas cloud temperature T_g will be unknown, and therefore should also be retrieved simultaneously with Q_g from the experimental measurements.

Thus, measurements of $\mathcal{L}_r(\nu)$ and $\mathcal{L}_m(\nu)$ over a spectral range rather than at a single ν will be necessary to provide a set of equations, but even so it is not possible to solve Equations (4) and (1) simultaneously for T_g and Q_g, because both parameters are coupled in the Lambert–Beer expression of transmittance (1), where the absorptivity a depends on T_g in a nontrivial way. Instead, they will be determined by a fitting process: we will calculate theoretical spectra for $\mathcal{L}_r(\nu)$ and $\mathcal{L}_m(\nu)$, divide them to obtain a theoretical nominal transmittance $\tau_{nom}^{th}(\nu)$ and assign to each pixel the column density and temperature values which provide the best fit to the experimental spectra $\tau_{nom}(\nu)$.

In summary, the final results of our method are a "column density image" and a "temperature image" with values of, respectively, Q_g and T_g at each point in the field of view, obtained by iteratively fitting the experimental nominal transmittance spectra with theoretical spectra generated according to the radiative model of Figure 1, through the Equations (1)–(4).

2.3. Theoretical Spectra and Fitting Procedure

The spectral positions and intensities of the emission/absorption lines have been obtained from the HITRAN database [12]. For methane and nitrous oxide, the HAPI [13] Python-based interface to HITRAN has been used to download the respective absorption coefficients. However, in this free-access database, there is no detailed information about propane. Absorption coefficients for it have been obtained from the absorption cross sections at an atmospheric pressure of 1 atm and three temperatures (278.15 K, 298.15 K and 323.15 K) available on the webpage of HITRAN online [14]. With this information, it is possible to calculate the absorption coefficients by multiplying the cross-section data by the number of molecules per volume unit at ambient conditions.

Theoretical spectra have been generated by summing up the standard linehapes of single absorption lines ("line-by-line method"). The dependence of a on temperature, due to variation of absorption cross sections with T, has been fitted by seventh-order polynomial functions with a spectral resolution of 0.01 cm^{-1} [10]. With this parametrization it is easy to construct theoretical transmittance spectra $\tau_g^{th}(\nu)$ for arbitrary values of T_g and Q_g, using Equation (1) and, in turn, theoretical $\tau_{nom}^{th}(\nu)$ spectra with (2)–(4).

In order to compare these spectra to the measured ones, the effect of finite instrument resolution must be accounted for. In our case, a triangular apodization was used, so that the instrumental lineshape function (ILS) is a squared *sinc* function [15].

However, when calculating the theoretical transmittance spectrum, it is not correct to simply convolve the ideal spectrum with the ILS. The reason is that the experimental nominal transmittance spectrum $\tau_{nom}(\nu)$ is not measured directly, but rather as a ratio (Equation (4)) of two radiance spectra measured by our instrument, $\mathcal{L}_m$ and $\mathcal{L}_r$. Therefore, the correct theoretical spectrum $\tau_{nom}^{th}(\nu)$ must be calculated as a ratio of widened radiances:

$$\tau_{nom}^{th}(\nu) = \frac{\int [\mathcal{L}^B(\nu', T_b) \cdot \varepsilon_b \cdot \tau_{a_1}(\nu') \cdot \tau_g(\nu') \cdot \tau_{a_2}(\nu') + \mathcal{L}^B(\nu', T_g) \cdot (1 - \tau_g(\nu')) \cdot \tau_{a_2}(\nu')] \cdot ILS(\nu - \nu')d\nu'}{\int \mathcal{L}^B(\nu', T_b) \cdot \varepsilon_b \cdot \tau_{a_1}(\nu') \cdot \tau_{a_2}(\nu') \cdot ILS(\nu - \nu')d\nu'} \tag{5}$$

where $\tau_g(\nu)$ and $\tau_{a_1}(\nu)$, $\tau_{a_2}(\nu)$ stand for the ideal transmittance spectra of the gas cloud and first and second atmospheric paths, respectively, as provided by HITRAN. They are functions (not explicitly displayed) of the temperatures (T_g, T_a) and column densities of the gas cloud (Q_g) and the atmospheric gases. In this work, it has been assumed that $\tau_{a_1} \approx \tau_{a_2} \approx 1$, which is a very good approximation for the measurement configuration and the spectral regions involved.

At each pixel, the fitting procedure is as follows (a single gas will be assumed; for each additional gas the procedure is the same but there is an additional unknown value of column density to be determined). We start by assuming a value for the couple (Q_g, T_g). The theoretical transmittance spectrum $\tau_{nom}^{th}(\nu)$ is calculated with Equations (1) and (5) at the points of the wavenumber axis of the experimental spectra. The differences with $\tau_{nom}(\nu)$ for each wavenumber are added up in quadrature to get the sum of squared errors (SSE). The Nelder–Mead minimization algorithm, as implemented in MATLAB software, is used then to find the value of (Q_g, T_g) for the next iteration, until convergence is reached. This iterative process is repeated for each pixel to obtain the images of column density and temperature.

3. Experimental Measurements

The experimental setup reproduces the scheme of Figure 1, but with the gas to be measured confined to a gas cell in order to know precisely the optical path (see Figure 2). The three main elements are: a blackbody radiator as a temperature controlled background, a gas cell for the pollutant to be characterized and the imaging Fourier transform spectrometer (IFTS) that captures both spectral and spatial information of the scene.

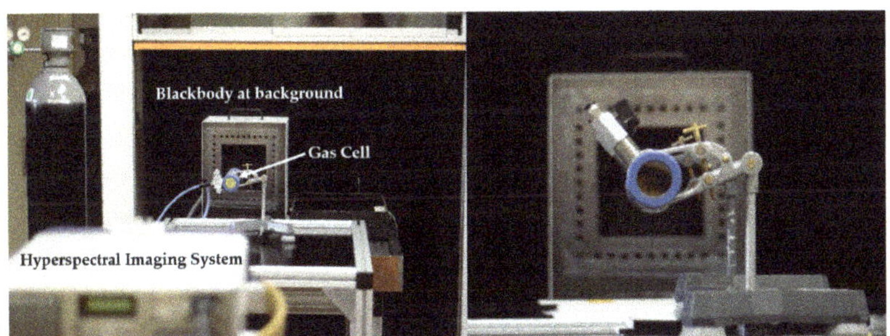

Figure 2. (**Left**) Overall view of the experimental setup. (**Right**) A close-up view of the gas cell without the gas supply tubes.

Specifically, an extended area (15 × 15 cm) blackbody radiator from Santa Barbara Infrared, Inc., with nominal emissivity of 0.9 was placed as uniform background, and a 43 cm long gas cell made of stainless steel with two 38 mm diameter sapphire optical windows was used to enclose the gas under test. This cell has two valves separated by a distance of 20 cm for gas input and output.

The experimental spectra have been acquired with a Telops FIRST-MW Hypercam IFTS [16,17] placed at a distance of two meters from the blackbody radiator, with the 43 cm metallic gas cell in-between. In this instrument, the incoming radiance is modulated by a Michelson interferometer, and then is detected by an InSb 320 × 256 focal plane array (IFOV = 0.35 mrad), sensitive in the mid-infrared (1850 to 6667 cm^{-1}). Interferograms are acquired for each pixel, which, after processing, can provide spectra with a maximum resolution of 0.25 cm^{-1}.

In order to reduce acquisition time to ≈ 1 min, in this work the spectral resolution of the measurements was set at 1 cm^{-1} and a spatial sub-windowing of 256×160 pixels was used. Integration time was 10 μs. Four interferograms were acquired for each measurement, and the dataset was pre-processed by calculating its median and then Fourier-transformed to obtain the radiance spectra. Processing of the interferograms includes triangular apodization, zero-padding to obtain experimental spectra with same wavenumbers as the theoretical ones, as well as off-axis correction [18]. All the processing steps have been described in [10].

Radiance spectra were obtained for both reference (with gas cell filled with N$_2$) and pollutant gas and divided according to Equation (4) to get a nominal transmittance spectrum.

Measurements have been carried out with the gas at ambient temperature and the blackbody background at 350 °C, for methane (CH$_4$), nitrous oxide (N$_2$O) and propane (C$_3$H$_8$) at the concentrations and in the spectral regions detailed in Table 1. The bottles were prepared by the Spanish Metrology Institute (CEM, Centro Español de Metrología), ensuring high accuracy in the concentration values.

Table 1. Air pollutants under test.

Pollutant Gas	Concentration (ppm)	Column Density (ppm·m)	Bandwidth (cm^{-1})
CH$_4$	600	258	2700–3200
N$_2$O	250	107.5	2100–2300
C$_3$H$_8$	500	215	2700–3200

4. Noise Filtering by Principal Component Analysis

Experimental radiance spectra for the three gases studied are shown in the left-hand graphs of Figure 3. These spectra, divided by the reference spectrum obtained with the gas cell full of N$_2$, give the transmittance spectra of the right-hand side. The best fitting by theoretical spectra (achieved with the iterative algorithm as explained in Section 2.3) is also shown.

It is well known that when two noisy spectra are divided, the signal to noise ratio (SNR) decreases greatly. Therefore, it would be very convenient to reduce the noise level of radiance spectra before calculating transmittance. This can be performed by acquiring more interferograms, at the cost of increasing measuring time, or by averaging over neighboring pixels, thus decreasing spatial resolution.

There is, however, a better solution provided by principal components analysis (PCA) [19]. This is a well-known statistical technique used to reduce the dimensionality of sets of multivariate data. If we have n measurements, each of m variables, the data can be interpreted as a cloud of n points in a m-dimensional *variable space*. PCA generates a new orthogonal basis in this space, optimally adapted to the data in the sense that (a) its origin coincides with the center of mass of the points and (b) the new (sometimes called "main") axes are oriented so that the projections of data on them are uncorrelated (i.e., in the new axes, the covariance matrix of the data is diagonal). The unit vectors corresponding to these axes are the eigenvectors of the covariance matrix, and PCA provides them in decreasing order of the associated eigenvalue. This means that the first principal direction is that along which the variance of the data is a maximum; the second principal component is, among the subset of vectors perpendicular to the first, the one whose direction contains the largest variance, and so on. The coordinates of a point in the spectral space with respect to the new basis are called principal components (PCs) or sometimes scores, and are obtained by subtracting the coordinates of the center of mass and then projecting on the basis of eigenvectors.

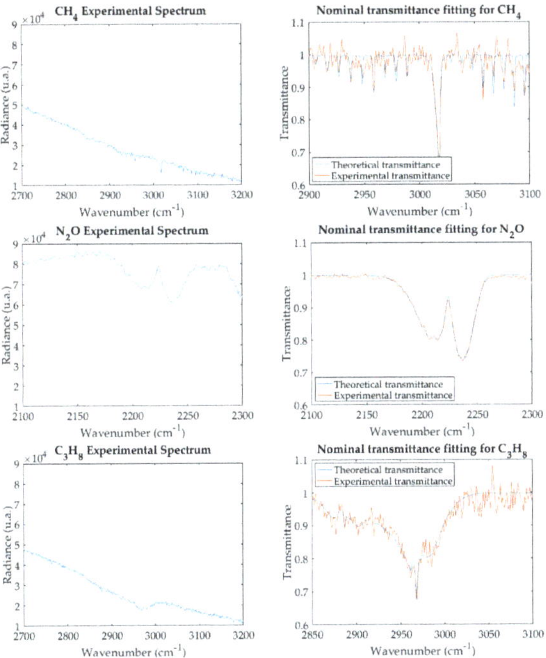

Figure 3. Experimental spectra of air pollutants: radiance (**left**) and nominal transmittance, with best fit (**right**).

Since most of the variance of the data is found in the first principal components, a good approximation to the original data set can be made by considering only a small number of principal components, say p. This is equivalent to projecting the data set in the p-dimensional sub-space built from the first p main axes, and achieves a reduction in the dimensionality of the data set from m to p.

In our case, the original data are the spectra (each one with m wavenumbers, $m \sim 15.000$ for 1 cm^{-1} resolution) from a region of n pixels corresponding to the gas cell. Since the spectra depend on two variables, T and Q, we can conjecture that the data should have an intrinsic dimensionality close to two. They should all, therefore, lie very close to a surface in the variable space, although this surface will not be a plane, since transmittance is not linear with Q or T. However, if the range of variation of T and Q in the data is relatively small, the corresponding surface region will be approximately flat, so that two principal components should be enough to describe with good approximation all the variability of the original data ($p = 2$). When T and Q have a wider variation, it will be necessary to take $p > 2$, but in any case, the principal components of large order will contain mainly noise. In summary, selecting the subspace spanned by the first major components not only dramatically reduces data volume, but also results in efficient noise filtering [20,21].

To apply PCA to our experimental data, a preliminary scene classification is performed by a standard k-means algorithm [22,23] to select the region of the image that corresponds to the gas in the cell. After applying PCA to the radiance spectra in that region, it is found that eigenvalues decrease sharply (Figure 4), so that for all the gases studied the first two account for more than 99.95% of the trace of the covariance matrix (i.e., the total variance of the data). This confirms our conjecture and suggests that a good spectrum reconstruction should be obtained with only two principal components. Indeed, Figure 5 (left-hand side) shows that the reconstructed radiance spectra reproduce with high fidelity the original ones (shown in Figure 3), but with noise filtered out; as expected, the effect is stronger in transmittance (Figure 5, right-hand side). The results of iterative fitting of these spectra are shown also in the right-hand side of Figure 5.

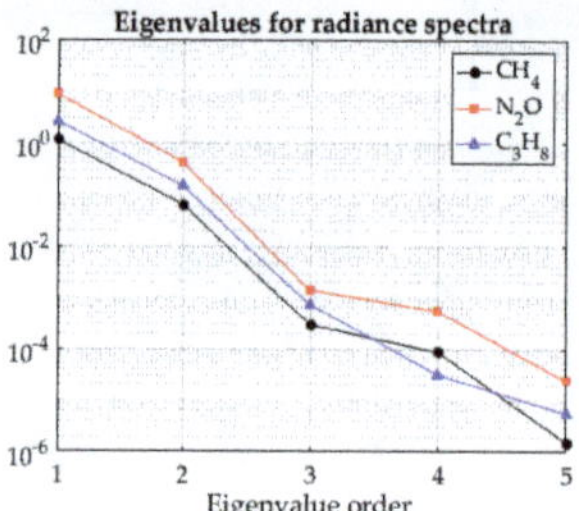

Figure 4. Values of the first 5 eigenvalues for the covariance matrix of the radiance spectra of the three gases studied.

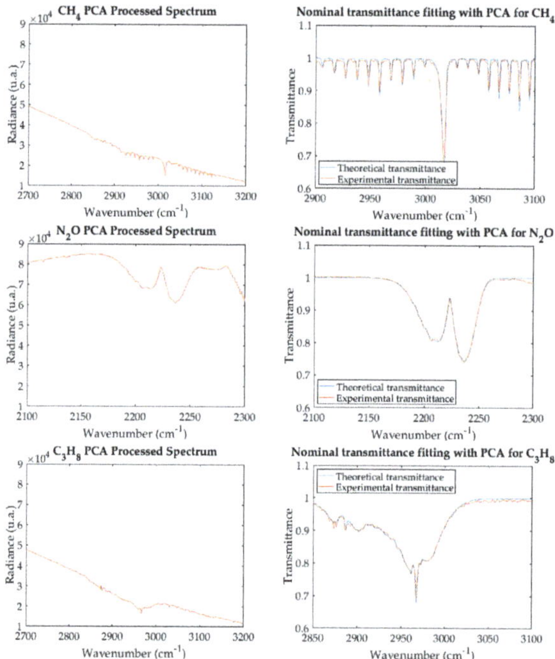

Figure 5. PCA–processed experimental spectra of air pollutants: radiance (**left**) and nominal transmittance, with best fit (**right**).

By fitting spectra over the whole field of view of the instrument, a map of retrieved Q is created. Figure 6 compares the C_3H_8 maps obtained from unprocessed spectra (left) and PCA-filtered spectra (right). As expected, only the round cell window regions have meaningful values, and they are quite similar in both cases, although the PCA-processed map is more uniform.

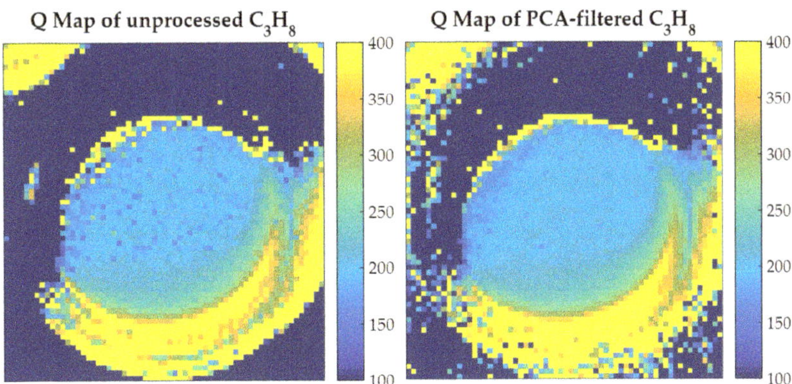

Figure 6. Maps of Q values retrieved by iterative fitting from τ_{nom}, unprocessed (**left**) and PCA-filtered (**right**). The scale is in ppm·m; the size of the field of view is 5.5 cm × 5.5 cm. Retrieved values of Q only have physical meaning in the central round region that corresponds to the gas cell window; it is clear that PCA filtering improves uniformity in that region.

Retrieved Q values are summarized in Table 2, both for PCA-filtered (Figure 5) and unfiltered spectra (Figure 3). Values are the mean ± the standard deviation in a square of 7×7 pixels at the center of the gas cell. Signal to noise ratios measured in dB are also tabulated. PCA increases SNR in all cases, and the effect is larger the noisier is the original spectrum: the dB value is multiplied by 3.2 for CH_4, by 2.1 for C_3H_8, and by 1.1 for N_2O. It must be pointed out that this improvement does not come at the expense of spatial resolution (which is not degraded) or acquisition time (which is not increased), since no spatial or time averaging is involved.

Comparison of the retrieved Q values with the nominal ones gives relative errors of -2.6% for CH_4, $+4.8\%$ for N_2O and -9.2% for C_3H_8 for non-PCA-processed spectra and similar values for the PCA-processed, except for a slightly better value for C_3H_8 (relative error -7.1%). These results, however, do not mean that PCA does not improve the measurement of Q. Since they have been obtained by spatially averaging over a uniform region, the most relevant parameter here is standard deviation, which is much smaller for PCA-filtered spectra. The conclusion to be extracted is that the main effect of PCA processing has been to improve the precision of retrieval rather than its accuracy.

Regarding the retrieved temperatures, for a room $T_g \approx 302$ K, results for CH_4, N_2O and C_3H_8 were, respectively, 310.6 ± 25.6 K, 305.0 ± 2.2 K and 312.6 ± 16.7 K for non-PCA-processed spectra, and 306.7 ± 2.4 K, 305.4 ± 1.4 K and 312.5 ± 8.7 K for the PCA-processed. These values show a similar behavior to those of Q: PCA processing has only improved slightly the value of T for CH_4 but has achieved an important reduction in standard deviations, i.e., gives better results regarding uniformity.

Table 2. Column density values retrieved and signal to noise ratio for air pollutants in a 7×7 square at the center of the gas cell. Values obtained by iterative search using as-measured experimental spectra and PCA-processed experimental spectra.

Gas	Nominal Q (ppm·m)	Retrieved Q w/o PCA (ppm·m)	Retrieved Q with PCA (ppm·m)	SNR w/o PCA (dB)	SNR with PCA (dB)
CH_4	258	251.2 ± 33.7	250.0 ± 9.8	5.5 ± 0.8	17.7 ± 0.1
N_2O	107.5	112.7 ± 1.6	112.3 ± 1.4	24.2 ± 0.8	26.1 ± 3.3
C_3H_8	215	195.2 ± 8.6	199.7 ± 3.8	12.6 ± 0.5	26.1 ± 1.1

5. Dimensionality Reduction by Principal Component Analysis

Up to now, PCA has been applied to a datacube of experimental nominal transmittance spectra and has been used only to filter out noise in those spectra by reconstructing them with a small number p of PCs (in the cases studied here, $p = 2$). Q_g and T_g have been retrieved by iterative fitting of the filtered spectra.

However, since filtered spectra are characterized by only $p \sim 2$ PCs, it seems that it is very inefficient to perform fitting in the full spectral space (where our objects are vectors of $m \sim 15.000$ components) instead of the subspace spanned by the relevant eigenvectors (where our objects are vectors of p components; we call this space "PC space").

The reason for this procedure is that simulation of spectra is based on the physics of absorption/emission and generates them line by line. So the spectra on which the iterative algorithm operates belong to the spectral space and have m components. If we want to operate in the PC space, they could be projected onto the p first eigenvectors obtained with PCA; then, the error between experiment and simulation could be calculated for the PCs. However, the bulk of the computation time is spent on the line-by-line simulation of the spectra and, once they are calculated, calculation of error is relatively straightforward. Thus, there is no appreciable efficiency gain in projecting the spectra on eigenvectors during iterative fitting and calculate errors in the PC space.

5.1. Retrieval by Search on Pre-Calculated Datacube

The previous observation underlines that the bottleneck of the retrieval process is the iterative generation of simulated spectra during fitting. Thus, a great improvement in efficiency could, in principle, be achieved by avoiding that process. This can be achieved if spectra are pre-calculated, as follows:

1. For a specific scene, a matrix of (T_g, Q_g) values can be defined, such that the ranges of T_g and Q_g cover the expected values in the scene. Nominal transmittance spectra $\tau_{nom}(\nu, Q_g, T_g)$ can be calculated for all the (T_g, Q) values of the matrix (for a given background temperature T_b). A simulated spectra datacube is thus obtained.
2. A experimental spectrum can now be compared to all the spectra of this datacube; the (T_g, Q_g) couple retrieved is the one that gives the smaller error (this can be measured as the sum of squared errors, SSE, or as the absolute error).

To test this procedure, simulated spectra datacubes with a spectral resolution of $1\ \mathrm{cm}^{-1}$ and $T_b = 350\ °\mathrm{C}$ were calculated for each of the three pollutant gases studied. Gas temperatures varied between $0\ °\mathrm{C} \le T_g < 69\ °\mathrm{C}$ with a step $\Delta T_g = 1\ °\mathrm{C}$, and the range of column densities was 70 ppm·m, centered for each gas at its expected column density, with $\Delta Q_g = 1$ ppm·m.

Results are shown in Table 3, under the heading SSD (simulated spectra datacube). Comparison with nominal values gives relative errors of -7.8% for CH_4, $+5.1\%$ for N_2O and -7.4% for C_3H_8, similar to those of the iterative fitting method except for a larger value in CH_4. Standard deviations are of the same order of those obtained previously with PCA-processed spectra.

Table 3. Column density retrieved for air pollutants in a 7×7 square at the center of the gas cell. Values obtained by search in simulated spectra datacube (SSD) and in simulated PC datacube (SPCD).

Gas	Nominal Q (ppm·m)	Retrieved Q SSD (ppm·m)	Retrieved Q SPCD (ppm·m)
CH_4	258	237.9 ± 11.7	253.2 ± 7.6
N_2O	107.5	113 ± 1.4	110.4 ± 3.4
C_3H_8	215	199.1 ± 3.3	218.1 ± 7.4

Generation of each simulated spectra datacube took 23.2 s of CPU time in an Intel i7 processor based computer at 3.2 GHz, with six cores and 64 GB of RAM. Then, the realization of a column density map over a region of 70×70 pixels took 5630 s of CPU time. This result was unexpected, since it is longer than the 1460 s of CPU time for the same task if completed by pixel-by-pixel iterative fitting.

The explanation is that in order to find the (T_g, Q_g) couple at each pixel an exhaustive search was used, i.e., the SSE was calculated between the experimental spectrum and *all* the spectra in the simulated datacube. This is a very inefficient strategy, and time can be reduced at least by an order of magnitude if a gradient search algorithm is used. Clearly, time will also be shorter if the simulated spectra datacube is made smaller, either by increasing the steps $(\Delta T_g, \Delta Q_g)$ or by reducing the range of (T_g, Q_g). No attempt of improvement along these lines has been made, however, since the approach based in PCA described in the following section is much more powerful.

5.2. Simulated PC Datacube

The retrieval strategy just described above compares experimental spectra as measured (i.e., in the spectral space) with the simulated ones. However, it can be enhanced by the use of principal components to make it faster.

If a PCA is performed on the simulated spectra datacube, its z dimension can be drastically reduced. The datacube thus obtained will be called the *simulated PC datacube*. Now, the number p of PCs needed may be larger than 2, since spectra in the simulated datacube have a larger variability than those of gas cell, because of the much wider interval of temperatures and column densities involved. However, the absence of noise reduces the variance of the simulated spectra, and, in our case, $p = 2$ is still enough to account for more than 99.95% of the total variance.

Now, to retrieve the values of T_g and Q_g for a pixel, the experimental spectrum is projected onto the first p eigenvectors of the simulated spectra datacube, in order to obtain its PCs (scores), and these p numbers are compared by a simple exhaustive search with those in the simulated PC datacube to find the (T_g, Q_g) couple with optimal agreement. It is important, however, not to make the direct comparison of the scores, but rather to multiply them by the magnitude of the corresponding eigenvector so as to to calculate correctly the distance between the experimental and the simulated spectra in the PC space.

Retrieval of Q and T is dramatically faster with this procedure. Generation of the simulated PC datacube from the simulated spectra datacube took 2.3 s of CPU. Then, creation of a map of Q over the same 70×70 region as above took only 1.0 s of CPU.

Results are shown in Table 3, under the heading SPCD (simulated PC datacube). Relative errors as compared to nominal values are now much smaller than previously: -1.9% for CH_4, $+2.7\%$ for N_2O and 1.4% for C_3H_8. Standard deviations are of the same order, being somewhat smaller for CH_4 and larger for C_3H_8.

Retrieved temperatures are also more accurate, and nearly identical for the three gases: 305.1 ± 2.7 K for CH_4, 305.7 ± 1.5 K for N_2O and 304.7 ± 5.1 K for C_3H_8.

A point worth noting is that, since this approach is based on a PCA performed on simulated spectra rather than on experimental ones, it can be applied as well to non-imaging spectrometers.

5.3. Retrieval of Q and T by Polynomial Fitting of Principal Components

One appealing aspect of the approach developed here is that the temperature and column density of the pollutant gas can be retrieved even without the ability to perform the complex process of spectrum simulation explained in Section 2.3. Rather, for a specific measurement conditions, with known T_b and expected ranges of T_g and Q_g, the user can be provided with the mean spectrum and the first p eigenvectors of the relevant simulated spectra datacube. Then, the components on the PC base of the experimental spectra can be written by subtracting the mean spectrum and projecting onto the eigenvectors.

In the previous section, T_g and Q_g for a pixel were obtained by an exhaustive search in the simulated PC datacube, to find the best agreement with those components. However, this can be further simplified for the user if explicit functions can be found, $T_g = T_g(PC_1, \ldots PC_p)$ and $Q = Q(PC_1, \ldots PC_p)$, that fit the dependence of T_g and Q_g from the PCs, as defined in the simulated PC datacube.

This has been perfomed for the three gases under study in this work, using the function package `polyfitn` available for use in MATLAB. It has been found that second-degree polynomial functions can provide values for T_g and Q_g as functions of (PC_1, PC_2), with very small errors. As an example (Figure 7), the error of the Q_g values furnished by the polynomial function is smaller than ±0.7 ppm·m for CH_4, ±1.7 ppm·m for N_2O, and ±5.5 ppm·m for C_3H_8 for most of the (T, Q) values of the pre-calculated datacube.

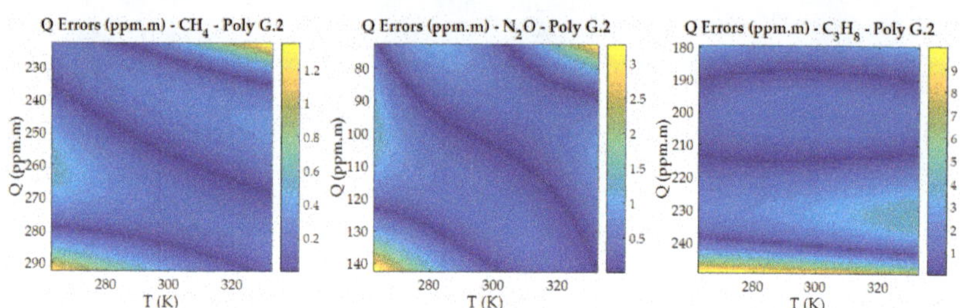

Figure 7. Absolute errors in the Q values obtained as second-degree polynomial functions of T_g (horizontal axis) and Q_g (vertical axis) for each of the gases studied. Errors are very small except for the cases when (T, Q) values are either very large or very small (for CH_4 and N_2O) and only for the very small values of Q (for C_3H_8).

6. Summary and Conclusions

The only way to improve signal-to-noise ratio (SNR) in a specific measurement condition with a non-imaging spectrometer was to average many spectra. In imaging spectrometers, averaging can be made over neighbouring pixels. In both cases, SNR improvement comes at a cost: time averaging degrades time resolution, and spatial averaging degrades spatial resolution.

Imaging spectroscopy, however, makes possible a better strategy: to apply principal component analysis to the datacube of experimental radiance spectra, and then reconstruct the spectra using only a reduced number of principal components. The reconstructed spectra have noise filtered out without losing spatial resolution.

In this work, this strategy has been applied to optimize measurements of column density (Q, concentration·path product) and temperature (T) of pollutant gases, specifically, methane, nitrous oxide, and propane.

A radiometric model that takes into account radiation emission and absorption, as well as instrumental lineshape, has been defined and applied to generate line-by-line theoretical spectra using the spectroscopic parameters of the HITRAN database. These spectra are compared to experimental spectra measured for the pollutant gases in order to retrieve their Q and T values. With an extended blackbody as background, two radiance spectra are acquired for each pixel: one with the gas cell full of pollutant at the prescribed concentration, the other with nitrogen as a reference, non-absorbing gas.

After PCA-processing, the increase in SNR, measured in dB, has been $\times1.1$ for N_2O, $\times2.1$ for C_3H_8, and $\times3.2$ for CH_4. These PCA-processed spectra have been used to obtain the nominal transmittance spectra whose comparison to theoretical spectra provides the retrieved Q and T values.

The more straightforward way to make that comparison is to generate theoretical spectra, and to compare them iteratively, wavenumber by wavenumber, to the experimental ones until the sum of squared errors is minimized. It has been found that the retrieved values of Q had a typical error of $\sim$ 7% both for unprocessed and PCA-processed spectra, although the latter provided better uniformity, with smaller standard deviations.

The strategy just described is, however, very slow and computing-intensive. PCA can be used also to speed up this process if the theoretical nominal transmittance spectra are pre-calculated for a range of T and Q appropriate to the expected values of the gas, and then PCA is applied to this *simulated spectra datacube*. Then, the comparison between experimental and theoretical spectra can be made in the PC space, whose dimension is drastically smaller than that of the spectra (in our case, two PCs versus $\sim$ 15.000 wavenumbers). Thus, a very significant reduction in calculation time (a factor larger than one thousand) is achieved. Accuracy of the retrieved Q and T values is also substantially improved: typical errors in retrieved Q values have been found to be $\sim$2%.

This procedure can be further simplified when the measurement conditions are repetitive, with known background temperature and gas T and Q within specific ranges. The user can be supplied with the results of the PCA applied to the relevant simulated datacube (mean spectrum and first eigenvectors), and can use them to obtain the first PCs of the experimental spectra. Then, if the ranges of T and Q are not too wide (e.g., 70 °C and 70 ppm·m in this work), explicit polynomic functions can be fitted to the simulated PC datacube that directly provides Q and T as functions of the first two PCs of the spectra. In this approach, the user only needs to measure the experimental nominal transmittance spectra, with no need to calculate simulated spectra or perform iterative fittings, and without significant loss of accuracy in the results.

Author Contributions: Conceptualization, J.M.; methodology, J.M.; software, G.G.; validation, G.G.; formal analysis, J.M.; investigation, J.M. and G.G.; writing—original draft preparation, J.M.; writing—review and editing, J.M. and G.G.; visualization, G.G.; supervision, J.M.; project administration, G.G. Both authors have read and agreed to the published version of the manuscript.

Funding: The authors would like to acknowledge the financial support from EURAMET through 16ENV08 IMPRESS 2 project. This project has received funding from the EMPIR programme co-financed by the Participating States and from the European Union's Horizon 2020 research and innovation programme.

Acknowledgments: The authors wish to thank Teresa Fernández and Enrique Santamaría from Centro Español de Metrología (CEM) for providing the gas mixtures studied in this work.

Conflicts of Interest: The authors declare no conflict of interest. The funders had no role in the design of the study; in the collection, analyses, or interpretation of data; in the writing of the manuscript, or in the decision to publish the results.

References

1. Héroux, M.-E.; Anderson, H.R.; Atkinson, R.; Brunekreef, B.; Cohen, A.; Forastiere, F.; Hurley, F.; Katsouyanni, K.; Krewski, D.; Krzyzanowski, M.; et al. Quantifying the healthimpacts of ambient air pollutants: Recommendations of a who/europe project. *Int. J. Public Health.* **2015**, *60*, 619–627. [CrossRef] [PubMed]
2. IMPRESS 2: Metrology for Air Pollutant Emissions. Available online: http://empir.npl.co.uk/impress/ (accessed on 14 September 2020).
3. Hodgkinson, J.; Tatam, R.P. Optical gas sensing: A review. *Meas. Sci. Technol.* **2012**, *24*, 012004. [CrossRef]
4. Prengle, H.W.; Morgan, C.A.; Fang, C.-S.; Huang, L.-K.; Campani, P.; Wu, W.W. Infrared remote sensing and determination of pollutants in gas plumes. *Environ. Sci. Technol.* **1973**, *7*, 417–423. [CrossRef] [PubMed]
5. Herget, W.F. Remote and cross-stack measurement of stack gas concentrations using a mobile FT-IR system. *Appl. Opt.* **1982**, *21*, 635–641 . [CrossRef] [PubMed]
6. Wormhoudt, J. (Ed.) *Infrared Methods for Gaseous Measurements: Theory and Practice*; Marcel Dekker: New York, NY, USA, 1985.
7. Manolakis, D.G.; Lockwood, R.B.; Cooley, T.W. *Hyperspectral Imaging Remote Sensing: Physics, Sensors, and Algorithms*; Cambridge University Press: Cambridge, UK, 2016.
8. Gross, K.C.; Bradley, K.C.; Perram, G.P. Remote Identification and Quantification of Industrial Smokestack Effluents via Imaging Fourier-Transform Spectroscopy. *Environ. Sci. Technol.* **2010**, *44*, 9390–9397. [CrossRef] [PubMed]
9. Harley, J.L.; Gross, K.C. Remote quantification of smokestack effluent mass flow rates using imaging Fourier transform spectrometry. *Proc. SPIE* **2011**, *8018*, 801813.

10. Rodríguez-Conejo, M.A.; Meléndez, J. Hyperspectral quantitative imaging of gas sources in the mid-infrared. *Appl. Opt.* **2015**, *54*, 141–149. [CrossRef] [PubMed]
11. Gålfalk, M.; Olofsson, G.; Bastviken, D. Approaches for hyperspectral remote flux quantification and visualization of GHGs in the environment. *Remote. Sens. Environ.* **2017**, *191*, 81–94. [CrossRef]
12. Gordon, I.E.; Rothman, L.S.; Hill, C.; Kochanov, R.V.; Tan, Y.; Bernath, P.F.; Birk, M.; Boudon, V.; Campargue, A.; Chance, K.V.; et al. The HITRAN2016 molecular spectroscopic database. *J. Quant. Spectrosc. Radiat. Transf.* **2017**, *203*, 3–69. [CrossRef]
13. Kochanov, R.V.; Gordon, I.E.; Rothman, L.S.; Wcisło, P.; Hill, C.; Wilzewski, J.S. HITRAN Application Programming Interface (HAPI): A comprehensive approach to working with spectroscopic data. *J. Quant. Spectrosc. Radiat. Transf.* **2016**, *177*, 15–30. [CrossRef]
14. HITRANonline. Available online: https://hitran.org/ (accessed on 26 February 2020).
15. Griffiths, P.R.; De Haseth, J.A. *Fourier Transform Infrared Spectrometry*, 2nd ed.; Wiley: Hoboken, NJ, USA, 2007.
16. Chamberland, M.; Farley, V.; Vallieres, A.; Villemaire, A.; Belhumeur, L.; Giroux, J.; Legault, J.-F. High-performance fieldportable imaging radiometric spectrometer technology for hyperspectral imaging applications. *Proc. SPIE* **2005**, *5994*, 59940N.
17. Gagnon, J.; Habte, Z.; George, J.; Farley, V.; Tremblay, P.; Chamberland, M.; Romano, J.; Rosario, D. Hyper-Cam automated calibration method for continuous hyperspectral imaging measurements. *Proc. SPIE* **2010**, *7687*, 76870E.
18. Gross, K.C.; Tremblay, P.; Bradley, K.C.; Chamberland, M.; Farley, V.; Perram, G.P. Instrument calibration and lineshape modeling for ultraspectral imagery measurements of industrial smokestack emissions. *Proc. SPIE* **2010**, *7695*, 769516.
19. Shlens, J. A tutorial on principal component analysis. *arXiv* **2014**, arXiv:1404.1100.
20. Natarajan, B.; Konstantinides, K.; Herley, C. Occam filters for stochastic sources with application to digital images. *IEEE Trans. Signal Process.* **1998**, *46*, 1434–1438. [CrossRef]
21. Antonelli, P.; Revercomb, H.E.; Sromovsky, L.A.; Smith, W.L.; Knuteson, R.O.; Tobin, D.C.; Garcia, R.K.; Howell, H.B.; Huang, H.-L.; Best, F.A. A principal component noise filter for high spectral resolution infrared measurements. *J. Geophys. Res. Atmos.* **2004**, *109*, D23102. [CrossRef]
22. Kanungo, T.; Mount, D.M.; Netanyahu, N.S.; Piatko, C.D.; Silverman, R.; Wu, A.Y. An efficient k-means clustering algorithm: Analysis and implementation. *IEEE Trans. Pattern Anal. Mach. Intell.* **2002**, *24*, 881–892. [CrossRef]
23. Duda, R.O.; Hart, P.E.; Stork, D.G. *Pattern Classification*, 2nd ed.; A Wiley-Interscience Publication; Wiley: Hoboken, NJ, USA, 2000.

Article

Analytic Optimization of Cantilevers for Photoacoustic Gas Sensor with Capacitive Transduction

Wioletta Trzpil, Nicolas Maurin, Roman Rousseau, Diba Ayache, Aurore Vicet and Michael Bahriz *

IES, University Montpellier, CNRS, 34095 Montpellier, France; wioletta.trzpil@umontpellier.fr (W.T.);
nicolas.maurin@ies.univ-montp2.fr (N.M.); roman.rousseau@ies.univ-montp2.fr (R.R.);
diba.ayache@ies.univ-montp2.fr (D.A.); aurore.vicet@umontpellier.fr (A.V.)
* Correspondence: michael.bahriz@umontpellier.fr

Abstract: We propose a new concept of photoacoustic gas sensing based on capacitive transduction which allows full integration while conserving the required characteristics of the sensor. For the sensor's performance optimization, trial and error method is not feasible due to economic and time constrains. Therefore, we focus on a theoretical optimization of the sensor reinforced by computational methods implemented in a Python programming environment. We present an analytic model to optimize the geometry of a cantilever used as a capacitive transducer in photoacoustic spectroscopy. We describe all the physical parameters which have to be considered for this optimization (photoacoustic force, damping, mechanical susceptibility, capacitive transduction, etc.). These parameters are characterized by opposite trends. They are studied and compared to obtain geometric values for which the signal output and signal-to-noise ratio are maximized.

Keywords: MEMS; gas sensor; photoacoustics; cantilever; capacitive detection; analytic model

Citation: Trzpil, W.; Maurin, N.; Rousseau, R.; Ayache, D.; Vicet, A.; Bahriz, M. Analytic Optimization of Cantilevers for Photoacoustic Gas Sensor with Capacitive Transduction. *Sensors* **2021**, *21*, 1489. https://doi.org/10.3390/s21041489

Academic Editor: Krzysztof M. Abramski

Received: 20 January 2021
Accepted: 18 February 2021
Published: 21 February 2021

Publisher's Note: MDPI stays neutral with regard to jurisdictional clai-ms in published maps and institutio-nal affiliations.

1. Introduction

The market for gas sensors was estimated to be 2.23 billion USD in 2020 and is expected to reach 4.49 billion USD in 2028 [1]. The growing interest in gas sensors is driven by various field of applications, e.g., medicine [2], air quality [3], food processing [4], or security and defense [5], that address legislative (e.g., EU's air quality directives), National Ambient Air Quality Standards) and/or individual needs. Sensors commonly used in the market, according to the highest percentage contribution into the gas sensor market income, are electrochemical, semiconductor, and infrared sensors [1]. Electrochemical sensor principles are based on creation of an electrical signal after reaction with a target gas. Semiconductor sensors are made of heated metal oxides which in the presence of the gas change their resistivity. Infrared gas sensors are based on electromagnetic signal conversion into electrical signal [6]. Characteristics of these sensors are presented in Table 1 [7].

Gas sensors for real-life applications [7], e.g., air quality, toxic gasses, medicine, food processing, are required to be selective (perfectly distinguish one species among others), sensitive (able to detect few particles per million in volume (ppmv)), reliable (stable, suffer from small drift), and compact. Infrared gas sensors, like the ones based on tunable diode laser spectroscopy (TDLS), can perfectly discriminate the spectral signature of a gas species among others, thus providing an excellent selectivity, combined with a high sensitivity (sub-ppb detection) [8] (Table 1). The main drawbacks of infrared detection are: lack of absorption line in infrared spectrum for some gasses, poor selectivity for gasses with absorption line at the same wavelength, and lack of compactness.

Photoacoustic spectroscopy, an evolution of TDLS, permits reducing the size of the gas sensor while maintaining equivalent performances. In TDLS the detected signal is proportional to the length of the optical path while in photoacoustic spectroscopy, it is related to the laser emitted power, which allows keeping a high sensitivity even in a compact gas cells.

149

Table 1. Characteristics for various types of gas sensors based on [7].

Parameters	Electrochemical	Semiconductor	Infrared
sensitivity	g	e	e
stability	b	g	g
selectivity	g	p	e
compactnesss	p	e	b
cost	e	g	p
application	air purity [9]	Industrial applications and civil use [10]	(a) Remote air quality monitoring; (b) Gas leak detection systems; (c) High-end market applications. [10]

e—excellent, g—good, p—poor, b—bad.

In photoacoustic spectroscopy, a modulated laser emitting at a wavelength corresponding to the absorption line of a targeted gas species is focused into a gas chamber. The measurement is performed by detecting the acoustic pressure generated by the local warming induced by molecular relaxation following optical absorption. The local temperature rise is a result of non-radiative vibrational–translational (V-T) relaxation processes occurring between excited molecules. At atmospheric pressure, the laser emission linewidth ($\sim$MHz) is much smaller than the gas linewidth ($\sim$GHz), which gives a perfect selectivity to this method.

The acoustic wave can be measured using a microphone [11] or a mechanical resonator such as a tuning fork [12]. The use of a mechanical resonator with high quality factor (around 10,000 for a quartz tuning fork (QTF)) improves the signal-to-noise (SNR) ratio and avoids the use of a resonant acoustic chamber.

Commercial QTF allows reaching very good sensing performances in Quartz Enhanced Photoacoustic Spectroscopy (QEPAS) [13] even if they were developed for the electronics market, and not for sensing purposes. As a consequence, the QTF is not optimized for photoacoustic spectroscopy and its potential integration in a compact system is limited compared to other mechanical resonators based on silicon materials. Silicon would offer several advantages such as its technological maturity, its design flexibility and its lower production costs. However, its best advantage lies in the feasibility of integration in complex CMOS electronics [14,15]. Recent progress in laser sources integration on silicon [16] makes it possible to consider fully integrated compact sensors. For these reasons, silicon-based micro-resonator seems to be the best choice for the future development of very compact gas sensors integrated on the same chip with electronics, a laser, and a mechanical resonator.

We study here the realization of a silicon-based micro resonator sensor, a cantilever, dedicated to photoacoustic sensing. This sensor, specifically designed for acoustic sensing purposes, would be an efficient transducer for sound wave detection.

The most common transduction methods in silicon-based micro-electromechanical systems (MEMS) are based on capacitive, piezoresistive, and piezoelectric effects. The capacitive transduction mechanism constitutes a more convenient method than piezoelectric [17] or piezoresistive [18] detection. It avoids any material deposition or implantation on the mechanical resonator, which may reduce the quality factor and make the fabrication process more complex. Capacitive detection employed in MEMS technology allows reaching high sensitivity. For example, the capacitive accuracy for accelerometers or position sensors is about a few ppm of their nominal capacitance [19], leading to a sub-femto-farrad resolution [20]. To improve a capacitive signal, it is advantageous to increase the capacitor surface which leads to a rise in viscous damping and abbreviates the devices performances. Undoubtedly, for parameters characterized by opposite trends, an optimization based on a theoretical model would be the first step towards sensor performance improvement.

The working principle of a gas sensor based on photoacoustic spectroscopy using a cantilever as a capacitive transducer is schematically presented in Figure 1. The acoustic pressure generated by laser light absorption applies a force on the cantilever and sets it in motion. To maximize the displacement, the acoustic wave is generated at the resonance frequency of the cantilever via laser wavelength modulation. The silicon cantilever is electrically insulated from the back silicon, forming a capacitor. One of the electrodes of the capacitor is the cantilever itself. The displacements of the cantilever cause the capacitance variations. Depending on the excitation frequency, the capacitance variations can be converted into a current or a voltage signal.

Performing solely a trial and error method for the sensor's performance optimization is not feasible due to economic and time constraints. Therefore, a computational method is the most reasonable choice. The sensing scheme imposes multi-physics problems in different domains and can be divided in four parts: (1) acoustic force, (2) damping mechanisms, (3) mechanical displacement, and (4) output signal. Many of these problems are not directly coupled and others are characterized by opposite trends in terms of geometry optimization. The main novelty in our approach is a simultaneous multi-physics optimization. This optimization aims to determine the geometrical parameters of the cantilever (length L, width b, thickness h, gap d (Figure 1)) and its resonance frequency, which would maximize the output electrical signal and the signal-to-noise ratio. For this, the cantilever has to be sized to maximize its displacement under acoustic wave exposition while exhibiting a strong capacitance variation.

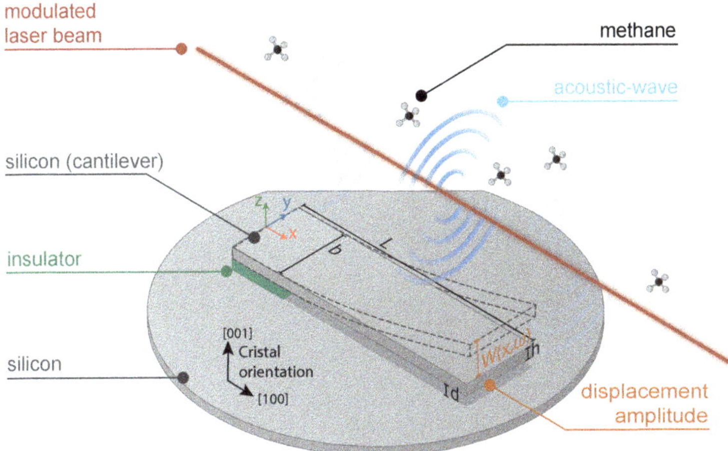

Figure 1. Sensing scheme of a silicon cantilever-based sensor for photoacoustic gas detection with capacitive transduction mechanisms.

The paper is divided as follows:

(1) The acoustic force part describes the generation of the photoacoustic wave and its interaction with the cantilever. As in TDLS, the acoustic wave is obtained by wavelength modulation technique at the pulsation ω [21]. The photoacoustic effect is related to heat production rate, depending on the non-radiative relaxation time of the target gas. The non-radiative relaxation strongly relies on the molecular species, the gas concentration, temperature, pressure, and the gas mixture. To obtain realistic values, our numerical estimation is performed on one specific gas: CH_4 diluted in N_2. However, the model is compatible with any gas mixture once the relaxation time is known.

(2) The damping mechanism part describes the following mechanisms of losses: viscous, thermoelastic, support, and acoustic damping.

(3) The mechanical part describes how the cantilever is set in motion by the acoustic wave, taking into account its susceptibility. The displacement amplitude of the cantilever is

described by $W(x, \omega)$ presented in Figure 1. Further terms describing the cantilever displacement refer to the fundamental vibration mode presented in Figure 1.

(4) The output signal part represents the energy conversion from mechanical motion to an electrical signal. It gives the relation between the cantilever deflection $W(x, \omega)$ and the electrical signal output $V_{out}(\omega)$.

(5) The *SNR* part illustrates the thermal noise $W_{noise}(x, \omega)$.

2. Acoustic Force

The purpose of this section is to study the cantilever dimensions (length L, width b, thickness h) and its resonance frequency in order to maximize the acoustic force. This part evaluates the photoacoustic pressure generation and the photoacoustic force applied to the cantilever. The source of the photoacoustic wave generation lies in periodic gas absorption induced by a modulated laser beam. This method is called wavelength modulation spectroscopy [21]. We consider a Gaussian laser beam propagating along the x-axis at an altitude $z = z_L$ and centered with respect to y-axis at $y = y_L$ (Figure 2).

The distribution of the light intensity $I(x, y, z)$ is related to the laser power P_L:

$$I(x, y, z) = P_L g(x, y, z) \tag{1}$$

$$g(x, y, z) = \frac{2}{\pi w_L(x)^2} \exp\left(-2\frac{(z - z_L)^2 + (y - y_L)^2}{w_L(x)^2}\right)$$

where $g(x, y, z)$ is a normalized Gaussian profile and $w_L(x) = w_L(x_L)\sqrt{\left(1 + \frac{(x - x_L)^2}{x_R^2}\right)}$ is the laser radius which depends on the Rayleigh length $x_R = \frac{\pi w_L(x_L)}{\lambda_L}$, with λ_L the laser emission wavelength.

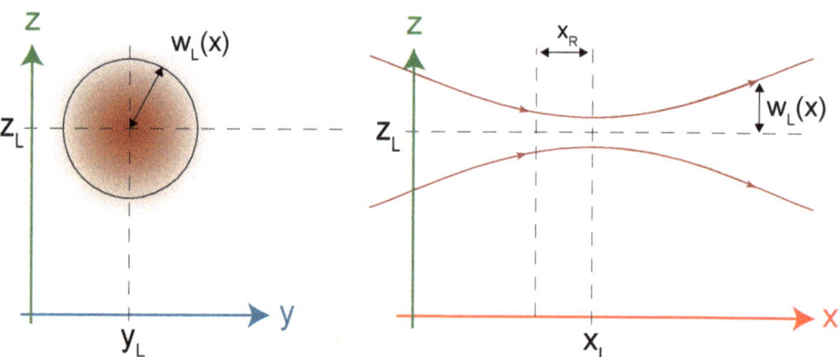

Figure 2. Gaussian beam profile and its position on the axis in relation to the cantilever microbeam. $w_L(x_L) = 100$ µm, $\lambda_L = 1.65$ µm, $x_L = 0.725L$, $y_L = 0$, $z_L = 150$ mm.

The theoretical model used to describe the pressure and force of the acoustic wave generated by molecular absorption is based on the model developed by Petra et al. [22]. However, our model takes into account the variation of the laser beam radius $w_L(x)$ along the optical axis (x-axis) and the effects of the gas relaxation time constant. The assumptions used in the model are:

1. The wavelength modulation is performed without modulation of the laser power.
2. Sommerfeld radiation conditions: no reflection from any walls of a gas cell and photoacoustic energy fading at infinity.
3. The photoacoustic pressure is unaltered by the presence of the cantilever.
4. The laser beam radius is smaller than the distance between the cantilever and the optical axis.

To fulfill the third assumption, the acoustic wave wavelength λ_a must be at least one order of magnitude larger than the thickness and width of the cantilever: $\lambda_a \sim 3.5$ cm at $\nu = 10$ kHz ($\lambda_a = \frac{c_s}{\nu}$, where c_s is the speed of sound).

The absorption of the modulated light causes periodic heat changes and subsequently an acoustic wave. The heat production rate is given by:

$$H(x,y,z,t) = \frac{C_f(\omega)g(x,y,z)}{\sqrt{1+(\omega\tau)^2}}e^{i(\omega t-\arctan(\omega\tau))} \tag{2}$$

where ω is the laser modulation frequency, τ is the target gas relaxation time $C_f(\omega)$ is the effective absorption coefficient. The absorption and transmission line shapes can be ideally described with a Lorentzian line shape function. The laser emission wavelength scan the absorption line and is modulated around the central wavelength λ_c. The modulated wavelength can be expressed as: $\lambda(t) = \lambda_c + \lambda_{amp}sin(\omega t)$, where λ_{amp} is the modulation amplitude. When the laser wavelength is modulated the power remains constant and equal to P_L, we can write $C_f(\omega) = 0.50\alpha(\omega)P_L$, where $\alpha(\omega)$ is the absorption coefficient of the gas. The 0.5 factor is obtained by expansion of the absorption function in Fourier series. We consider only the first Fourier component ($a_1 = 0.5$) for the $1f$ detection method. The second Fourier components would result in a coefficient of 0.35 ($a_2 = 0.35$) [22].

The photoacoustic wave generation is related to the heat production due to the light absorption. The expression of photoacoustic pressure $P(x,y,z)$ is given by the wave equation:

$$\frac{\partial^2 P(x,y,z,t)}{\partial t^2} - c_s^2\Delta P(x,y,z,t) = (\gamma - 1)\frac{\partial H(x,y,z,t)}{\partial t} \tag{3}$$

where $c_s = 347.276$ m/s is the sound velocity in air and $\gamma = \frac{C_p}{C_v}$ the adiabatic gas coefficient or heat capacity ratio equal to the fraction ratio between heat capacities at constant pressure and volume.

Equation (3) is an inhomogeneous equation with time. By substituting $P(x,y,z,t) = p(x,y,z)e^{i\omega t}$ and $H(x,y,z,t) = h(x,y,z)e^{i\omega t}$ and imposing Sommerfeld radiation boundary conditions, Petra et al. [22] showed that the pressure equation takes the following form:

$$p(x,y,z) = -\frac{\pi A}{2c_s^2 k_s^2}(Y_0(k_s r) + iJ_0(k_s r))\int_0^{+\infty} uJ_0(u)\exp\left(\frac{-2u^2}{k_s^2 w_L(x)^2}\right)du \tag{4}$$

where J_0, Y_0 are the zero-order Bessel functions of the first and the second kind, respectively. $A = -(\gamma-1)\omega H(x,y,z)\frac{2}{\pi w_L(x)^2}$ represents the amplitude of photoacoustic pressure, $k_s = \omega/c_s$ is the wave number, and $r = \sqrt{(z-z_L)^2 + (y-y_L)^2}$ is the distance between the laser beam and the cantilever.

The photoacoustic force F_{PA} applied on the cantilever is defined as the difference of pressure between the top and bottom surfaces of the cantilever.

$$F_{PA} = \int_0^L \int_{-b/2}^{b/2} (p(x,y,z) - p(x,y,z-h))\phi_n(x)\,dx\,dy \tag{5}$$

$\phi_n(x)$ describes the one-dimension shape of the cantilever mechanical mode n. It gives the cantilever deflection and can be found analytically by solving an eigenvalue problem of the Euler–Bernoulli equation. The mode shape for a clamped-free cantilever is given by [23]:

$$\phi_n(x) = \cosh\left(\alpha_n\frac{x}{L}\right) - \cos\left(\alpha_n\frac{x}{L}\right) - \frac{\sinh(\alpha_n) - \sin(\alpha_n)}{\cosh(\alpha_n) + \cos(\alpha_n)}\left(\sinh\left(\alpha_n\frac{x}{L}\right) - \sin\left(\alpha_n\frac{x}{L}\right)\right) \tag{6}$$

The acoustic force acting on the cantilever is frequency-modulated at the wavelength modulation frequency of the laser source. For a first harmonic detection (1f detection) it is adjusted to the cantilever mode frequency. In our model the cantilever vibrates at its fundamental mode-first harmonic $n = 1$ which corresponds to a mode constant $\alpha_1 = 1.875$.

Results and Discussion

The parameters used in the simulation are detailed in Table 2. We chose a laser emitting at 1.65 μm to target a strong methane CH_4 absorption line. Based on our numerical simulation presented in Appendix A.1, Figure A1 illustrates how x_L and y_L coordinates maximize the acoustic force, while $z_L = 250$ μm conserves the assumption that laser light does not interfere with the cantilever.

Table 2. Parameters used to describe the laser source and the acoustic wave.

Parameter	Description	Value
λ_L	laser wavelength	1.65 μm
$w_L(x_L)$	laser waist (experimental value)	100 μm
x_L, y_L, z_L	laser beam waist radius position	0.725 L, 0, 250 mm
P_L	laser power	50 mW
C_{gas}	CH_4 concentration in N_2	1%
α	absorption coefficient	CH_4 0.38 m^{-1}
τ	target gas relaxation time	CH_4 11.5 μs [24]
c_s	speed of sound	Air 347.276 m/s
γ	heat ratio capacity	Air 1.4

Due to the thermal relaxation time, the modulation frequency strongly affects the heat production rate (Equation (2)) and subsequently the acoustic force. Indeed, to allow the molecules to thermalize efficiently, the laser modulation needs to be lower than the molecules relaxation time.

Each molecule exhibits a different relaxation time. To maximize the photoacoustic force, the optimisation needs to be made with respect to one type of gas. We chose CH_4 diluted in nitrogen N_2 for which the relaxation time is equal to 11.5 μs [24]. However, the relaxation time between the molecules might differ by several orders of magnitude.

Figure 3 presents the acoustic pressure and force for CH_4 diluted in N_2, 1% and 0.5%, respectively. Only the acoustic force depends on cantilever geometry. To maintain a fixed frequency, the cantilever length is adjusted with the following equation:

$$f_n = \frac{\omega_n}{2\pi} = \frac{\alpha_n^2}{2\pi\sqrt{12}} \frac{h}{L^2} \sqrt{\frac{E}{\rho_b}} \tag{7}$$

where f_n is the resonance frequency of a clamped-free cantilever, $\rho_b = 2330$ kg/m^3 is the silicon density and $E = 130$ GPa is Young's modulus for silicon in [100] direction [25].

The values of the acoustic force and pressure clearly depend on the modulation frequency as it is presented in Figure 3. For each concentration, they increase with the frequency until reaching a maximum around 20 kHz for the acoustic pressure and around 11 kHz for the acoustic force. This maximum is related to CH_4 relaxation time value. The maximum shift to lower frequency between the acoustic pressure and the acoustic force is due to the cantilever length which appears only in the acoustic force, Equation (5). According to Equation (7), the length of the cantilever is longer for lower frequencies. Therefore, the surface exposed to the acoustic pressure is larger, which subsequently increases the acoustic force at low frequencies.

The maximum value of the acoustic force is at 11 kHz. To maximize the force applied on the cantilever, this frequency is used in the following numerical simulations of the cantilever geometry (width b, thickness h, and length L). However, the model is adaptable to any frequency with respect to the assumptions.

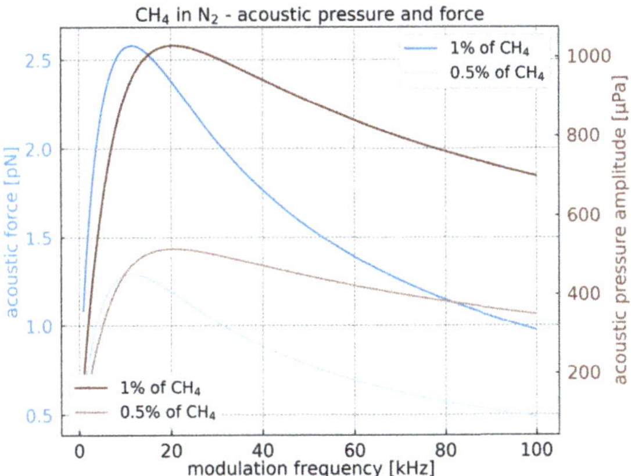

Figure 3. Acoustic force and acoustic pressure dependency on the modulation frequency for diluted CH_4 at 1% and 0.5% in nitrogen. Cantilever width $b = 25$ μm and thickness $h = 100$ μm.

Figure 4 represents the total photoacoustic force applied on cantilever for different cantilever geometries. It shows two general trends. Firstly, the photoacoustic force increases with the width b and the thickness h. Indeed, the surface enlargement increases the energy collection from the acoustic wave. Secondly, the thickness increment increases the pressure difference between the top and bottom sides of the cantilever, which enhances the acoustic force. For a fixed cantilever frequency, the increase of the thickness causes the length increment and enlarges the total surface (Equation (7)). The results presented in Figure 4 would change while using different gases, different volume mixing ratios, or different frequencies (Figure 3). Nevertheless, the general trend would remain constant.

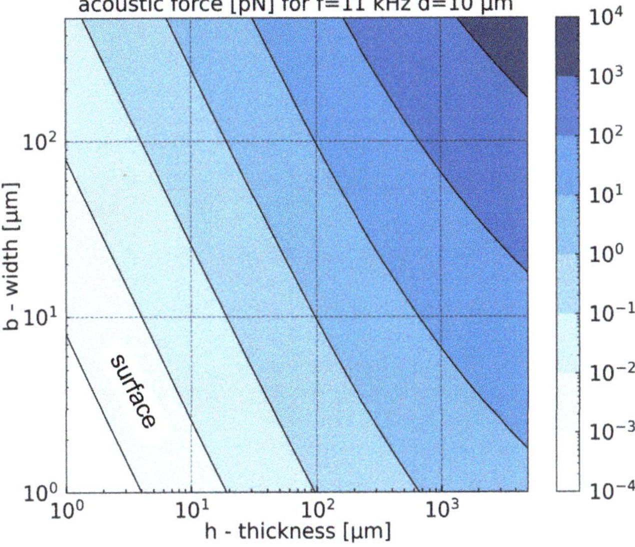

Figure 4. Acoustic force for 1% of CH_4 in N_2 as a function of width b and thickness h of the cantilever. For different thickness, the length is adjusted to maintain constant frequency: 11 kHz. This frequency was chosen to maximize acoustic force. The area with the weakest acoustic force corresponds to cantilevers with the smallest surface.

The results presented in the following sections are further taken into the calculation to get the optimized geometry of the cantilever with regard to electrical signals and signal-to-noise ratio.

3. Damping Mechanism—Quality Factor

The quality factor Q is a dimensionless number which describes the energy losses in the system. It can be expressed as the ratio between the energy stored in a cycle of vibration E_{stored} and the energy dissipated in a cycle of vibration $E_{dissipated}$

$$Q = 2\pi \frac{E_{stored}}{E_{dissipated}} \tag{8}$$

There are two main mechanisms where the cantilever can lose energy: through internal energy dissipation, like thermoelastic losses, and external dissipation, like viscous damping, support losses or acoustic losses. The total quality factor consists of quality factors originating from different losses and can be calculated using the following equation:

$$\frac{1}{Q_{total}} = \frac{1}{Q_{viscous}} + \frac{1}{Q_{thermo}} + \frac{1}{Q_{support}} + \frac{1}{Q_{acoustic}} \tag{9}$$

3.1. Thermoelastic Losses

Thermoelastic damping (TED) is a loss mechanism due to the irreversible heat flow in vibrating structures. A temperature gradient occurs between regions under tension (where the temperature drops) and regions under compression (where the temperature rises).

We use an analytical model proposed by Lifshitz [26,27], where the thermoelastic quality factor is given by:

$$Q_{thermo} = \frac{C_p}{E\alpha_T^2 T}\left(\frac{6}{\xi^2} - \frac{6}{\xi^3}\left(\frac{\sinh(\xi) + \sin(\xi)}{\cosh(\xi) + \cos(\xi)}\right)\right)^{-1} \tag{10}$$

ω, C_p, α_T, T, E are the pulsation, specific heat capacity, linear thermal expansion coefficient, temperature, Silicon Young's modulus, respectively. $\xi = h\sqrt{\frac{\omega \rho_b C_p}{2K}}$ represents a dimensionless number where K is the thermal conductivity. The values of all these parameters can be found in Table 3. The maximum of thermoelastic damping [26] occurs for $\xi = 2.225$. This value corresponds to a transition frequency $f_t = \frac{\pi}{2}\frac{K}{\rho_b C_p h^2}$. For a cantilever frequency f_n lower than the transition frequency f_t ($f_n < f_t$), the beam is permanently in thermal equilibrium. In this case the vibration is called isothermal. On the other hand, when $f_n > f_t$ the cantilever frequency is higher than the transition frequency, the beam does not have enough time to thermally equilibrate and this vibration is called adiabatic. In both cases, the energy dissipation is low. However, the Q_{thermo} quality factor is higher in isothermal than in adiabatic regime [28]. In case of constant-frequency regime, one needs to calculate the thickness that gives the maximal damping. Based on the f_t expression, the isothermal zone corresponds to the thin cantilever thickness and the adiabatic zone to the large thickness. For $f_n = 11$ kHz, the maximal thermoelastic damping, i.e., the lowest $Q_{thermo} = 12,500$, corresponds to a cantilever with thickness $h = 90$ µm. Therefore, for frequency of 11 kHz, thermoelastic damping is not a limiting factor.

Table 3. Parameters used to describe the damping mechanism.

Parameter	Description	Value
ρ_b	mechanical resonator density	Si 2330 kg/m^3
ρ_f	fluid density	Air 1.177 kg/m^3
μ_f	fluid dynamic viscosity	Air 1.85 $\times$ 10^{-5} kg/m/s
d	air gap	10 µm
P_a	pressure	101,325 Pa
T	temperature	300 K
E	Young's modulus	Si$_{<100>}$130 GPa
ν	Poisson's ratio	Si$_{<100>}$ 0.28
α_T	thermal expansion coefficient	Si 2.6 $\times$ 10^{-6} 1/K
C_p	specific heat at constant pressure	Si 700 J/(kgK)
K	thermal conductivity	Si 90 W/m/K

3.2. Acoustic Losses

Acoustic losses refer to losses caused by a vibrating structure being a source of acoustic wave radiation. A good approximation of these losses can be expressed with an analytical model for cantilever with elliptical cross-section [29–31]. In this approach the quality factor related to acoustic losses is given by the following equation:

$$Q_{acoustic} = \frac{256}{\pi} \frac{\rho_b}{\rho_f} \frac{1}{(k_s b)^3} \frac{h \int_0^L \phi_n^2(x)\,dx}{\int_{\varphi=0}^{\pi} sin^3\varphi \left| \int_0^L \phi_n(x) \exp(-ik_s x cos(\varphi))\,dx \right|^2 d\varphi} \tag{11}$$

where ρ_f is a fluid density, $k_s = \omega/c_s$ the acoustic wave number and c_s is the speed of sound. Numerical calculations using Equation (11) show that the losses due to acoustic radiation become important when the cantilever length is comparable to the acoustic wavelength λ_a. It is less significant at low frequencies. Moreover, acoustic losses increase quickly as the width increases and the thickness decreases (for constant-frequency regime). For instance, for $b = 5000$ µm and $h = 1$ µm $Q_{acoustic} \simeq 605$.

3.3. Support Losses

The cantilever presented in Figure 1 is held by a support. During the cantilever movement a part of the energy is dissipated into the support. This dissipation is described by the support quality factor. An analytical solution for support losses in case of a clamped-free cantilever was proposed by Hao [32] and takes the following form:

$$Q_{support} = \left(\frac{0.24(1-\nu)}{(1+\nu)\Psi} \right) \frac{1}{(\frac{\alpha_n}{\pi}\chi_n)^2} \left(\frac{L}{h} \right)^3 \tag{12}$$

where ν, α_n, χ_n is the Poisson's ratio, a mode constant, and a mode shape factor, respectively. For the clamped-free cantilever fundamental mode $n = 1$ and $\alpha_1 = 1.875$, the mode shape factor $\chi_1 = \frac{sin(\alpha_1)-sinh(\alpha_1)}{cos(\alpha_1)+cosh(\alpha_1)}$ and $\Psi = 0.336$. It can be seen from Equation (12) that the energy dissipation from the support is inversely proportional to $(L/h)^3$. If we look at a fixed frequency, without considering the length of the cantilever, then the quality factor of the support is $Q_{support} \propto \frac{1}{\sqrt{\omega_n^3 h^3}}$.

3.4. Viscous Damping

Viscous damping originates from the fluid resistance. It is considered to be the most significant damping mechanism in MEMS operating in ambient conditions.

During the beam movement in fluid, an additional force related to the medium appears. The quality factor due to viscous damping can be analytically expressed using a normalized time-independent function called hydrodynamic function Γ_{hydro}:

$$Q_{viscous} = \frac{\frac{4\rho_b h}{\pi \rho_f b} + \Gamma^R_{hydro}(\omega)}{\Gamma^I_{hydro}(\omega)} \tag{13}$$

where $\rho_b, \rho_f, \Gamma^R_{hydro}, \Gamma^I_{hydro}$ are the density of the beam, the density of the fluid, and the real and imaginary parts of the hydrodynamic function, respectively. The total hydrodynamic function originates from the linearized Navier–Stokes equation. Thus, it can be represented as a linear combination of hydrodynamic functions originating from each sidewall of the beam cross-section [33]. Pictorially, it is presented in Figure 5, while mathematically it is expressed as:

$$\Gamma_{hydro} = \frac{1}{2}\Gamma_{tb} + \frac{1}{2}\Gamma_{tb} + \Gamma_{sq} + \frac{1}{2}\Gamma_{lr} + \frac{1}{2}\Gamma_{lr} \tag{14}$$

where $\Gamma_{tb}, \Gamma_{sq}, \Gamma_{lr}$ are hydrodynamic functions originating from the top and bottom side of the cantilever, squeeze film, and the left and right side of the cantilever, respectively.

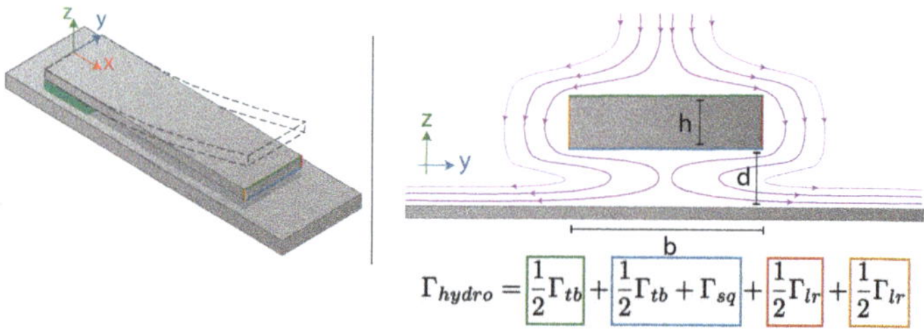

Figure 5. Scheme of streamlines acting on the cross-section sidewalls of the cantilever oscillating in its first mode of vibration with corresponding hydrodynamic functions. Γ_{tb} is used to describe the forces applied at the top and the bottom of the cantilever, Γ_{lr} relates to the left and right sides of the cantilever, while Γ_{sq} is a hydrodynamic force originating from squeeze film effect.

3.4.1. Viscous Damping on the Top and the Bottom

Γ_{tb} describes the viscous damping on the front and the back of the cantilever.

Sader [34] used the exact analytic solution for a circular-cross section cantilever. Then he used a multiplicative correction function Ω_{sader} in order to provide a more precise result in case of infinitely thin rectangular beams. Correction function Ω_{sader} depends on the Reynolds number and therefore on the width and frequency of the cantilever. The expression of Γ_{tb} is given in Equation (15) and the Ω_{sader} expression in [34].

$$\Gamma_{tb}(\omega) = \left(1 + \frac{4iK_1(-i\sqrt{iR_e})}{\sqrt{iR_e}K_0(-i\sqrt{iR_e})}\right)\Omega_{sader}(\omega) \tag{15}$$

K_0, K_1 are modified Bessel functions of the second kind, $Re = \frac{\rho_f \omega b^2}{4\mu_f}$ is the Reynolds number, ρ_f is the density of the fluid, and μ_f is the dynamic viscosity.

3.4.2. Viscous Damping on the Left and on the Right

The theoretical approach of Γ_{lr} can be found in [33] and takes the following form:

$$\Gamma_{lr}(\omega) = \frac{2\sqrt{2}h}{\pi b\sqrt{R_e}}(1 + i) \tag{16}$$

As explained in [35], this expression neglects edge and thickness effects, but remains a sufficient approximation in our configuration since it has been tested and compared to experimental results in [33].

3.4.3. Viscous Damping Due to the Squeeze Film Effect

On the cantilever sidewall, where the gas is trapped between substrate and cantilever, there exists an additional counter reactive force originating from squeeze film action. A mathematical description of squeeze film was given by Bao et al. [36]:

$$\Gamma_{sq}(\omega) = \frac{-4P_a}{\pi d b \rho_f \omega^2} \left(f_e(\sigma) - i f_d(\sigma) \right) \tag{17}$$

where P_a, d are the surrounding pressure and air gap shown in Figure 5, σ is a squeeze number [36] given by the following equation:

$$\sigma = \frac{12 \mu_f \omega L^2}{P_a d^2} \tag{18}$$

and $f_e(\sigma)$ and $f_d(\sigma)$ are functions introduced by Langlois [37] with the following form:

$$\begin{pmatrix} f_e(\sigma) &=& 1 - \sqrt{\dfrac{2}{\sigma}} \dfrac{\sinh\left(\sqrt{\frac{\sigma}{2}}\right) + \sin\left(\sqrt{\frac{\sigma}{2}}\right)}{\cosh\left(\sqrt{\frac{\sigma}{2}}\right) + \cos\left(\sqrt{\frac{\sigma}{2}}\right)} \\[2ex] f_d(\sigma) &=& \sqrt{\dfrac{2}{\sigma}} \dfrac{\sinh\left(\sqrt{\frac{\sigma}{2}}\right) - \sin\left(\sqrt{\frac{\sigma}{2}}\right)}{\cosh\left(\sqrt{\frac{\sigma}{2}}\right) - \cos\left(\sqrt{\frac{\sigma}{2}}\right)} \end{pmatrix} \tag{19}$$

3.5. Results and Discussion

Despite the lack of a general trend for the quality factor optimization in terms of all losses, it is possible to find the optimal value of the quality factor in terms of geometry for a given frequency. This optimum is presented in Figure 6. It takes into account all damping mechanisms presented in the previous subsections. The values of parameters used in this numerical simulation are presented in Table 3. Simulations have been realized at 11 kHz where the acoustic force is maximal, and for comparison at 60 kHz. An optimum has been found for laser modulation frequency at 11 kHz. For other physical mechanisms like damping mechanism, other optimums in modulation frequency can be expected. Simulation at higher frequency illustrates the evolution of these physical parameters with the frequency. As it will be seen in Section 6, although the different frequency dependency of physical mechanisms, the optimum frequency for the gas sensor is the same as the one for the acoustic force, here 11 kHz. A complete frequency study is presented in Appendix A.4. For the present simulations, a gap value of $d = 10$ μm has been chosen as a good compromise between fabrication constraint and sensor performances. The extensive study of the gap is presented in Section 7. Simulations show that the maximum value is slightly larger at high frequencies, and the most significant effect is the shift of the optimum to the lowest thickness when the frequency increases. As we will detail below, this effect is due to the losses of the mechanical supports.

In the figure, we identified the areas which correspond to the main limiting mechanisms. As it was shown $Q_{support} \propto \frac{1}{\sqrt{\omega_n^3 h^3}}$, therefore the limitation for the quality factor with high thickness originates from the damping of the support. The term in ω^3 in this equation explains the shift of the optimum to the lowest thickness, when the frequency increases.

The effect of the squeeze film damping appears for the largest width when the gas is trapped under the cantilever. For the smallest width, where the inertial forces are smaller than the viscous forces, the total quality factor is limited by the viscous damping. This area corresponds to the lowest Reynolds number.

In this model, neither thermoelastic nor acoustic damping are limiting factors. For all geometries and frequencies, the two associated quality factors are at least one order of magnitude higher than the other damping mechanisms. For more detail on the individual limits of each quality factor, the reader can refer to Appendix A.2, Figures A2 and A3.

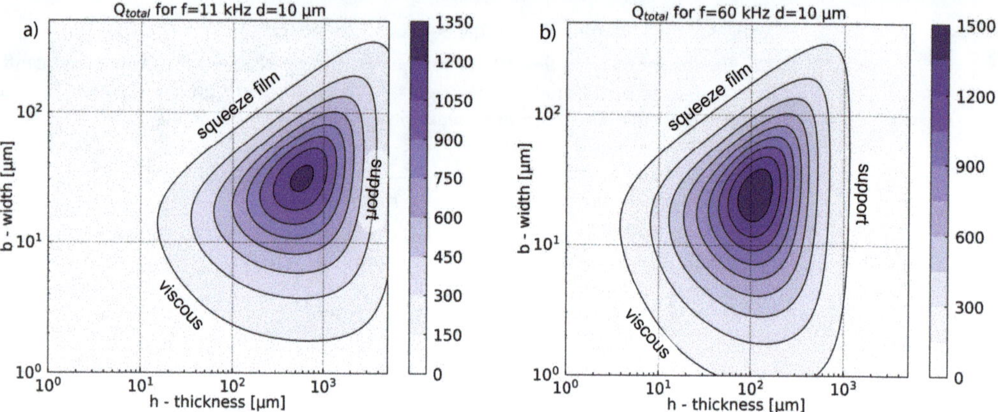

Figure 6. Total quality factor as a function of width and thickness for a cantilever of fundamental resonance frequencies equal to 11 kHz (**a**) and 60 kHz (**b**), for a gap between support and cantilever equal to $d = 10$ μm.

4. Displacement

The deflection $W_n(x, \omega)$ at the position x of the n-th mode of the beam under a photoacoustic driving force $F_{PA}(\omega)$ is given by:

$$W_n(x, \omega) = \chi_n(\omega) F_{PA}(\omega) \phi_n(x) = w_n(\omega) \phi_n(x) \tag{20}$$

where $\chi_n(\omega)$, $F_{PA}(\omega)$, $\phi_n(x)$, $w_n(\omega)$ are the mechanical susceptibility, photoacoustic driving force, mode shape function normalized with $max(\phi_m(x)) = 1$, and the maximal displacement, respectively. For the fundamental mode $w_n(\omega)$ denotes the displacement amplitude at the extremity of the beam. The susceptibility represents the frequency-dependent response of the cantilever under an external force and can be expressed as:

$$\chi_n(\omega) = \frac{1}{m_n(\omega_n^2 - \omega^2) + i(\frac{\omega_n \omega m_n}{Q_{total}})}$$

where Q_{total}, m_n are the total mechanical quality factor and the effective mass, respectively. The effective mass represents the part of structure actually involved in the movement.

The structure is subjected to two opposite forces: the photoacoustic force F_{PA} which is periodic and drives the beam into motion and the resistance caused by damping of the structure. The damping is given by the total quality factor Q_{total}. Both forces are presented in previous sections. The effective mass is described by:

$$m_n = \rho_b h b \int_0^L \phi_n^2(x)\, dx \tag{21}$$

It is related to the resistance of the resonator for motion changes. Consequently, it decreases the susceptibility and amplitude displacement of the resonator.

Results and Discussion

Figure 7 has been calculated with the mathematical expression of the previous section (Equation (4)). In this section, some approximations will be proposed to explain the shape of the graph.

The mechanical displacement $w_n(\omega_n)$ presented in Figure 7 is the product between the photoacoustic force and the mechanical susceptibility χ_n. At the resonance frequency $2\pi f_n = \omega_n$, the susceptibility can be approximated as $\chi_n = Q/(m_n\omega_n^2)$ and the displacement as $w_n(\omega_n) = Q_{total}F_{PA}/(\omega_n^2 m_n)$.

For the fundamental mode of the cantilever, we can write the acoustic force as $F_{PA} \simeq 0.39bL\Delta p(b,h)$, where $\Delta p(b,h)$ is the pressure difference between the top and the back of the cantilever. The function $\Delta p(b,h)$ increases with the thickness h and in this approximation remains quite constant for various widths b. The effective mass of the cantilever fundamental mode is $m_n \simeq 0.25\rho_b hbL$ (i.e., 25% of the total mass). The displacement can then be approximated by:

$$w_n(\omega_n) = \frac{Q_{total}F_{PA}}{\omega_n^2 m_n} \simeq 1.56\frac{Q_{total}}{\omega_n^2}\frac{\Delta p(b,h)}{\rho_b h} \tag{22}$$

The simulations show that the fraction $\frac{\Delta p(b,h)}{\rho_b h}$ remains quite constant for different widths and is inversely proportional to the cantilever thickness: $\frac{\Delta p(b,h)}{\rho_b h} \propto \frac{1}{h}$. Due to its homogeneity, this term is called "acceleration" in Figure 7. It is reducing the maximal displacement when the thickness increases. This region corresponds to weak acoustic force or/and heavy effective masses. Counterintuitively, the simplified Equation (22) shows that increasing the cantilever surface to collect more photoacoustic energy increases the effective mass, resulting in constant mechanical displacement. Indeed, a simplification by the surface bL appears between the term of the acoustic force and the effective mass.

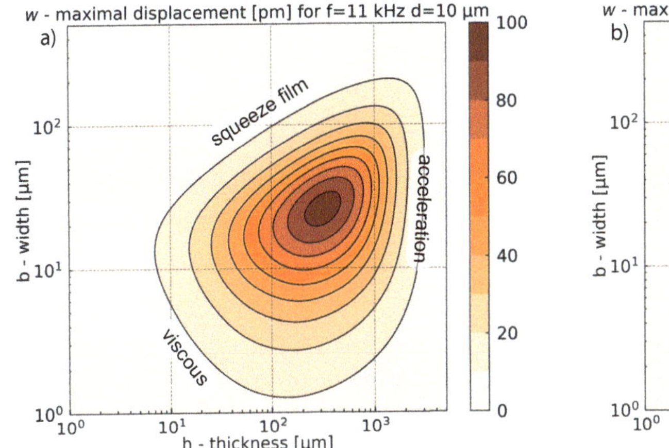

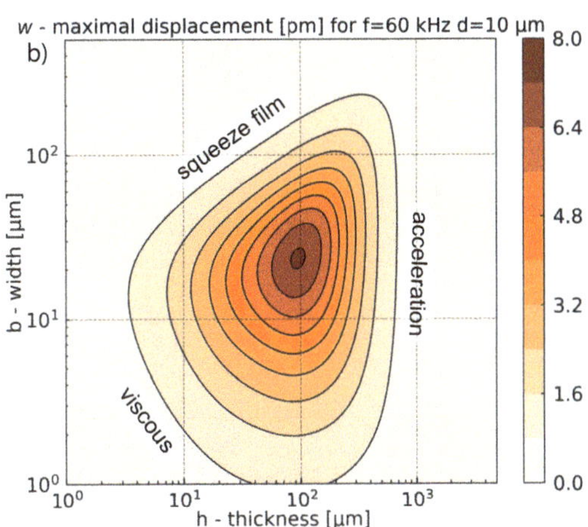

Figure 7. Total displacement versus width and thickness for a cantilever with fundamental resonance frequency equal to 11 kHz (**a**) and 60 kHz (**b**) for a gap between support and cantilever equal to $d = 10$ µm.

The other limitations come from the viscous damping introduced by the Q_{total} term, and are similar to those shown in the Figure 6.

Figure 7 shows a large difference in displacement amplitude between a modulation frequency of 11 kHz and 60 kHz. Despite the improvement of quality factor with increasing frequency, the acoustic force significantly drops at high frequency (Figure 3) and the susceptibility, as it is inversely proportional to ω_n^2. The results then show that photoacoustic force and susceptibility gain more importance with the change of frequency.

5. Electrical Part

This section focuses on maximizing the conversion between mechanical deflection and electrical signal.

The nominal capacitance $C_0 = Lb\varepsilon_0\varepsilon_r/d$ is the capacitance value without any displacement, where $\varepsilon_r, \varepsilon_0$ are the relative permittivity of the media (equal to unity in air) and vacuum, respectively. For the different geometries considered, C_0 may take values between 10^{-4} and 100 pF. The expression of nominal capacitance indicates that the change of the distance between two electrodes will cause a capacitance variation.

The dynamic capacitance caused by deflection of the cantilever is given by [38]:

$$C(t) = \int_0^L \frac{b\varepsilon_0\varepsilon_r}{d + W_n(x,\omega)}\, dx \exp(i\omega t) \tag{23}$$

The model applies a method called DC bias sensing [39] (Chapter 5). Figure 8 presents the sensing scheme. In an electromechanical system, a polarization voltage V_{dc} on the electrodes is required to generate an electrical signal related to the mechanical behavior of the moving electrode.

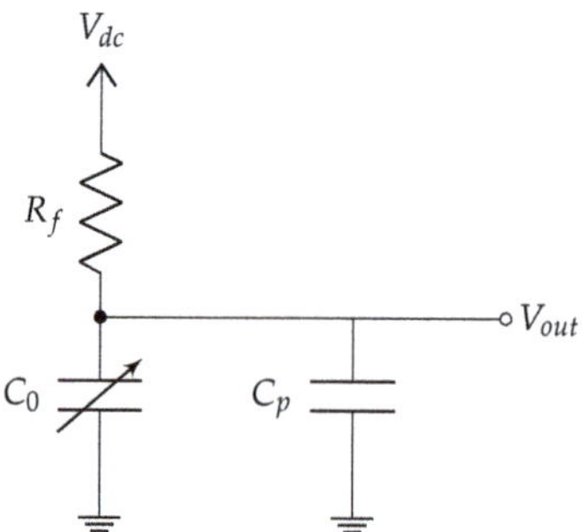

Figure 8. Sensor conditioning circuit, where: $C_0, C_p, R_f, V_{dc}, V_{out}$ are the capacitance of the cantilever, parasitic capacitance, resistance, polarization voltage, and output voltage, respectively.

The application of the force generated by the photoacoustic effect sets the movable electrode in motion. This movement causes the changes of the capacitance from the maximal value C_{max} to the minimal C_{min}. C_{max} and C_{min} correspond to the minimal and maximal distances between the cantilever and support, respectively.

The capacitance variation can take place at a constant charge or a constant voltage [40]. If the time constant $R_f C_0 >> 1/\omega_n$, the electric charge stored in the capacitor remains constant. $R_f \simeq 100\ G\Omega - 10\ T\Omega$ is the value of the resistor placed between the cantilever and the polarization voltage V_{dc}. In the constant charge regime, the voltage of the measured signal is given by $V_{out}(t) = C_0 V_{dc}/C(t)$ while its amplitude is given by:

$$V_{out} = V_{dc} \int_0^L \frac{W_n(x,\omega)}{Ld}\, dx \tag{24}$$

Results and Discussion

Figure 9 presents the results obtained for a bias $V_{dc} = 1$ V. The maximum values follow the tendencies given by the displacement. The values change with the gap d and frequency f_n. Indeed, according to Equation (22), Equation (24) can be simplified as follows:

$$V_{out} \simeq 0.39 V_{dc} \frac{w_n(\omega)}{d} \tag{25}$$

Decreasing the gap d between the two electrodes should lead to an output signal amplitude increase. However, simultaneously it increases the squeeze film damping and reduces the cantilever displacement. The optimization of this parameter will be discussed in the last section.

Equation (25) indicates that the signal output does not depend on the area of the capacitor as it would be expected based on Equation (23). The output signal amplitude is given for an open circuit, without any read-out circuit which can modify the signal. In a complete system, the signal is attenuated by a parasitic capacitance C_p, which is the sum of the parasitic capacitance of the resonator itself and the one which comes from read-out circuit. The output signal attenuation can be estimated with the ratio $C_0/(C_0 + C_p)$ [41].

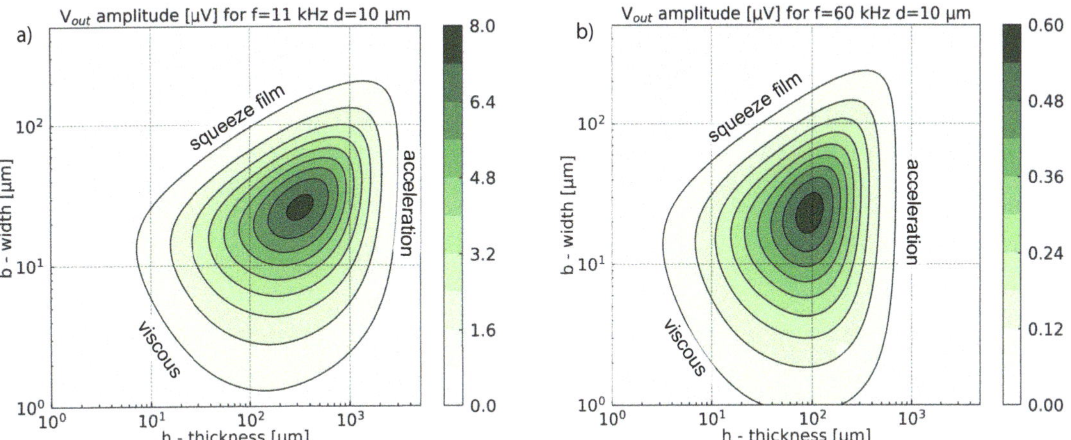

Figure 9. Amplitude of the output voltage versus width and thickness for a cantilever of fundamental resonance frequency equal to 11 kHz (**a**) and 60 kHz (**b**) for a gap between the support and the cantilever equal to $d = 10$ μm. The polarization voltage is $V_{dc} = 1$ V.

6. Thermal Noise

The sensor performance is limited by the unavoidable noise caused by the thermal fluctuations (Brownian movement) which set the resonator in motion. Therefore, it must be considered to construct a high-performance sensor. The maximum displacement of the cantilever caused by the Brownian noise $w_{noise}(\omega_n)$ [42] is given by the fluctuation-dissipation theorem:

$$w_{noise}(\omega_n) = \sqrt{\frac{4k_b T \Delta f Q}{\omega_n^3 m_n}} = \sqrt{4k_b T \Delta f} \sqrt{\frac{w_n}{\omega_n F_{PA}}} \tag{26}$$

where k_b, T, and Δf are the Boltzmann constant, cantilever temperature, and detection bandwidth, respectively. The plot for thermal noise as a function of cantilever geometry is presented in the Appendix A.3, Figure A4.

The signal-to-noise ratio at the resonance pulsation ω_n is given by:

$$SNR = \frac{w_n(\omega_n)}{w_{noise}(\omega_n)} = \sqrt{\frac{w_n \omega_n F_{PA}}{4k_b T \Delta f}} \tag{27}$$

Discussion about Signal-to-Noise Ratio

The optimum for the SNR presented in Figure 10 does not match with the highest output signal amplitude presented in Figure 9. The highest output signal corresponds to the highest mechanical displacement w_n (Equation (25)). The Brownian noise can be considered as a force acting on the cantilever. The optimal way to improve SNR is maximization of the photoacoustic force which comes down to increasing the surface area for photoacoustic pressure collection. Therefore, the optimum SNR is shifted to greater widths and thicknesses, where the acoustic force is greater (Figure 4), and the collected energy increases. As in the previous section, the increase of the surface collecting the photoacoustic energy will be limited to the larger widths by the squeeze film damping, and by the acceleration term for the larger thicknesses.

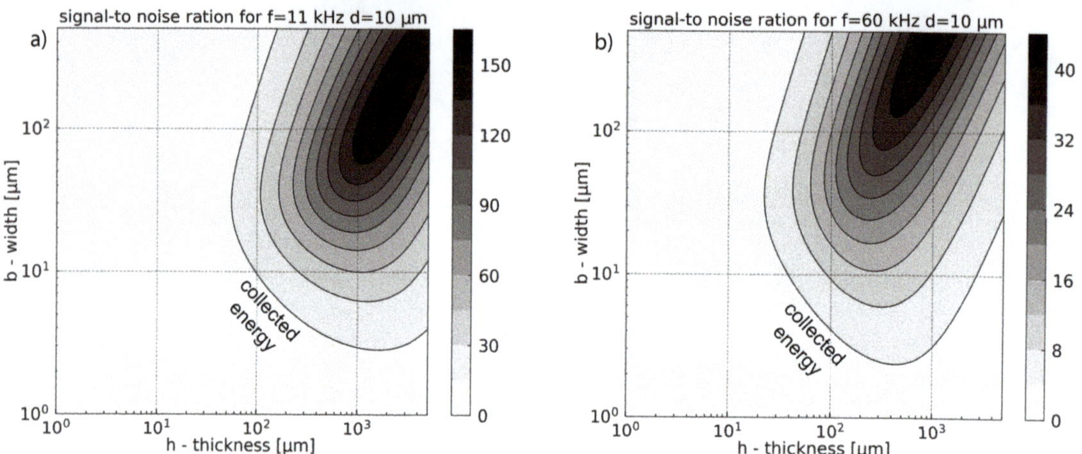

Figure 10. Signal-to noise ratio for a cantilever of fundamental resonance frequency equal to 11 kHz (**a**) and 60 kHz (**b**), a gap between support and cantilever equal to $d = 10$ µm.

The second significant parameter in Equation (27) is the pulsation ω_n. Unlike the previous result, ω_n reduces the difference between low and high frequencies. This is due to the fact that SNR is inversely proportional to the square root of frequency $SNR \propto \frac{1}{\sqrt{\omega_n}}$, while the amplitude of displacement $w_n(\omega_n)$ is inversely proportional to the frequency square power $w_n(\omega_n) \propto \frac{1}{\omega_n^2}$. For instance, the ratio between maximal value of SNR at 11 kHz and 60 kHz $\frac{max(SNR(11kHz))}{max(SNR(60kHz))} \simeq \frac{1}{22} \frac{max(w_n(11kHz))}{max(w_n(60kHz))}$ is around 22 times lower than the ratio of displacement. This indicates that for SNR the frequency term is less significant than in terms of displacement (voltage output).

7. Study of the Gap Effect

This section focuses on the effect of the distance between electrodes d (Figure 1) on the general sensor performance for a constant resonance frequency of 11 kHz. The analytic solutions previously presented (Equation (27)) were implemented in a Python programming environment to estimate an optimal value of signal-to-noise ratio for each value of the gap d. Subsequently, for each optimal SNR value, we get the geometrical parameters of the cantilever (width, length, thickness, Figure 11a) and their corresponding output voltages (Figure 11b).

When d increases, the signal-to-noise ratio increases as the displacement increases due to the decrease of squeeze film damping. At the same time, the signal output voltage decreases due to the increase of the distance between electrodes. This improvement on the SNR can also be explained by the optimized width, which increases with the gap d and which allows more energy collection. This increase in width is made possible by the decrease in the squeeze film damping for large gap d. Above $d \simeq 200$ µm the acoustic damping becomes dominant and some saturation appears on the width curve.

The variations on the thickness curve are less important than on the width. The length curve follows the thickness rise to satisfy the constant frequency condition. As for the curve of the width, we can identify on the curves for length and thickness two different regimes that are probably due to the transition where acoustic damping becomes more important than squeeze film damping.

By taking into account the fabrication process issues, a cantilever with a gap $d = 10$ μm can be realized on a silicon-on-oxide (SOI) wafer. In this case, the SNR ratio will reach 150 and the amplitude of the output signal should reach 0.9 μV. Depending on the possibilities of the fabrication process, size of the final design, and required performance of the device (SNR), Figure 11 can be used as a reference to create a cantilever for optimal photoacoustic gas detection with capacitive transduction mechanisms.

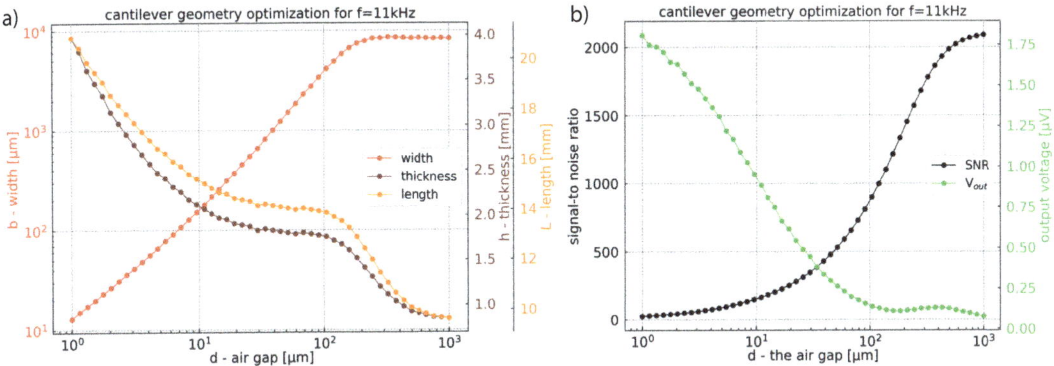

Figure 11. (**a**) Geometrical values giving the highest signal-to-noise ratio as a function of the gap d between support and cantilever. (**b**) Highest signal-to-noise ratio and its corresponding output amplitude signal as a function of the gap d between support and cantilever. For these simulations, the cantilever fundamental resonance frequency is equal to 11 kHz.

8. Conclusions

The presented sensor combines multiple physical phenomena and, therefore, its optimization is not straightforward. In this paper, we discussed multiple parameters which affect the sensor performances and we analyzed their contributions. Our model takes into account some optimization for (1) the acoustic force; (2) the system damping; (3) the mechanical displacement under photoacoustic force considering damping mechanism; (4) the electrical signal under capacitive transduction mechanism, and (5) the signal-to-noise ratio. This model provides a method to retrieve the optimized cantilever geometry depending on the required size of the sensor and other restrictions which might be imposed by the fabrication process or the operating conditions of the sensor.

(1) Our model includes the gas relaxation time. We proposed a cantilever geometry optimized for photoacoustic gas sensor with a capacitive transduction mechanism in the context of CH_4 absorption with a concentration of 1%. For the same temperature and pressure conditions, only the amplitude of the photoacoustic force will change with the concentration. For this reason, the geometric parameters of the cantilever will remain the same for all the concentrations. For different gasses, only the optimal laser modulation frequency and the absolute acoustic pressure will change; the trends presented in other sections will remain the same. Based on our model, the cantilever optimal geometry can be recalculated for any other gas by taking into account the relaxation time and absorption coefficient. However, the optimal sensor should be created for one specific gas.

(2) The study of the different damping mechanisms show that viscous damping, and particularly squeeze film effect, is fundamental. The impact of squeeze film effect is

visible in Figure 11 for d values up to 200 μm. For d above this value, the acoustic damping becomes the limiting damping mechanism.

(3)–(5) To improve the SNR, one can increase the gap d and/or the surface collecting the photoacoustic energy. However, increasing the gap d between the two electrodes will decrease the output signal amplitude. Depending on the read-out circuit and the parasitic capacitance it is possible to obtain a high SNR while maintaining a sufficient output signal. For example, with a gap of $d = 10$ μm which can be realized on a silicon-on-oxide (SOI) wafer, the signal-to-noise ratio will reach 150 and the amplitude of the output signal will be around 0.9 μV. For different values of the gap, one can use Figure 11.

Finally, despite a complex multiphysical problem, we have proposed a complete analytic model able to find the optimum geometric parameters of a cantilever for photoacoustic sensing with capacitive transduction. Beyond the simple optimization, this study is intended to provide all the tools allowing understanding of all the mechanisms of this complex problem. The variety of these physical mechanisms, often incompatible with each other during a finite element simulation, gives all its strength to our analytical approach to the problem. This paper demonstrates that a simple cantilever with capacitive transduction mechanism will not reach the same performance in terms of limit of detection as the best QEPAS technique or best standard photoacoustic technique using a microphone. However, besides being the optimization tool, this work is intended to be an educational tool allowing a mechanical resonator to be developed with more complex geometry and other transduction mechanisms. This study paves the way to develop new mechanical resonators for compact, integrated, and sensitive gas sensors.

Author Contributions: W.T. was in charge of the conceptualization, methodology and formal analysis under the supervision of M.B. She was assisted by N.M., R.R., and D.A. M.B. was in charge of the Python code writing. The article was written by W.T. and M.B., reviewed and edited by M.B. and A.V. All authors have read and agreed to the published version of the manuscript.

Funding: This research was financially supported by the French Ministry of Defense (DGA-MRIS), Region Languedoc Roussillon, European Community (FEDER) and Renatech, Agence Nationale de la Recherche (MULTIPAS Project No. [ANR-16-CE04-0012], and NOMADE Project No. [ANR-18-CE04-0002-01]).

Data Availability Statement: Data and Python code developed for this article is available by contacting the correspoding author.

Conflicts of Interest: The authors declare no conflict of interest.

Sample Availability: The Python code developed for this article is available by contacting the correspoding author.

Abbreviations

The following abbreviations are used in this manuscript:

MEMS	micro-electromechanical systems
PA	photoacoustic
QEPAS	quartz enhanced photoacoustic spectroscopy
QTF	quartz tuning fork
SNR	signal-to-noise ratio
SOI	silicon-on-oxide
TDLS	tunable diode laser spectroscopy
V-T	vibrational–translational

Appendix A

Appendix A.1. Optimal Beam Position

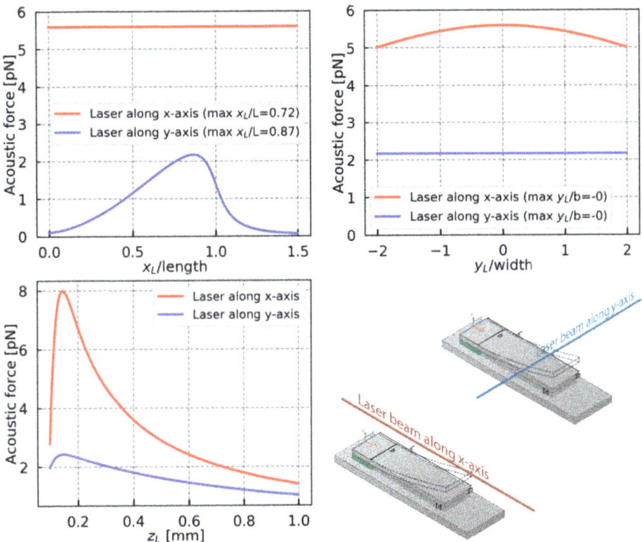

Figure A1. Amplitude of the photoacoustic force applied on a cantilever as a function of the laser beam position for two configurations: red—laser beam along the x-axis; blue—laser beam along the y-axis. Where $f_n = 11$ kHz, $b = 50$ μm, $h = 100$ μm, and $L = 3.5$ mm, respectively are the resonance frequency, the width, the thickness, and the length of the cantilever.

Appendix A.2. Quality Factor

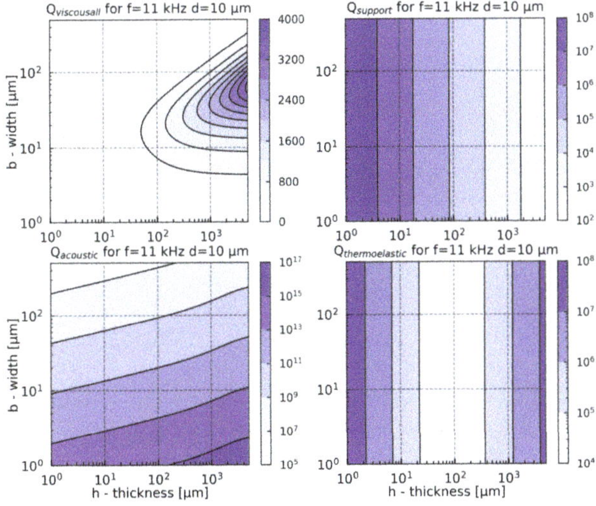

Figure A2. Quality factors for viscous damping with squeeze film (top-left panel), support damping (top-right panel), acoustic damping (bottom left panel), and thermoelastic damping (bottom-right panel) as a function of width and thickness of a cantilever. Cantilever's fundamental resonance frequency: 11 kHz, gap between support and cantilever: $d = 10$ μm.

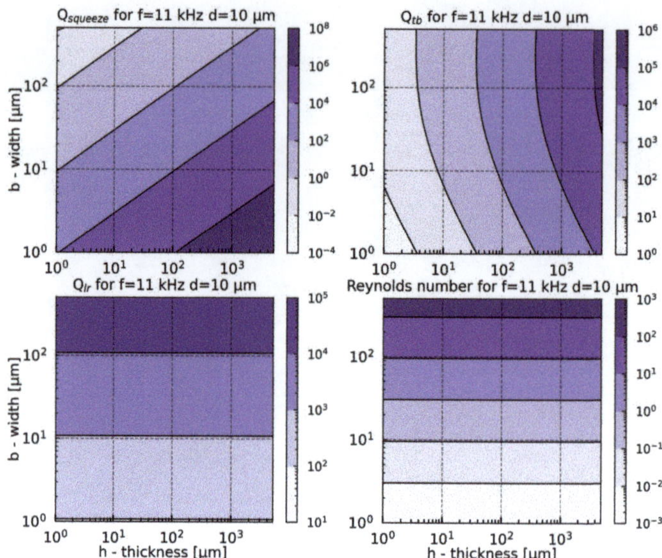

Figure A3. Quality factors for squeeze film (top-left panel), viscous damping originating from the top and bottom side of a cantilever (top-right panel), viscous damping originating from left and right side of a cantilever (bottom left panel) and Reynolds number (bottom-right panel) as a function of width and thickness of a cantilever. Cantilever's fundamental resonance frequency: 11 kHz, gap between support and cantilever: $d = 10$ μm.

Appendix A.3. Thermal Noise

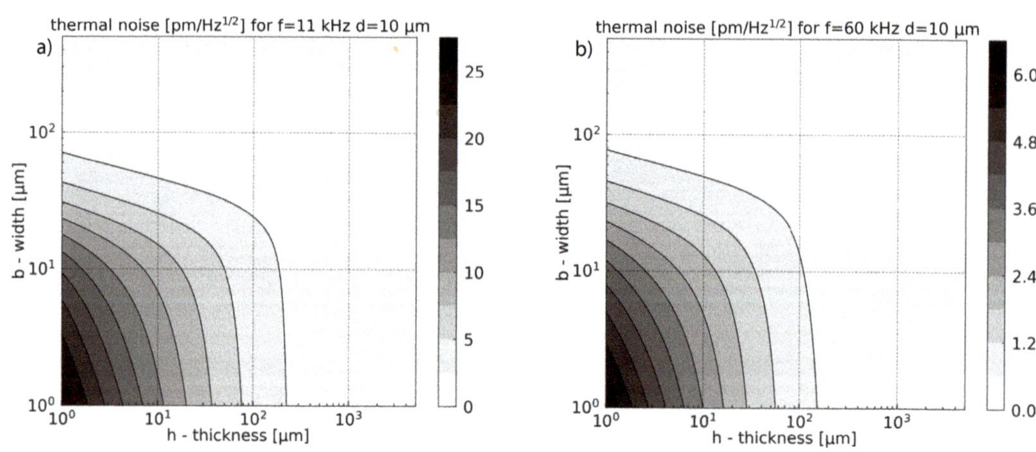

Figure A4. Thermal noise as a function of width and thickness of a cantilever. Cantilever's fundamental resonance frequency: 11 kHz (**a**) and 60 kHz (**b**), gap between support and cantilever: $d = 10$ μm.

Appendix A.4. Frequency Study

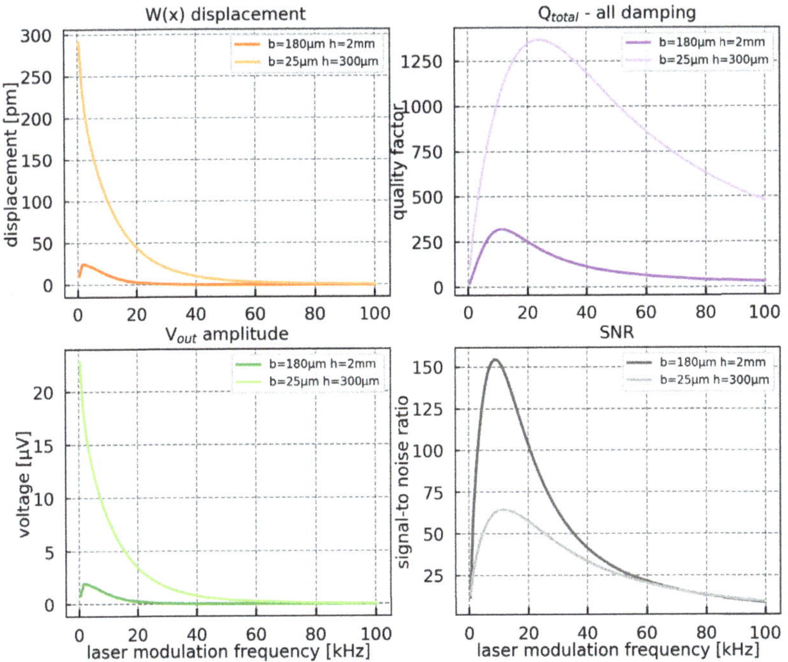

Figure A5. Frequency study for two geometries of cantilever with a distance between the electrodes $d = 10$ µm: geometry optimized for SNR ($b = 180$ µm, $h = 2$ mm) and geometry optimized for V_{out} ($b = 25$ µm, $h = 300$ µm).

References

1. Gran-View-Research. *Gas Sensor Market Size, Share and Trends Analysis Report*; Gran-View-Research: San Francisco, CA, USA, 2021.
2. Maurin, N.; Rousseau, R.; Trzpil, W.; Aoust, G.; Hayot, M.; Mercier, J.; Bahriz, M.; Gouzi, F.; Vicet, A. First clinical evaluation of a quartz enhanced photo-acoustic CO sensor for human breath analysis. *Sens. Actuators B Chem.* **2020**, *319*, 128247. [CrossRef]
3. Laj, P.; Klausen, J.; Bilde, M.; Plass-Duelmer, C.; Pappalardo, G.; Clerbaux, C.; Baltensperger, U.; Hjorth, J.; Simpson, D.; Reimann, S.; et al. Measuring atmospheric composition change. *Atmos. Environ.* **2009**, *43*, 5351–5414. [CrossRef]
4. Giglio, M.; Elefante, A.; Patimisco, P.; Sampaolo, A.; Sgobba, F.; Rossmadl, H.; Mackowiak, V.; Wu, H.; Tittel, F.K.; Dong, L.; et al. Quartz-enhanced photoacoustic sensor for ethylene detection implementing optimized custom tuning fork-based spectrophone. *Opt. Express* **2019**, *27*, 4271–4280. [CrossRef] [PubMed]
5. Karapuzikov, A.; Nabiev, S.S.; Nadezhdinskii, A.; Ponomarev, Y.N. Laser methods of detecting vapor explosives in the open atmosphere: Analytical possibilities for counteracting terrorist acts. *Atmos. Ocean. Opt.* **2011**, *24*, 133–143. [CrossRef]
6. Awang, Z. Gas sensors: A review. *Sens. Transducers* **2014**, *168*, 61–75.
7. Korotcenkov, G. Metal oxides for solid-state gas sensors: What determines our choice? *Mater. Sci. Eng. B* **2007**, *139*, 1–23. [CrossRef]
8. Namjou, K.; Cai, S.; Whittaker, E.A.; Faist, J.; Gmachl, C.; Capasso, F.; Sivco, D.L.; Cho, A.Y. Sensitive absorption spectroscopy with a room-temperature distributed-feedback quantum-cascade laser. *Opt. Lett.* **1998**, *23*, 219. [CrossRef] [PubMed]
9. Cretescu, I.; Lutic, D.; Manea, L.R. Electrochemical sensors for monitoring of indoor and outdoor air pollution. In *Electrochemical Sensors Technology*; IntechOpen: London, UK, 2017; p. 65.
10. Liu, X.; Cheng, S.; Liu, H.; Hu, S.; Zhang, D.; Ning, H. A survey on gas sensing technology. *Sensors* **2012**, *12*, 9635–9665. [CrossRef]
11. Castleden, S.L.; Kirkbright, G.F.; Spillane, D.E.M. Wavelength Modulation in Photo-Acoustic Spectroscopy. *Anal. Chem.* **1981**, *53*, 2228–2231. [CrossRef]
12. Kosterev, A.A.; Bakhirkin, Y.A.; Curl, R.F.; Tittel, F.K. Quartz-enhanced photoacoustic spectroscopy. *Opt. Lett.* **2002**, *27*, 1902. [CrossRef]
13. Rousseau, R.; Loghmari, Z.; Bahriz, M.; Chamassi, K.; Teissier, R.; Baranov, A.N.; Vicet, A. Off-beam QEPAS sensor using an 11-µm DFB-QCL with an optimized acoustic resonator. *Opt. Express* **2019**, *27*, 7435–7446. [CrossRef]

14. Uranga, A.; Verd, J.; Barniol, N. CMOS–MEMS resonators: From devices to applications. *Microelectron. Eng.* **2015**, *132*, 58–73. [CrossRef]

15. Quintero, A.; Cardes, F.; Perez, C.; Buffa, C.; Wiesbauer, A.; Hernandez, L. A VCO-based CMOS readout circuit for capacitive MEMS microphones. *Sensors* **2019**, *19*, 4126. [CrossRef] [PubMed]

16. Rio Calvo, M.; Monge Bartolomé, L.; Bahriz, M.; Boissier, G.; Cerutti, L.; Rodriguez, J.-B.; Tournié, E. Mid-infrared laser diodes epitaxially grown on on-axis (001) silicon. *Optica* **2020**, *7*, 263–266. [CrossRef]

17. Tadigadapa, S.; Mateti, K. Piezoelectric MEMS sensors: State-of-the-art and perspectives. *Meas. Sci. Technol.* **2009**, *20*, 092001. [CrossRef]

18. Barlian, A.A.; Park, W.T.; Mallon, J.R.; Rastegar, A.J.; Pruitt, B.L. Semiconductor piezoresistance for microsystems. *Proc. IEEE* **2009**, *97*, 513–552. [CrossRef]

19. Lötters, J.C.; Olthuis, W.; Veltink, P.H.; Bergveld, P. A sensitive differential capacitance to voltage converter for sensor applications. *IEEE Trans. Instrum. Meas.* **1999**, *48*, 89–96. [CrossRef]

20. Ferlito, U.; Grasso, A.D.; Pennisi, S.; Vaiana, M.; Bruno, G. Sub-Femto-Farad Resolution Electronic Interfaces for Integrated Capacitive Sensors: A Review. *IEEE Access* **2020**, *8*, 153969–153980. [CrossRef]

21. Schilt, S.; Thevenaz, L.; Robert, P. Wavelength modulation spectroscopy: Combined frequency and intensity laser modulation. *Appl. Opt.* **2003**, *42*, 6728–6738. [CrossRef] [PubMed]

22. Petra, N.; Zweck, J.; Kosterev, A.A.; Minkoff, S.E.; Thomazy, D. Theoretical analysis of a quartz-enhanced photoacoustic spectroscopy sensor. *Appl. Phys. B* **2009**, *94*, 673–680. [CrossRef]

23. Yabuno, H.; Nayfeh, A.H. *Nonlinear Normal Modes of a Parametrically Excited Cantilever Beam*; Springer: Berlin/Heidelberg, Germany, 2001; pp. 65–77.

24. Schilt, S.; Besson, J.P.; Thévenaz, L. Near-infrared laser photoacoustic detection of methane: The impact of molecular relaxation. *Appl. Phys. B* **2006**, *82*, 319–328. [CrossRef]

25. Hopcroft, M.A.; Nix, W.D.; Kenny, T.W. What is the Young's modulus of silicon? *J. Microelectromech. Syst.* **2010**, *19*, 229–238. [CrossRef]

26. Lifshitz, R.; Roukes, M. Thermoelastic damping in micro- and nanomechanical systems. *Phys. Rev. B Condens. Matter Mater. Phys.* **2000**, *61*, 5600–5609. [CrossRef]

27. Lochon, F.; Dufour, I.; Rebière, D. A microcantilever chemical sensors optimization by taking into account losses. *Sens. Actuators B Chem.* **2006**, *118*, 292–296. [CrossRef]

28. Le Foulgoc, B.; Bourouina, T.; Le Traon, O.; Bosseboeuf, A.; Marty, F.; Breluzeau, C.; Grandchamp, J.P.; Masson, S. Highly decoupled single-crystal silicon resonators: An approach for the intrinsic quality factor. *J. Micromech. Microeng.* **2006**, *16*. [CrossRef]

29. Blake, W.K. The radiation from free-free beams in air and in water. *J. Sound Vib.* **1974**, *33*, 427–450. [CrossRef]

30. Johnston, R.A.; Barr, A.D.S. Acoustic and Internal Damping in Uniform Beams. *J. Mech. Eng. Sci.* **1969**, *11*, 117–127. [CrossRef]

31. Morse, P.M.; Feshbach, H. Methods of theoretical physics. *Am. J. Phys.* **1954**, *22*, 410–413. [CrossRef]

32. Hao, Z.; Erbil, A.; Ayazi, F. An analytical model for support loss in micromachined beam resonators with in-plane flexural vibrations. *Sens. Actuators A Phys.* **2003**, *109*, 156–164. [CrossRef]

33. Aoust, G.; Levy, R.; Bourgeteau, B.; Le Traon, O. Viscous damping on flexural mechanical resonators. *Sens. Actuators A Phys.* **2015**, *230*, 126–135. [CrossRef]

34. Sader, J.E. Frequency response of cantilever beams immersed in viscous fluids with applications to the atomic force microscope. *J. Appl. Phys.* **1998**, *84*, 64–76. [CrossRef]

35. Cox, R.; Josse, F.; Heinrich, S.M.; Brand, O.; Dufour, I. Characteristics of laterally vibrating resonant microcantilevers in viscous liquid media. *J. Appl. Phys.* **2012**, *111*, 014907. [CrossRef]

36. Bao, M.; Yang, H. Squeeze film air damping in MEMS. *Sens. Actuators A Phys.* **2007**, *136*, 3–27. [CrossRef]

37. Langlois, W. Isothermal squeeze films. *Q. Appl. Math.* **1962**, *20*, 131–150. [CrossRef]

38. Ijntema, D.J.; Tilmans, H.A. Static and dynamic aspects of an air-gap capacitor. *Sens. Actuators A Phys.* **1992**, *35*, 121–128. [CrossRef]

39. Bao, M. *Analysis and Design Principles of MEMS Devices*; Elsevier: Amsterdam, The Netherlands, 2005.

40. Meninger, S.; Mur-Miranda, J.O.; Amirtharajah, R.; Chandrakasan, A.; Lang, J.H. Vibration-to-electric energy conversion. *IEEE Trans. Very Large Scale Integr. (VLSI) Syst.* **2001**, *9*, 64–76. [CrossRef]

41. Scheeper, P.; Van der Donk, A.; Olthuis, W.; Bergveld, P. A review of silicon microphones. *Sens. Actuators A Phys.* **1994**, *44*, 1–11. [CrossRef]

42. Adamson, B.D.; Sader, J.E.; Bieske, E.J. Photoacoustic detection of gases using microcantilevers. *J. Appl. Phys.* **2009**, *106*, 114510. [CrossRef]

Article

Spectroscopic Techniques versus Pitot Tube for the Measurement of Flow Velocity in Narrow Ducts

Francesco D'Amato [1], Silvia Viciani [1,*], Alessio Montori [1], Marco Barucci [1], Carmen Morreale [2], Silvia Bertagna [2] and Gabriele Migliavacca [2]

[1] CNR-INO, Area CNR, Via Madonna del Piano 10, 50019 Sesto Fiorentino, Italy;
     francesco.damato@ino.cnr.it (F.D.); alessio.montori@ino.cnr.it (A.M.); marco.barucci@ino.cnr.it (M.B.)
[2] Innovhub Stazioni Sperimentali per l'Industria srl, Via G. Galilei 1, 20097 San Donato Milanese, Italy;
     carmen.morreale@mi.camcom.it (C.M.); silvia.bertagna@mi.camcom.it (S.B.);
     gabriele.migliavacca@mi.camcom.it (G.M.)
*   Correspondence: silvia.viciani@ino.cnr.it; Tel.: +39-055-522-6332

Received: 26 November 2020; Accepted: 17 December 2020; Published: 21 December 2020

Abstract: In order to assess the limits and applicability of Pitot tubes for the measurement of flow velocity in narrow ducts, e.g., biomass burning plants, an optical, dual function device was implemented. This sensor, based on spectroscopic techniques, targets a trace gas, injected inside the stack either in bursts, or continuously, so performing transit time or dilution measurements. A comparison of the two optical techniques with respect to Pitot readings was carried out in different flow conditions (speed, temperature, gas composition). The results of the two optical measurements are in agreement with each other and fit quite well the theoretical simulation of the flow field, while the results of the Pitot measurements show a remarkable dependence on position and inclination of the Pitot tube with respect to the duct axis. The implications for the metrology of small combustors' emissions are outlined.

Keywords: laser flow meter; Pitot tube; flow speed; time of flight; dilution method; flow simulation; flow turbulence; gas sensing applications

---

1. Introduction

Flow measurement in stacks is of fundamental importance in the assessment of pollutant emissions, because when combined with the measurement of the concentrations, it provides the mass flux of any emitted pollutant. Inside stacks flow is not laminar in most cases and the length of straight and unobstructed pipe sections, available for measurements, is often not enough to allow the full development of a regular velocity profile. Consequently, velocity and flow measurements are often affected by large uncertainty. Small ducts (inner diameter ≤ 50 cm) are more influenced than larger ones by perturbations, even due to the presence of sensors. Ducts can include curves, scrubbers, swirlers, extraction ports, probes and other devices, which can perturb the movement of the exhaust gases. Another issue is related to the presence of pollutants in the gas stream that can produce fouling or corrosion of the measuring devices. For example, particulate matter, often present in high concentration in many plant exhausts, such as biomass boilers, can obstruct the orifices for measuring the differential pressure or clog the moving parts of an anemometer. A recent review paper [1] addresses the origin of random and systematic errors for in-stack velocity measurements using Pitot tubes, how these problems are treated in the pertaining international standards and what effect they may have on the uncertainty of the measurements of pollutant emissions. From this survey, it emerges that, in the presence of cyclonic flows, the use of S-type Pitot tubes can result in errors of up to 12%, due to non-axial flows. A similar error can be produced by the misalignment of the Pitot tube during the measuring procedures. The errors associated with velocity measurements, using S-type

Pitot tubes in cyclonic flows, have been investigated in detail also in [2] by means of computational fluid dynamic (CFD) modelling. Different pipe configurations, producing different flow patterns have been studied and the associated errors have been analyzed. In particular, the authors pointed out that, in presence of an asymmetric velocity profile, typically with swirling flows, the maximum velocity in a cross-section of stack describes a spiral along the duct. In this case, the results of the measurements depend on the location of the measuring point, whose best position is unpredictable. Another source of error, associated with the use of S-type Pitot tubes and examined in [2], is due to the inclined gas velocity vectors in presence of non-axial velocity components: the contribution of this error, estimated under the specific conditions simulated in [2], is up to 2.5%. Measurements were carried out in a wind tunnel, in different experimental conditions and geometries in [3], yielding similar results.

Several techniques are available for measuring in-stack velocity and flow rate: most of them are described in two international technical standards: EN ISO 16911-1:2013 [4] and EN ISO 16911-2:2013, the former focuses on manual and the latter on automatic methods. The most commonly used approaches make use of Pitot tubes or vane anemometers to determine the flow velocity either on a single fixed point or on a grid of points on a measuring plane. The other methods described in [4] include a tracer gas dilution technique and a calculation procedure, based on the energy consumption of the combustion plants. Both of them provide directly the gas flow rate and indirectly the average velocity in the stack. The tracer dilution technique follows a very basic approach, because it determines the flow rate using just two physical quantities: the mass flow rate of the injected tracer and the concentration of this tracer in the gas stream. The energy consumption approach relays on the possibility to measure the flow rate and composition of the fuel and the oxygen content in the fumes, continuously and very accurately. The Transit Time (TT) tracer gas method is another possible approach, briefly described in EN ISO 16911-1. This method allows determining the average velocity of the flowing gas within a portion of a duct having a constant cross-section. Many different solutions may be adopted to put this method into practice, which make use of a tracer gas which follows the gas flow and which can be measured with a sufficiently high time rate. For this technique to be adopted, some constraints should be fulfilled. It should be possible to inject a tracer inside the stack, as homogeneously (with respect to the transverse section of the duct) as possible. The tracer should not be naturally present in the atmosphere, should not interfere with the normal operation of the plant, should be neither poisonous or toxic, nor environmentally detrimental. It should not be a byproduct of the plant itself. For an accurate measurement the maximum response time of the detection technique must be $\sim 10 \div 100$ ms, to be negligible compared to the rise time of the concentration of the tracer, which is in general $\sim$second. Short half-life radioactive tracers are often considered the most suitable, because they can be detected through the duct walls without the need for probes, ports or windows, but the use of these tracers is often restricted by national legislation. A detection technique that measures the average concentration of the tracer on a cross-section is to be preferred to a point measurement, because it reduces the effects of an improper mixing. Optical and laser techniques exist since long for the measurement of gas flows: Laser Two Focus [5], Doppler effect [6], Particle Imaging Velocimetry [7]. Laser Two Focus investigates very small regions (0.25 mm $\times$ 0.25 mm $\times$ 0.25 mm) at time, so requiring some time, and moving optics, to obtain a raster image. Laser Doppler Velocimetry provides information along a direction at an angle with the stack axis. In order to keep this angle as small as possible, inclined beams are required. In ref. [6] a setup with two measurement channels is shown. Particle Imaging Velocimetry requires the injection in the stack of particles with suitable dimensions, a pulsed, high power laser source, and a fast, high-resolution camera. Finally, a careful calibration procedure must be carried out. TT and Dilution techniques are much simpler, requiring low power laser sources and standard detectors, with straightforward analysis procedures. Spectroscopy is particularly suitable for the TT tracer gas method. In this case, the tracer should feature optical absorptions in wavelength regions where user-friendly laser sources are available, and these absorptions should be sufficiently strong to reduce the tracer concentration to very low levels ($\leq 1\permil$).

In the framework of the EMPIR Project IMPRESS II, a laser-based device, presented in this article, has been developed, which allows us to perform both dilution and TT measurements of the in-stack velocity. The two methods have been tested and cross-validated in a stack simulator, having a small section, under strongly cyclonic flow conditions. The flow pattern inside the test rig has been simulated through CFD modeling, using Ansys® Academic Fluent, Release 15.0, ANSYS, Inc., Canonsburg, PA, USA in order to get a better understanding of the experimental conditions. In parallel, fixed point measures, using an S-type Pitot tube, have been carried out and compared under conditions which are particularly critical for using a Pitot tube, in order to evaluate the possible bias between this conventional method and the other two optical techniques.

2. Optical Detection

In order to adopt any optical technique, we had to choose the tracer molecule. Following the constrains described above, we selected acetylene (C_2H_2). It has a strong absorption band around 1520 nm, which is one of the regions of optical telecommunications, and both laser sources and optical devices are available off-the-shelf. It is not toxic, and there is no risk of explosion at concentrations below 1‰ (Low Explosion Level for acetylene is 2.5% [8]).

A possible drawback is that it can be produced during incomplete combustion of methane [9]. However, most of our measurements were carried out when heating the air inside the stack by means of electrically driven resistors, so the risk of interferences was avoided. In a real case, other fuels than methane could be used and, even in the case of incomplete combustion of methane, the injection of an extra amount of acetylene can be easily detected. In order to verify which concentration should be used in the measurements, to have a sufficient Signal-to-Noise Ratio, we examined the behavior of the absorptions in the temperature range of 285–385 K.

Molecular absorptions are described by the Beer-Lambert law:

$$I_{out} = I_{in} \cdot e^{-\alpha L}, \tag{1}$$

where I_{out} and I_{in} are the powers of the light beam after and before crossing the sample, respectively; L is the length of the sample, and

$$\alpha = S \cdot g(v) \cdot n, \tag{2}$$

where S (cm/molecule) is the absorption strength, $g(v)$ (cm) the shape of the absorption (area normalized to 1), and n (molecule/cm^3) the density of the absorbing species. One of the consequences of this behavior is that the absorbance of a transition depends on temperature, as both linestrength S and lineshape $g(v)$ are sensitive to temperature. Figure 1 shows the comparison of the transmissions, in a narrow wavelength range, at 285 and 385 K, at P = 1 Bar, L = 0.6 m, of a simulated exhaust mixture including H_2O 10%, CO_2 10%, O_2 4% (green line), and the same mixture with addiction of C_2H_2 1‰ (blue) [10].

It is evident that linestrengths decrease with increasing temperature. The best choice with respect to interference from other molecules is the line close to 1.521 μm, even if it is not the most intense one. Its strength decreases by about 38% at 385 K (with respect to 285 K), nevertheless, the green line in Figure 1 is very close to transmission 1 (i.e., unperturbed transmission) at both temperature conditions. Finally, a concentration around 1‰ is sufficient to produce absorptions of a few %, easily detectable by any optical technique. Actually, we easily worked at levels about one order of magnitude less than this. Due to all these features, and because suitable laser sources are commercially available at this wavelength, the transition at 1.521 μm should be selected in real operation conditions. As a matter of fact, in many of the present measurements we used hot air, or room temperature air, rather than exhaust gases, so the concentrations of water and carbon dioxide were the atmospheric values. For this reason in our tests we used either the absorption at 1.521 μm, or the more intense one at 1.520 μm.

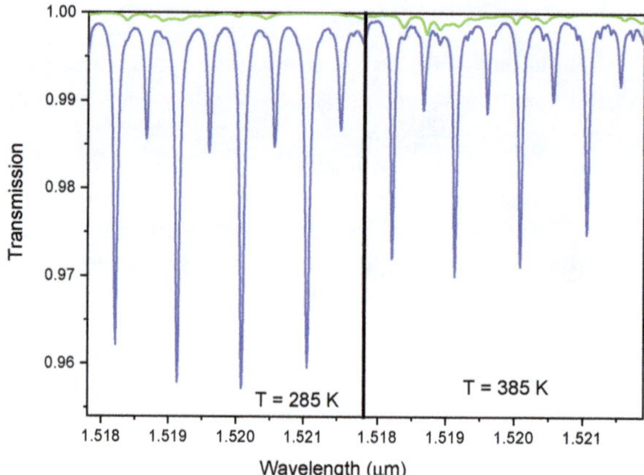

Figure 1. Comparison of the transmission spectra of an exhaust including H_2O 10%, CO_2 10%, O_2 4% (green line), and the same mixture with addiction of C_2H_2 1‰ (blue) at P = 1 Bar, L = 0.6 m, T = 285 K (**left**) and T = 385 K (**right**).

As for the detection technique to be used, our choice was between Direct Absorption (DA) and Wavelength Modulation Spectroscopy (WMS) [11], depending on the kind of flow measurement. In DA the power of a light beam, transmitted across a sample, is converted by a detector into a voltage or current. When sweeping the laser wavelength, the absorption profile is retrieved. The analysis of this profile yields the absolute value of the concentration, provided that pressure, temperature, pathlength and molecular linestrength are known. In WMS, a modulation is applied to the laser wavelength at a period τ_m much lower than the sweep time τ_s (typically $\tau_m \leq 10^{-3} \cdot \tau_s$). The detector signal is demodulated at a multiple of the modulation frequency. The most used choice in commercial devices is a demodulation at twice the modulation frequency. In this case, the signal is approximately the second derivative of the absorption profile, and the central peak is proportional to the density of absorbing molecules. WMS exhibits a better signal-to-noise with respect to DA [12]. In a previous paper [13] we demonstrated the calibration problems of WMS when the composition of the sample varies, in particular because of large percentages of carbon dioxide and water, problems which do not occur with DA [14]. As a matter of fact, the two different optical flow measurements (TT and dilution) feature different calibration issues: TT is related to the time evolution of the concentration of the injected tracer and not to the absolute value of the mixing ratio. On the contrary, dilution requires an accurate measurement of the tracer mixing ratio. For these reasons we chose DA for dilution, and WMS for TT measurements.

3. Materials and Methods

3.1. Layout of the Test Rig

The measurements were carried out at the Innovhub premises in San Donato Milanese (MI), Italy. The testing facility, used to validate the proposed technique, is a small-scale stack simulator, consisting of a flue generator and a vertical duct, where two measuring planes are available. A detailed description of the plant is shown in Figure 2. Real exhaust gases can be produced using one of the boilers connected to the plant; in particular, a 300 kW gas boiler, a 150 kW fuel oil boiler and a 100 kW biomass boiler are available. Otherwise, simulated exhausts can be produced using a fan, an electric heating unit and a steam generator; the flue gases produced this way can be more easily modulated since each of the parameters (namely flow rate, temperature and moisture content) can be set within a relatively large range. On the contrary, when a boiler is used, the composition, temperature and flow rate of the exhausts are mostly fixed, depending of the power and feed of the boiler, just minor adjustments can be produced by acting on the boiler's settings. Both the real exhausts and the simulated fumes are conveyed through horizontal ducts to the same vertical stack, having a constant circular section of 300 mm of internal diameter (ID) and being 12 m high totally. The stack is a double wall stainless steel duct with a 25 mm thick thermal insulation. The gas is conveyed into the stack through a 45° inclined insertion duct; a coaxial section, consisting of a 250 mm inner duct, allows straightening the velocity field. Downstream, a screw-shaped mixing section is placed, in order to force the stream coming from the boilers to mix with the one coming from the fan, when both are necessary at the same time. The first measuring plane is located about 3 diameters downstream of this section. The second one is placed 3.54 m higher (i.e., 12 ID downstream). Both measurement points are equipped with two opposite flanged ports, housing the optical windows. About 5 m higher the fumes are released in the atmosphere. Additional ports for inserting probes used for the analysis of the gas composition and for the measurement of point velocity are located close to the two main measuring planes. The Pitot tubes used in the present work were placed 0.3 m above the upper measuring point. The injection point of the tracer was placed as far as possible from the measuring planes, so that the better possible mixing could be obtained. A small perforated pipe, inserted into the main duct close to the flue generator, was used to distribute the tracer in the mainstream. A gas bottle, containing the pure tracer, was connected to the injection point through a flexible PTFE tube; a pressure regulator was used to set the pressure of the tracer at the injection point. A quick connection hose was used to manually connect/disconnect the tracer carrying tube with the injection pipe at the beginning of each TT test. On the other hand, a calibrated mass-flow meter was used to feed a known amount of tracer during the dilution tests. The tracer is mixed with the fumes in the straight horizontal duct (about 10 m long), a further and more intense mixing is produced in the mixing section of the stack, previously described.

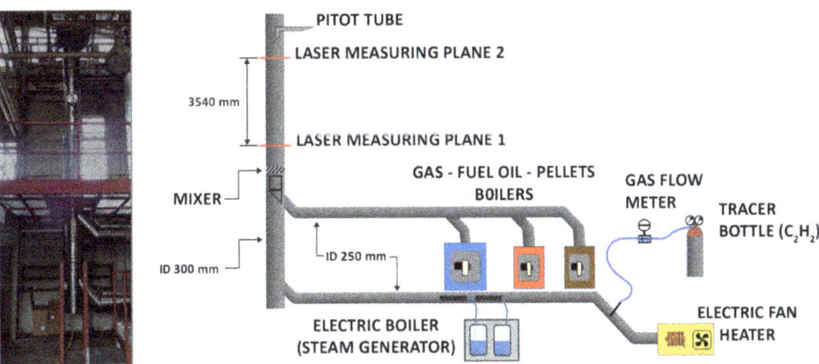

Figure 2. General layout of the stack simulation facility used in this study.

3.2. Optical Platform

The optical platform, based on the first version described in [11], was developed in order to adapt it to provide both DA and WMS signals for two measurement points inside the stack and for a reference arm.

3.2.1. Optics

The optical scheme is described in Figure 3. The laser source is a DFB fiber laser with an emission wavelength at 1521 nm (Eblana Photonics EP1521-0-DM-B01-FA, butterfly package, pigtail output). Immediately out of the laser, a beam splitter (Thorlabs TW1550R1A2) takes 1% of the beam to a fiber-coupled reference cell (Wavelength References C2H2-12-H(5.5)-50-FCAPC, 50 Torr, 5.5 cm optical path) and home-made BK7 etalon, length 30.75 mm, Free Spectral Range 3.25 GHz, to obtain a relative frequency scale. The remaining 99% is 50:50 split (Thorlabs TW1550R5A2) into two beams, which are sent to the measurement points. As shown in Figure 4 Center, the launcher is very close (2 cm) to the detector. The beam crosses the stack and is reflected by a plane mirror, directly onto the detector. As the stack diameter is 30 cm, and inside the stack the distance between the two beams is less than 2 cm, we can affirm with a good approximation that we are probing the tracer concentration along a diameter of the stack. It's worth noting that both optical measurement channels are necessary for the TT measurement only, as dilution is measured in steady-state conditions, and one channel only could be used. Yet, once the experimental apparatus is completely installed, the presence of two optical lines makes it possible to have a redundancy in the dilution measurements.

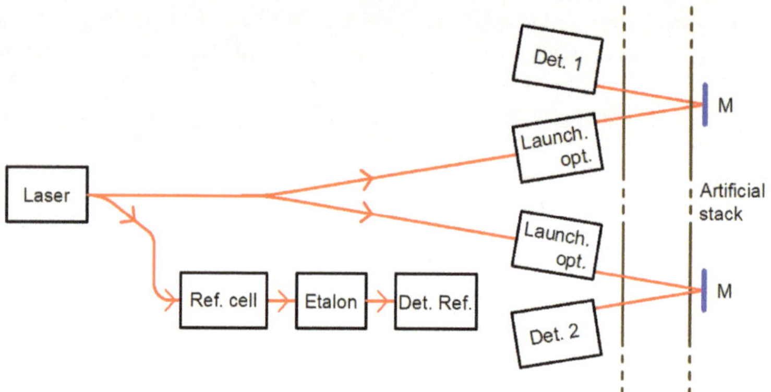

Figure 3. Block diagram of the optics. M: flat Mirror. The angle between the two couples of beams inside the stack is exaggerated for the sake of clarity.

3.2.2. Mechanics

Mechanics is divided into three parts. The first is an aluminum breadboard that hosts the electronics, the laser and the reference optics (Figure 4 Left). The breadboard is framed into a structure that fits a plastic suitcase for easy transportation. The other two parts of the mechanics are mounted onto the stack. The launching/receiving optics can be seen in Figure 4 Center. The laser beam is shot across the stack, onto a mirror (Figure 4 Right), which directly reflects the beam onto the detector, for a total of two passes inside the stack. The mechanical stability of the system relies on the flanges protruding out the stack. All the optical ports are sealed by 1″ anti-reflection coated BK7 windows (Thorlabs WG11050-C).

Figure 4. (**left**): main plate of the mechanics, together with the waveform generator and the laptop; (**center**): launching optics, driven by the yellow optic fiber, and detector; (**right**): steering mirror, which reflects the laser beam onto the detector.

3.2.3. Electronics

A block diagram of the electronics is shown in Figure 5. The laser is current-supplied and temperature stabilized by a commercial driver (ppqSense QubeCL). A National Instruments crate (cRIO 9067) hosts: two 4-channels, 20 MSamples/s acquisition rate each, 14 bits vertical resolution ADC plug-in's (NI-9775); a programmable, 4-channels digital I/O plug-in (NI-9402); and a 4-channels PT100 reader (NI-9217) for housekeeping. An important feature of this crate is its FPGA, Xilinx Zynq-7000, with 85.000 logic cells and 106.400 flip-flops. A Tektronix double output Arbitrary Function Generator 3022 provided: the trigger (1 kHz) for the QubeCL to start the sawtooth ramp to sweep the laser frequency across an absorption profile, the TTL output signal for the digital I/O plug-in which starts the acquisition on the rising edge of the TTL, and the high frequency (1 MHz) sinusoid used for WMS. The latter signal was converted by the QubeCL into current, and added, together with the ramp, to the bias current of the laser. We used three detectors, Hamamatsu InGaAs mod. G12180-210A, 1 mm diameter, 2-stage Peltier cooling, 40 MHz cutoff frequency, two for the detection points and the third for the reference arm. Each detector is equipped with two outputs, low pass ($<$100 kHz) and high pass ($>$500 kHz) filtered. The outputs of the detectors are acquired by six channels of the NI-9775 modules. The low pass signals are used for DA and the high pass signals are used for WMS.

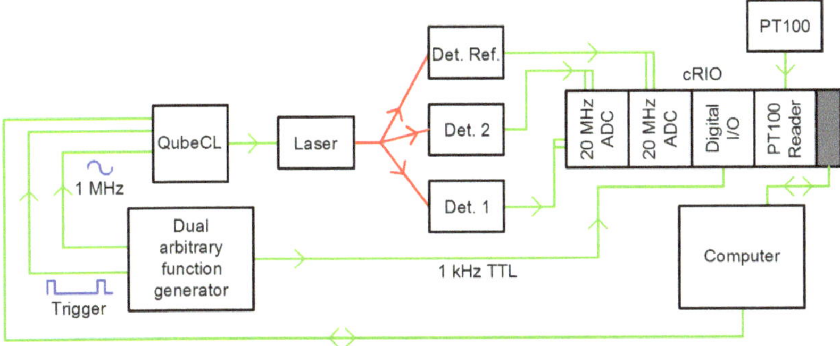

Figure 5. Block diagram of the electronics.

3.2.4. Software

Once set the function generator and the laser driver, the procedure of synchronization, acquisition and storage is carried out by the FPGA of the NI cRIO. We wrote a dedicated LabVIEW program for data acquisition and for the implementation of a digital lock-in amplifier, according to well-known principles and routines [15–17]. The lock-in can operate in single and dual-phase mode. The modulated signals are acquired by the 20 MS/s ADCs for a time corresponding to an integer multiple N of the modulation period. The higher N, the higher the integration time. The actual value of N is chosen taking into account the rise/fall time of the concentration peak. The deconvolution procedure is carried out by the FPGA: the signals are multiplied by a reference sinusoid and then integrated during the N periods. In single-phase mode, the phase of the reference signal can be adjusted, in order to maximize the output. In dual phase-mode, a quadrature detection can be performed, multiplying the acquired signals by two reference signals, 90° out of phase with respect to each other, and then taking the module and the sign.

The software allows all these choices, and stores data in the USB memory stick connected to the cRIO.

3.3. Optical Methods

3.3.1. Direct Absorption and Dilution Method

The dilution measurements consist of a repeated laser wavelength scan (without the sinusoidal modulation) across the selected C_2H_2 absorption line. Each scan is acquired as a set of 20,000 points for each channel, in 1 ms. A binning over 100 points reduces the points per ramp to 200. Binning, and average of the scans, are performed by the FPGA, producing a spectrum every 28 ms. These spectra are saved in the computer. The final waveform is fitted by using the theoretical Voigt absorption profile. According to the procedure described in [18], the mixing ratio can be inferred by the fitting parameters, provided that temperature, pressure, molecular linestrength and path-length are known. Pressure and temperature are measured by sensors included in the Pitot probe (accuracy 0.05% and 0.3%, respectively), the linestrength is reported in HITRAN molecular database [10] with accuracy 1% and the pathlength was measured to be 580 ± 1.5 mm.

In order to verify the quality of our data, and to decide the best integration time for our measurements, we performed the Allan-Werle Variance [19,20] on mixing ratio measurements. Figure 6 shows the results obtained for two different speeds of the fan (expressed as % of the maximum speed). The best integration time results to be 8 s, corresponding to an Allan-Werle Variance σ_{AWV} of 1.5 ppm at 30% and 0.9 at 75%. The slight difference can be due to the different measurement conditions. It is worth noting that this is the Allan-Werle test of the whole system, formed by the fan, the mass-flow meter and our optical apparatus. Dilution measurements are stationary, so, according to the Allan–Werle analysis, we can further average over 285 scans, for 8 s total measurement time. Figure 7 shows the averaged absorption profile, with the fit curve and the residual.

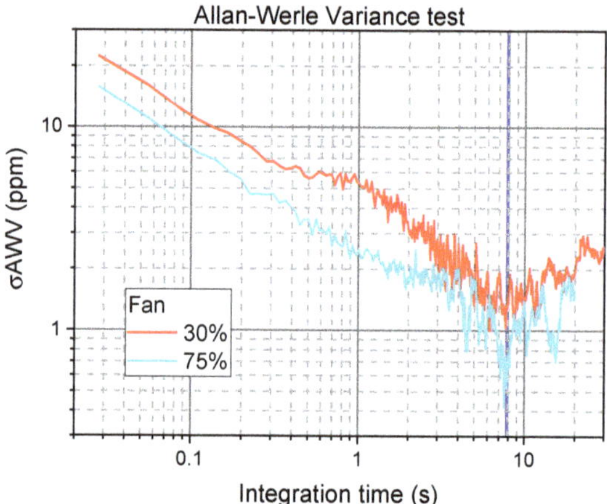

Figure 6. Allan-Werle test of the system at two different fan speeds, with respect to maximum.

The calculation of the flow speed is performed by measuring the steady-state concentration of the tracer, when the amount of injected tracer per unit time and the transverse section of the duct are known, according to the formula:

$$V = \frac{T}{CS} \tag{3}$$

where V is the velocity (m/s), T is the flow (m^3/s) of the injected tracer, C is the mixing ratio measured downstream, and S (m^2) is the transverse area of the duct.

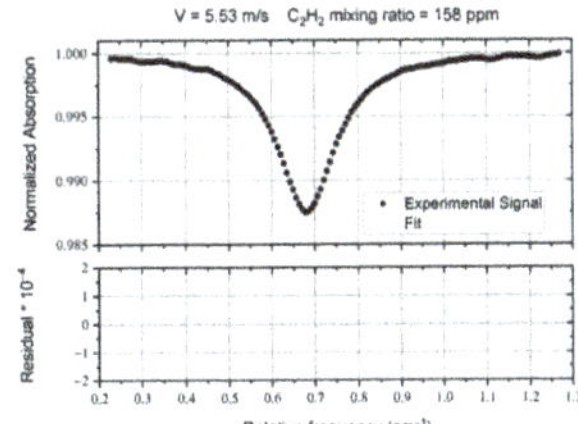

Figure 7. Plot of a direct absorption acquisition and its analysis.

3.3.2. Wavelength Modulation Spectroscopy and TT Method

To perform Second Derivative WMS, the fast modulation (1 MHz) is added to the laser sweep (1 kHz) and the acquired signal is deconvolved at 2 MHz. An example of WMS signals from the reference cell and from one measuring point along the stack is shown in Figure 8, where the different shapes of the two waveforms are due to the different pressures inside the cell and the stack. The peak values of the profiles in Figure 8 are proportional to the concentration. In the TT method, the time profile of the peak heights is acquired at two different measuring points (reference is only used to check that the laser wavelength is not drifting), when a concentration burst of the tracer is inserted inside the stack. In order to have the maximum acquisition rate, the laser wavelength sweep is stopped and the laser is set at the peak of the absorption. In principle, it would be possible to implement an active stabilization routine, based on an odd derivative (demodulation at an odd multiple of the modulation frequency). As a matter of fact, due to the stability of the laser driver, active stabilization was not necessary, and emission stability was maintained for a time larger than the 10 s of the measurement duration. With this procedure, the acquisition rate is 1 kHz, sufficient to follow the time profile of C_2H_2 inside the stack.

To obtain the concentration burst, the tracer gas line was closed and opened with a delay as short as possible. An example of the TT measurements, at two different points of the stack, at three different flow speeds, is reported in Figure 9. The velocity can be obtained by calculating the time delay between two corresponding timestamps of the red and blue curves of Figure 9. The selection of the timestamps is not a minor issue and it will be analyzed in the next section. The geometrical distance between the two measuring planes (3.54 m) divided by the delay yields the average gas velocity.

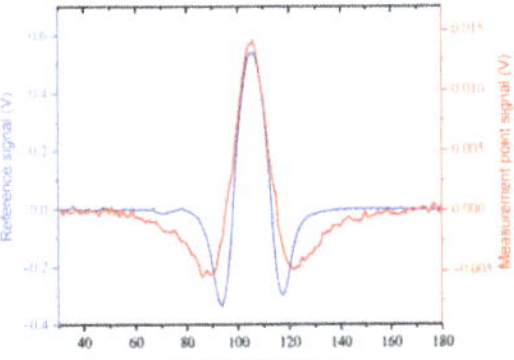

Figure 8. Wavelength Modulation Spectroscopy (WMS) signals from the reference cell (blue) and one of the measurement points along the stack (red).

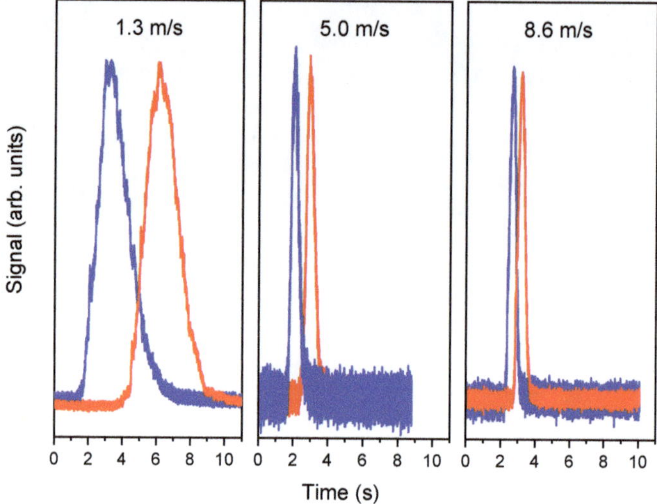

Figure 9. Set of Transit Time (TT) measurements at three different flow speeds, at the lower (blue) and upper (red) measurement points.

3.4. CFD Modeling

CFD calculations were used to simulate the flow field inside the stack, comparing the model with the experimental results, and driving further measurements. The commercial software Ansys® Academic Fluent, Release 15.0 has been used for this calculation. It solves conservation equations for mass and momentum, according to an Eulerian approach, while an additional equation for energy conservation is solved to account for heat transfer processes. For flows involving species mixing or reactions, a species conservation equation is solved for each one and additional transport equations are also solved when the flow is turbulent.

In this model, the turbulent viscosity μ_t, assumed as isotropic, is computed by combining the turbulent kinetic energy k and its dissipation rate ϵ as follows:

$$\mu_t = \rho C_\mu \frac{k^2}{\epsilon},$$ (4)

where ρ is the density and C_μ is a constant.

A tetrahedral unstructured grid, consisting of 2.6 M cells, has been set up; the standard $k\epsilon$ model of turbulence has been adopted; the time-dependent simulations have been carried out using a fixed time step of 0.05 s. The composition, velocity and temperature of the gas stream have been set at the inlet, as boundary conditions for the numerical solver of the model equations. A homogeneous distribution of the tracer has been assumed on the inlet section at time zero conditions. Then, the average concentrations of the tracer on two straight lines corresponding, in the model grid, to the laser beam path lines in the real plant, have been computed for each time step, by averaging the tracer concentration calculated in each cell laying on these lines. These average concentrations are directly comparable with the detected signals.

3.4.1. Simulation of the Tracer Concentration Profile

By using CFD simulation, the time profiles of the concentration of the tracer were calculated at any point along the stack, in order to make a comparison with experimental data and to select the best method to calculate the gas velocity during TT measurements. The simulated concentration profile varies along the vertical axis, as a consequence of the progressive mixing of the tracer. This makes the curves, at the two measuring planes, not perfectly overlapping and a specific criterion has to be defined in order to select the two reference timestamps (one for each curve) to be compared to measure the time delay for velocity calculation. Different choices, for these points, produce slightly different outcomes in terms of velocities, which means that this choice is an essential aspect of the method. The expected profiles of the concentration of the tracer, along the vertical axis of the stack, as calculated through CFD simulations, are reported in Figure 10, for an average gas velocity of 6 m/s and a temperature of 105 °C, as well as the corresponding bi-dimensional distributions, on a vertical plane, at 0.75, 1.25 and 1.75 s, as explained in the caption.

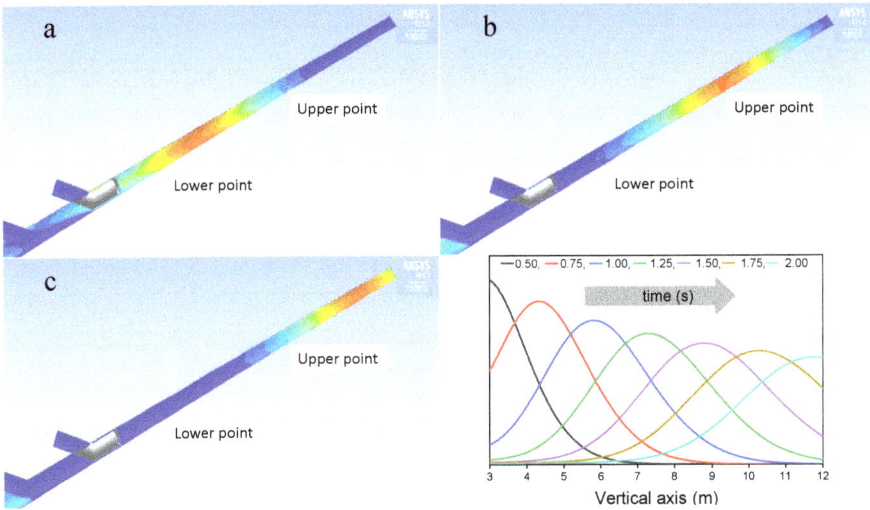

Figure 10. Computational fluid dynamic (CFD) tracer concentration profile along the stack axis (average velocity 6 m/s, T = 105 °C). Time origin is the transit time of the tracer at the upper junction (grey in the figures). (**a**) 0.75 s; (**b**) 1.25 s; (**c**) 1.75 s.

It can be observed that the increase/decay curve spans for a long distance along the vertical axis; from this point of view, these conditions are quite different from an ideal plug flow. The shape of the curve actually changes along the axis: its slope decreases and the peak broadens, as expected. In Figure 11 experimental and model results are compared: the signals of TT measurements collected at the lower and upper measuring planes are plotted together with the results of CFD simulation (circles) at the same test conditions. The broadening effect is produced by the combination of the bulk velocity of the flow and the diffusion velocity of the tracer molecules. The second peak in the simulation shows a larger broadening effect compared to the experimental results. It can be assumed that the adopted model of turbulence overestimated the turbulent diffusivity of the system. More advanced modeling, for instance including Large Eddy Simulation (LES) could have produced a better matching, but a fully detailed simulation is outside the scope of this work.

Molecular and turbulent diffusion produces an apparent velocity, which gives incorrect results if the reference timestamps of each curve are not selected properly. As far as an isotropic diffusivity is assumed, each molecule diffuses randomly in any direction: molecules mainly diffusing in the flow direction move faster than average fluid-dynamic velocity, while molecules diffusing mainly against the flow direction move slower than average fluid-dynamic velocity. Consequently, these points are not suitable for measuring the bulk velocity of the flow. The centroid of the tracer spike should be identified in order to get a correct result from the TT technique; EN ISO 16911-1 states that the best choice is the median of the concentration distribution. A specific analysis has been carried out to compare different possible criteria. The simulated concentration distributions can be used to compare the results one can expect by applying different approaches of calculation on the same data. In Figure 12 the average gas velocity was estimated from three different sets of points, on the simulated curves: the peak of the curve, the median, and the mid-point between the maximum slope at both sides of the peak (the inset shows the location of these points). These calculations have been repeated for several axial distances, in order to evaluate the effect of the separation between the two measuring points. The best matching with the known average actual velocity is produced using the peaks. The velocities calculated based on the median are equally in good agreement with the expected result, except for very short distances when velocities would be overestimated. The sloping approach produces more erratic results, probably due to the fluctuations produced by the numerical derivation of the curves. It is noteworthy, anyway, that the three points, selected on each curve, are quite close to each other, because of the symmetry of these curves, while many different results may be expected in more asymmetric conditions.

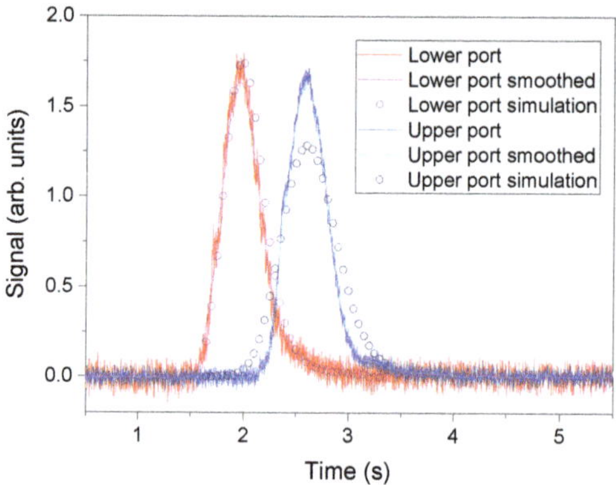

Figure 11. Comparison between simulation and experimental data for concentration profiles at the average velocity 6 m/s, T = 105 °C).

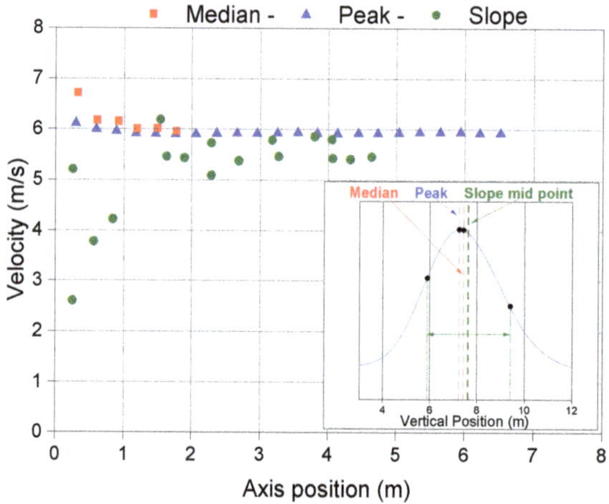

Figure 12. Calculated velocities using different points on the simulated tracer concentration curves at different times (average velocity 6 m/s, T = 105 °C).

3.4.2. Simulation of Velocity Field Inside the Stack

A simulation of the velocity field inside the stack was carried out in order to understand how critical is the positioning of the Pitot tube, due to the fact that the flow field inside the stack does not feature a cylindrical symmetry, but a complex behavior, as reported in literature [3,21]. Figure 13 shows the calculated components of the gas velocity along the stack. The velocity vectors on the horizontal section where the Pitot probe was installed are shown in detail in Figure 14: the flow field is strongly asymmetric and the vectors are inclined in a quite complex configuration. As a consequence, the location of the Pitot probe can critically affect the velocity measurements and deserves further investigation.

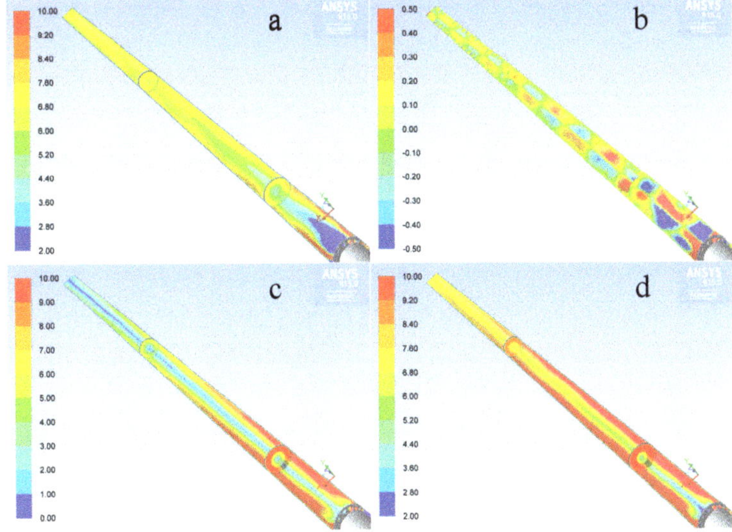

Figure 13. Simulated velocity field inside the stack simulator in all its components: (**a**) total, (**b**) axial, (**c**) radial and (**d**) tangential velocity. Units are m/s.

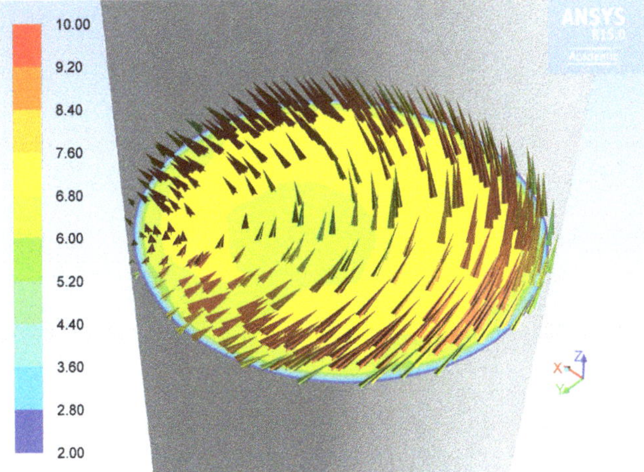

Figure 14. Simulation of the flow field along a horizontal section of the duct, corresponding to the Pitot port position. Units are m/s.

4. Measurements and Results

The experimental tests were carried out under different conditions of gas velocity, temperature and composition, in order to check the response of the three techniques and the effect of each parameter.

4.1. Pitot Probe Test

We do not describe the working principle of Pitot tubes, it is enough to say that an S-type Pitot, connected to a micromanometric device, specifically designed for in-stack velocity and flow rate measurements was used in the first series of intercomparison tests. This device is designed to be integral with the probe and has an integrated inclinometer. The Pitot tube was placed 0.3 m downstream the upper measuring plane and velocity measurements were carried at a fixed point on the central axis of the stack, in accordance with EN 15259 for ducts having an internal diameter smaller than 0.35 m. Despite this, it is not possible to be sure that this is the best choice in the specific case of a cyclonic and asymmetrical flow, such as the one studied in the previous section. Hence, we decided to investigate this feature in deeper detail. We performed a series of measurements, using the Pitot tube only, testing a flow of air at room conditions. The Pitot probe was inserted inside the stack at different distances from the duct axis (negative values mean between the port and the duct axis, positive values mean beyond the duct axis), and at different angles with respect to the vertical direction. The results of this test are shown in Figure 15. It is evident that the velocities are not constant along a diameter. Moreover, when tilting the Pitot probe, even by a few degrees, the readings change. From this investigation, we concluded that the reading at the central position is approximately equal to the average velocity on that line and only if the probe is strictly vertical. Just a little displacement along the axis or a small rotation is enough to read a quite different velocity.

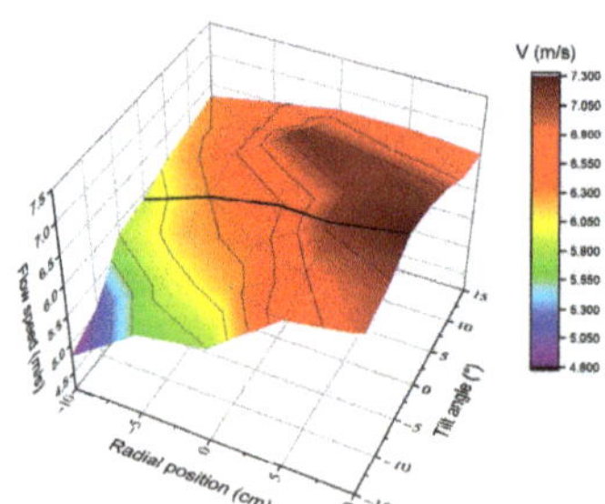

Figure 15. Plot of the flow field, along a diameter of the stack, with different Pitot angles with respect to the duct axis. Units are m/s.

4.2. First Set of Measurements

The first set of measurements was carried out with ambient air at room temperature, in order to have a set of intercomparison data: for each flow fan-speed we recorded the velocity readings obtained by the Pitot, and the results of the dilution and the TT techniques. As dilution and TT measurements cannot be carried out at the same time, we used the fan speed (% with respect to maximum speed) as a reference for different series of measurements.

In order to perform dilution measurements, the injection was constant, measured by using a thermal-mass-flowmeter (Bronkhorst High-Tech model EL-FLOW, accuracy 3.5%), at the level of $3.7 \div 3.9$ L/min ($6.17 \times 10^{-5} \div 6.50 \times 10^{-5}$ m^3/s). The concentration readings were in the range $136 \div 845$ ppm, so well below 1‰, as explained in Section 2. In order to measure the time delay to calculate the velocity, we took the time stamp of the transit of a peak in different ways, described in Section 3.4.1: the median of the distribution of the signal, the peak and the mid-point of the maximum peak slopes. Figure 16 shows the readings of dilution and different time stamps of TT, vs. Pitot readings, for different fan speeds. For TT method each point is the mean value of a set of measurements and the error bars are the corresponding standard deviations. For the dilution method, each point is calculated according to (3), and the error bars are the accuracy of each measurement.

There is a very good correlation between the Pitot tube and the two optical techniques, whatever the analysis of the TT signals. In each graph of Figure 16 the first point is not taken into account for the linear fit. The reason is that the Pitot tube shouldn't be used around 1 m/s, as its linearity is poor in this range. In fact, the average flow speed at the first point of each graph, read with optical techniques is 1.04 m/s, while the Pitot reading is 1.21 m/s, which is the largest discrepancy among Pitot and optical readings. In Table 1 the different slopes of the fit curves (i.e., the proportion coefficient with respect to Pitot tube) are reported. We can note that all the slopes are equal, within their uncertainties, and slightly less than unity. As the dilution technique is not affected by any fluid-dynamic effect, its readings can be assumed as the reference values and the very small difference with the TT results proves the reliability of the TT technique, and its independence of the selection of the line of sight.

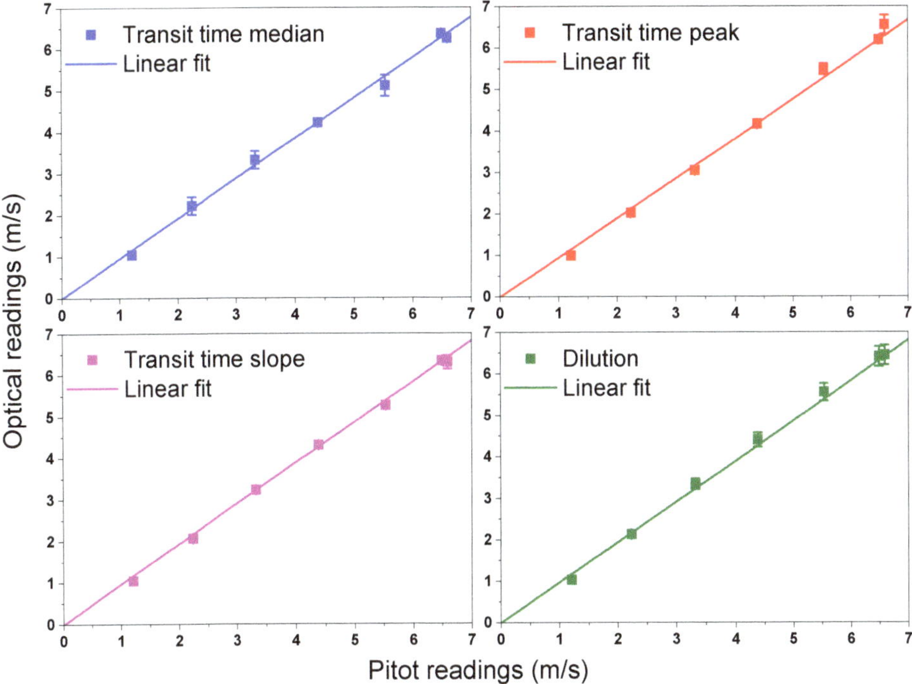

Figure 16. First set of flow velocity measurements, comparing Pitot tube with three different analysis of the TT signals and with dilution. Points are related to different fan velocities.

Table 1. Table of the proportionality coefficients between optical techniques and Pitot tube for different analysis procedures.

Technique	Ratio	Uncertainty
TT median	0.968	0.007
TT peak	0.97	0.01
TT slope	0.976	0.005
Dilution	0.97	0.01

4.3. Second Set of Measurements

We carried out a second run of measurements, in order to investigate a wider range of experimental conditions, and the use of a conventional Pitot probe, adopted for in-field isokinetic sampling, manually aligned, both vertically and with respect to the duct axis, as routinely performed in real-life periodic measurements. This time the fluid was room air, heated using electric resistors. At the highest temperature, steam was added to room air, at a rate depending on the flow speed. For each temperature, the fan speed was set at different levels. In this set of measurements, the Pitot readings were compared with TT only. The results show very good linearity of the TT readings, at any temperature and gas composition, with respect to Pitot data, whatever the criterion for the time stamp of each peak (Figure 17). On the other side, it is evident that there is a systematic deviation between the TT readings and the Pitot results, whose ratio was measured to be 0.822 ± 0.002. This result is not surprising for a manually aligned probe, according to previous literature, and following our simulations and experimental tests. The amount of the discrepancy is in this case above 20%, which means that the most unfavorable conditions of the positioning of the Pitot probe were encountered here.

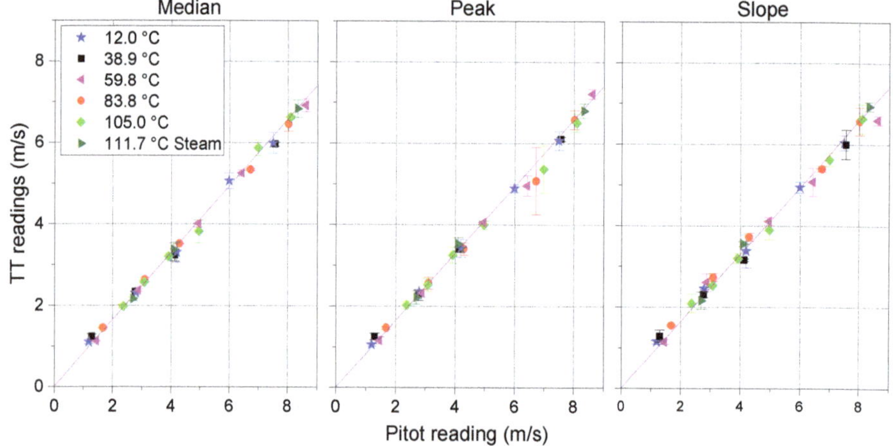

Figure 17. TT readings, with different time stamps, vs. Pitot readings, with manual setting of the Pitot probe for different temperatures and fan speeds. Steam flow rate at 111.7 °C: 60 kg/h @ 2.7 m/s; 31 kg/h @ 4.1 m/s; 18 kg/h @ 8.3 m/s.

5. Discussion and Perspectives

We have applied two spectroscopic detection techniques to the measurement of flow in narrow ducts, and to the calibration of standard sensors and methods. Our multipurpose device was deployed in a stack simulator, proving to fulfill all the requirements for the above task. It is already stated in EN ISO 16911-1 that the dilution method is a valuable reference method, which is intrinsically free from wall effects, or from any perturbation due to the geometry of the duct, or to any objects inserted in the duct, inducing turbulence. On the contrary, the more turbulent the flow, the more homogeneous the distribution of the tracer. TT method, in particular the one implemented in this work, based on optical techniques, compares very closely with dilution.

The conventional procedure adopted for calibrating Pitot tubes is carried out ex-situ in a standardized wind tunnel using a primary reference device. It generally provides either a single calibration factor or a set of calibration factors for the different velocity ranges, used to calculate the point velocity from the readings of the differential pressure at the Pitot ends. Calibrated Pitot tubes are then used both to measure in stack velocity directly and to periodically calibrate other automated measuring systems, installed on the stacks to measure the flow rate on line. This study shows that this approach can produce inaccurate results when cyclonic flows in small ducts are involved, in particular:

1. the conditions of 5 internal diameters of straight duct upstream and 3 downstream of the measuring section, without elements disturbing the flow, may not be a sufficient guarantee to provide a regular velocity profile at the measuring plain;
2. single-point measurements may be non-representative of the average velocity, a multipoint technique should be required even for ducts having an internal diameter smaller than 0.35 m;
3. manual alignment of a Pitot tube is not accurate and stable enough, large deviations of the measured values may be produced by very small variations in the placement of the probe;
4. techniques relying on the bulk properties of the flow, such as the concentration of a tracer injected in the gas stream, are largely insensitive to the flow pattern and the local disturbances, producing linear and accurate results.

As a consequence of these considerations, the pertaining international standards, namely UNI EN 15259:2008, EN ISO 16911-1:2013 and EN ISO 16911-2:2013, should include specific warnings for small ducts, suggesting the use of reliable and robust techniques for the in-site calibration of the automated measurement systems, such as dilution based and transit time methods. Moreover, apart from periodical checks, the necessity to repeat the calibration every time a modification occurs in the duct, upstream the sensor should be introduced.

Table 2 compares the requirements of the different techniques used in this work. Pitot tubes are undoubtedly the simplest technique for flow velocity measurements, as they require an insertion port only, and no consumables. TT requires four optical ports. Dilution can be implemented either across the stack, or after gas extraction. In the first case, two optical ports are necessary, in the second case an extraction port (downstream the Pitot, to avoid any interference) and a heated line must be used. In both the latter cases, a tracer is required, which means consumables and an injection point. Despite the higher complexity, we proved that in narrow ducts it is necessary to add ports, or injection/extraction points, suitable for the application of more complicated techniques, for intercomparison and calibration. As a final remark, the described optical techniques are not so much time consuming, as they require one workday for set-up, measurement and packing.

Table 2. Comparison of Pitot tube, TT and dilution methods.

	Affected by Position	On-Site Calibration	Traces Gas	Optical Ports	Extraction
Pitot tube	YES	Compulsory	NO	NO	NO
Dilution	NO	NO	YES	YES/NO	NO/YES
TT	NO	NO	YES	YES	NO

Author Contributions: Conceptualization, F.D. and G.M.; methodology, F.D., S.V., M.B. and G.M.; software, S.V. and A.M.; formal analysis, S.V. and G.M.; investigation, F.D., S.V., A.M., M.B., C.M., S.B. and G.M.; resources, A.M. and M.B.; writing—original draft preparation, F.D., S.V. and G.M.; writing—review and editing, F.D., S.V. and G.M.; visualization, F.D., S.V. and G.M.; project administration, F.D. and G.M.; funding acquisition, F.D. and G.M. All authors have read and agreed to the published version of the manuscript.

Funding: The project IMPRESS 2 has received funding from the EMPIR programme co-financed by the Participating States and from the European Union's Horizon 2020 research and innovation programme (Project Number: 16ENV08).

Acknowledgments: The authors want to thank Massimo D'Uva (CNR-INO) for the mechanical components, and Stefano Ariazzi and Massimiliano Siviero (Innovhub) for technical assistance during measurements.

Conflicts of Interest: The authors declare no conflict of interest. The funders had no role in the design of the study; in the collection, analyses, or interpretation of data; in the writing of the manuscript, or in the decision to publish the results.

References

1. Dimopoulos, C.; Robinson, R.A.; Coleman, M.D. Mass emissions and carbon trading: A critical review of available reference methods for industrial stack flow measurement. *Accred. Qual. Assur.* **2017**, *22*, 161–165. [CrossRef]

2. Geršl, J.; Knotek, S.; Belligoli, Z.; Dwight, R.P.; Robinson, R.A.; Coleman, M.D. Flow rate measurement in stacks with cyclonic flow–Error estimations using CFD modelling. *Measurement* **2018**, *129*, 167–183. [CrossRef]

3. Kang, W.; Trang, N.D.; Lee, S.H.; Choi, H.M.; Shim, J.S.; Jang, H.S.; Choi, Y.M. Experimental and numerical investigations of the factors affecting the S-type Pitot tube coefficients. *Flow Meas. Instrum.* **2015**, *44*, 11–18. [CrossRef]

4. *EN ISO 16911-1: Stationary Source Emissions–Manual and Automatic Determination of Velocity and Volume Flow Rate in Ducts—Part 1: Manual Reference Method*; International Organization for Standardization: Geneva, Switzerland, 2013.

5. Selbach, H. Mapping of velocity flow fields using the laser-two-focus technique. *ICALEO* **1989**, *1989*, 48–56. [CrossRef]

6. Allen, M.G. Diode laser absorption sensors for gas-dynamic and combustion flows. *Meas. Sci. Technol.* **1998**, *9*, 545–562. [CrossRef] [PubMed]

7. Atkins, M.D. Chapter 5—Velocity Field Measurement Using Particle Image Velocimetry (PIV). In *Application of Thermo-Fluidic Measurement Techniques*; Kim, T., Lu, T.J., Song, S.J., Eds.; Butterworth-Heinemann (Elsevier): Oxford, UK, 2016; pp. 125–166, ISBN 9780128097311. [CrossRef]

8. Air Liquide Safety Data Sheet for Acetylene. Available online: https://encyclopedia.airliquide.com/acetylene (accessed on 19 December 2020).

9. Watt, L.J. The Production of Acetylene from Methane by Partial Oxydation. Master's Thesis, Department of Chemistry, University of British Columbia, Vancouver, BC, Canada, 1951.

10. Gordon, I.E.; Rothman, L.S.; Hill, C.; Kochanov, R.V.; Tan, Y.; Bernath, P.F.; Birk, M.; Boudon, V.; Campargue, A.; Chance, K.V.; et al. The HITRAN2016 molecular spectroscopic database. *J. Quant. Spectrosc. Radiat. Transf.* **2017**, *203*, 3–69. [CrossRef]

11. Silver, J.A.; Bomse, D.S.; Stanton, A.C. Diode laser measurements of trace concentrations of ammonia in an entrained-flow coal reactor. *Sensors* **1991**, *30*, 1505–1511. [CrossRef] [PubMed]

12. Duffin, K.; McGettrick, A.J.; Johnstone, W.; Stewart, G.; Moodie, D.G. Tunable Diode-Laser Spectroscopy with Wavelength Modulation: A Calibration-Free Approach to the Recovery of Absolute Gas Absorption Line Shapes. *J. Lightwave Technol.* **2007**, *25*, 3114–3125. [CrossRef]

13. D'Amato, F.; Viciani, S.; Montori, A.; Lapini, A.; Fraboulet, I.; Poulleau, J. Optical Detection of Ammonia Inside a Stack: Comparison of Different Techniques. *Measurement* **2020**, *159*, 107746. [CrossRef]

14. Viciani, S.; Montori, A.; Chiarugi, A.; D'Amato, F. A Portable Quantum Cascade Laser Spectrometer for Atmospheric Measurements of Carbon Monoxide. *Sensors* **2018**, *18*, 2380. [CrossRef] [PubMed]

15. Cifuentes, A.; Marín, E. Implementation of a field programmable gate array-based lock-in amplifier. *Measurement* **2015**, *69*, 31–41. [CrossRef]

16. Wang, J.; Wang, H.; Liu, X. A Portable Laser Photoacoustic Methane Sensor Based on FPGA. *Sensors* **2016**, *16*, 1551. [CrossRef] [PubMed]

17. Giaconia, G.C.; Greco, G.; Mistretta, L.; Rizzo, R. Exploring FPGA-Based Lock-In Techniques for Brain Monitoring Applications. *Electronics* **2017**, *6*, 18. [CrossRef]

18. Viciani, S.; D'Amato, F.; Mazzinghi, P.; Castagnoli, F.; Toci, G.; Werle, P. A cryogenically operated laser diode spectrometer for airborne measurement of stratospheric trace gases. *Appl. Phys. B* **2008**, *90*, 581–592. [CrossRef]

19. Allan, D.W. Statistics of atomic frequency standards. *Proc. IEEE* **1966**, *54*, 221–230. [CrossRef]

20. Werle, P.; Mücke, R.; Slemr, F. The limits of signal averaging in atmospheric trace-gas monitoring by tunable diode-laser absorption spectroscopy (TDLAS). *Appl. Phys. B* **1993**, *57*, 131–139. [CrossRef]

21. Węcel, D.; Chmielniak, T.; Kotowicz, J. Experimental and numerical investigations of the averaging Pitot tube and analysis of installation effects on the flow coefficient. *Flow Meas. Instrum.* **2008**, *19*, 301–306. [CrossRef]

Publisher's Note: MDPI stays neutral with regard to jurisdictional claims in published maps and institutional affiliations.

 *sensors*

MDPI

Article

Anodically Bonded Photoacoustic Transducer: An Approach towards Wafer-Level Optical Gas Sensors

Simon Gassner [1,2,*], Rainer Schaller [2], Matthias Eberl [2], Carsten von Koblinski [3], Simon Essing [4], Mohammadamir Ghaderi [2], Katrin Schmitt [1,5] and Jürgen Wöllenstein [1,5]

[1] Department of Microsystems Engineering (IMTEK), Albert-Ludwigs-Universität Freiburg, 79085 Freiburg im Breisgau, Germany; katrin.schmitt@imtek.uni-freiburg.de (K.S.); juergen.woellenstein@imtek.uni-freiburg.de (J.W.)
[2] Infineon Technologies AG, 81549 Neubiberg, Germany; rainer.schaller@infineon.com (R.S.); matthias.eberl@infineon.com (M.E.); Amir.Ghaderi@infineon.com (M.G.)
[3] Infineon Technologies Austria AG, 9500 Villach, Austria; Carsten.Koblinski@infineon.com
[4] Department of Electrical and Computer Engineering, Technical University of Munich, 80333 München, Germany; simon.essing@tum.de
[5] Fraunhofer IPM, 79110 Freiburg, Germany
* Correspondence: simon.gassner@infineon.com

Abstract: We present a concept for a wafer-level manufactured photoacoustic transducer, suitable to be used in consumer-grade gas sensors. The transducer consists of an anodically bonded two-layer stack of a blank silicon wafer and an 11 μm membrane, which was wet-etched from a borosilicate wafer. The membrane separates two cavities; one of which was hermetically sealed and filled with CO_2 during the anodic bonding and acts as an infrared absorber. The second cavity was designed to be connected to a standard MEMS microphone on PCB-level forming an infrared-sensitive photoacoustic detector. CO_2 sensors consisting of the detector and a MEMS infrared emitter were built up and characterized towards their sensitivity and noise levels at six different component distance ranging from 3.0 mm to 15.5 mm. The signal response for the sample with the longest absorption path ranged from a decrease of 8.3 % at a CO_2 concentration of 9400 ppm to a decrease of 0.8 % at a concentration of 560 ppm. A standard deviation of the measured values of 18 ppm was determined when the sensor was exposed to 1000 ppm CO_2.

Keywords: gas sensor; photoacoustic; pressure transducer; wafer-level; CO_2

Citation: Gassner, S.; Schaller, R.; Eberl, M.; von Koblinski, C.; Essing, S.; Ghaderi, M.; Schmitt, K.; Wöllenstein, J. Anodically Bonded Photoacoustic Transducer: An Approach towards Wafer-Level Optical Gas Sensors. *Sensors* **2022**, *22*, 685. https://doi.org/10.3390/s22020685

Academic Editor: Krzysztof M. Abramski and Piotr Jaworski

Received: 28 December 2021
Accepted: 14 January 2022
Published: 17 January 2022

Publisher's Note: MDPI stays neutral with regard to jurisdictional claims in published maps and institutional affiliations.

1. Introduction

Over the last decade, the public interest in air pollution measurement has gradually increased, giving rise to high demand for air quality sensors [1,2]. There are various different approaches to classify the air quality based on the concentration measurement of air pollutants including either a specific or a mix of volatile organic compounds (VOC), total suspended particles (TSP), or relative humidity (RH) [3–6]. Moreover, the concentration of carbon dioxide (CO_2) has been shown to be one of the most applicable indicators for indoor air quality. CO_2 is a major indoor pollutant which, even at slightly elevated gas concentrations, leads to declining work performance and diminishing focus capacity in the human organism [7]. Monitoring CO_2 levels indoors is, therefore, essential to achieve an optimal balance between maximizing human performance and minimizing the need for energy-expensive ventilation of indoor spaces. In automotive environments, CO_2 can reach increased concentration levels with undesirable physiological effects in a very short period of time, due to exhaust gases entering the vehicle cabin or due to metabolically induced CO_2 emissions by the passengers. Typical adverse effects are fatigue, drowsiness, and lethargy which can lead to a higher risk of traffic accidents [8].

Although the CO_2 concentration limits inducing adverse effects in humans are not well defined, the standardly used value dates back to Max von Pettenkofer in 1858, who defined a hygienic limit of 1000 ppm CO_2 [9]. The consensus is that monitoring and keeping the CO_2 concentration as low as possible is beneficial for indoor air quality.

To date, various different gas sensing technologies are available to determine the CO_2 content in the air, including, but not limited to, non-dispersive infrared (NDIR) spectrometry, photoacoustic spectroscopy (PAS), quartz crystal microbalance (QCM) sensors, and several technologies based on chemical interactions of gas with a sensing material [10–15]. Most sensors currently available on the market for consumer electronics are based on either NDIR or PAS, as they are the only mature techniques with high selectivity to CO_2, which are still cheap and durable enough to be relevant. Both are based on the absorption of electromagnetic waves with a wavelength tailored to match the specific CO_2 absorption wavelengths.

NDIR sensors directly measure gas quantities by detecting slight changes in the transmittance of infrared light in an absorption path between a source and a detector. For selectivity towards a target analyte, optical filters are used. However, these filters used in NDIR sensors can only be designed for optical bandwidths rather than for the absorption spectrum of a molecule. Thus, the sensor may not only be selective towards its target gas, but also towards interfering gases, such as humidity. Furthermore, these filters are prone to temperature variations as the optical properties of such filters change with temperature. The change of intensity is described by the Beer-Lambert-Law and is strongly dependent on the concentration of absorbing molecules in the path and the traveled distance of the light. NDIR sensors typically use pyroelectric or bolometer infrared detectors, which naturally have a relatively low signal-to-noise ratio (SNR). In order to get significant signal levels, these sensors use long absorption paths achieved by multiple reflections inside the housing and narrow optical filtering. Although there were successful optimizations of reflector geometries for the design of compact and still optically efficient NDIR sensors, these reflectors are rather costly to manufacture [16].

PAS sensors, on the other hand, make use of the photoacoustic effect, discovered by Bell and Roentgen in 1881, and sense CO_2 indirectly by measuring sound [17,18]. These sensors detect gas concentrations by introducing periodically modulated light of a specific wavelength into a detection chamber in which a target gas absorbs precisely that wavelength. The molecules of the target gas get excited to higher energy levels by the absorption and generate thermal energy by colliding with molecules of a carrier gas. The thermal energy change in the closed chamber is translated into a pressure change, which is sensed by a microphone. The signal level of photoacoustic sensors correlates with the concentration of CO_2 inside the sensor. However, as these sensors use microphones to detect the photoacoustic signal, optimized gas port geometries need to be used in order to prevent acoustical interference such as banging doors or loud conversations. These gas inlets cause an increase in the response time of PAS CO_2 sensors [19,20].

Here, we investigate a new sensing concept which combines elements of both, NDIR and PAS sensors and includes a manufacturing approach for a detector module, which offers potential for miniaturized and very selective gas measurements. In contrast to NDIR detectors, the presented photoacoustic transducer is highly selective towards CO_2 as only its CO_2 filling absorbs infrared light. The detectors are simple to manufacture and therefore offer great potential for fast and low-cost production of miniaturized CO_2 sensors. In addition, the approach also resolves the issue of acoustic interference by concept, as the microphone is embedded within a closed cavity, without increasing the response time of the sensor.

2. Materials and Methods

2.1. Sensor Concept

The core of the presented concept is the development of a photoacoustic transducer chip that allows the spatial separation of the acoustic sensor from the photoacoustic cell. The transducer consists of a gas-tight and gas-filled cavity, acting as an optical absorption layer that transfers the irradiated infrared light energy to acoustical pulses and a thin, flexible membrane that can pass these pulses to a second cavity containing a microphone.

Figure 1 illustrates a conceptual sketch of the sensing mechanism. A MEMS hotplate emits infrared (IR) pulses of defined frequency and pulse length, which may be filtered by means of an optical filter with a transmission spectrum matching the characteristic absorption band of CO_2 at $\lambda = 4.26\,\mu m$. Although this filter is not necessarily needed, it can prevent wavelengths different from those of the CO_2 absorption spectrum to get absorbed in Silicon or glass parts of the transducer and thus increasing the performance. The pulses of IR light pass an optical absorption path, where they get partly absorbed by potentially present CO_2 molecules.

After the absorption path, the light pulses enter a hermetically encapsulated absorption cell through an optically transparent and rigid window. The absorption cell is filled with CO_2, which acts as a gas-specific absorption medium and therefore absorbs the IR light pulses with the characteristic absorption wavelengths of CO_2. The absorbed energy gets transformed into acoustic waves by means of the photoacoustic effect. The photoacoustic effect itself consists of three different steps: The radiative absorption by the gas molecules, the successive relaxation of the molecules while increasing the total thermal energy in the gas, and—by means of the ideal gas law in a closed system—the resulting emergence of a pressure signal [20]. As the IR light pulses are modulated with a defined frequency, the pressure signal also follows this frequency. The more energy is absorbed in the absorption path, the lower the resulting intensity of the acoustic waves generated in the cell is.

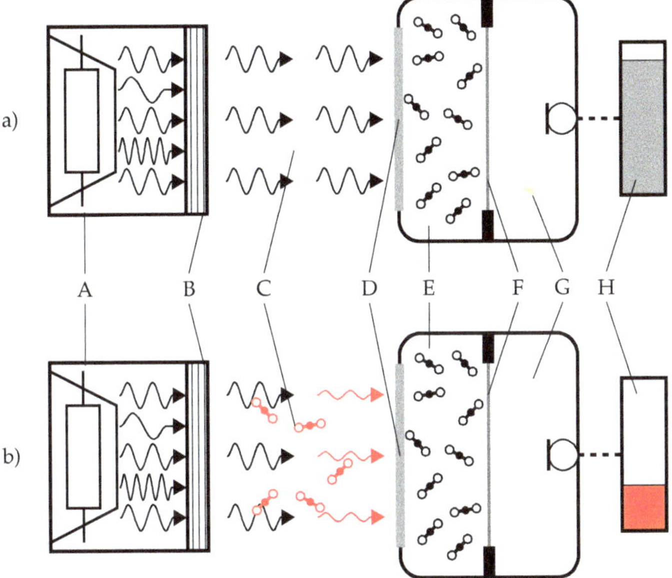

Figure 1. Conceptual sketch of the proposed sensing mechanism comprising: (A) pulsed IR-emitter (B) optical filter (optional) (C) optical absorption path (D) IR-transparent entrance window (E) hermetic cell with encapsulated CO_2 (F) flexible membrane (G) acoustically tight cavity with microphone (H) sensor signal. (**a**) sensor without CO_2 (**b**) sensor in presence of CO_2.

One wall of the absorption cell consists of a thin and flexible, but gas-tight membrane, which slightly deforms with the pressure fluctuations and therefore allows the generated pressure pulses to be transmitted to the other side while maintaining the CO_2 atmosphere inside the absorption cell. This way, the acoustic pulses can enter the second cavity, which needs to be only acoustically-tight, but not necessarily gas-tight. Inside this second cell, a MEMS microphone is placed, which detects the arriving acoustic pulses. The amplitude of the acoustic pulses is indirectly proportional to the number of CO_2 molecules present in the absorption path.

The concept of using a slightly deforming membrane or diaphragm for gas sensing was first described by Golay as part of an infrared active pneumatic detector [21]. In contrast to the concept presented here, the membrane was utilized as a deforming mirror in an optical modulation circuit rather than as an acoustic transducer.

2.2. Manufacturing

Figure 2 depicts a schematic cross-section of the photoacoustic transducer. Its center-piece is a thin, homogeneous glass membrane that has been wet-etched from a borosilicate glass wafer with a thickness of approximately 500 µm. A silicon nitride (SiN) hard mask was applied from the top and the bottom side to the glass wafer and was structured by lithography. At the openings of the hard mask, an anisotropic etch process symmetrically created a cavity on both sides of the wafer, leaving residual thin membranes with a diameter of approximately 3.6 mm. Defined by the duration of the etching, thicknesses between about 10 µm and 70 µm could be achieved. Within each membrane, the thickness tolerance was found to be ±1 µm. For investigation of the concept, we focused on samples with the thinnest membrane thickness, as they are the most flexible and have the highest mechanical compliance.

After etching, the glass wafer was anodically bonded to a blank silicon wafer, which was dry-polished in order to achieve the highest bond quality. Furthermore, for the purpose of maximizing the infrared light transmission, the used Si wafer was low doped and backside-grounded to a thickness of 250 µm. The anodic bond was formed at 300 °C while applying a CO_2 bias atmosphere of 2 bar. Thus, after cooling down to ambient temperature, a CO_2 filling of about 1 bar was captured in the gas-tight cavity between the glass membrane and silicon lid. After bonding, the wafer was mechanically sawed to form transducers of the size 5.5 mm by 5.5 mm.

Using Fourier transformation infrared (FTIR) spectroscopy, an absorbing behavior at the CO_2 absorption bands was confirmed as shown in the transmission spectrum (Figure A1). The characteristic shape of the CO_2 absorption at 4.26 µm was clearly visible with a relative absorption of about 15 % to 20 % compared to the background. The spectrum was limited at the lower end by the natural transparency of silicon and the capabilities of the used FTIR spectrometer and on the upper end by the borosilicate glass, which is nontransparent for light with wavelengths above 6.6 µm.

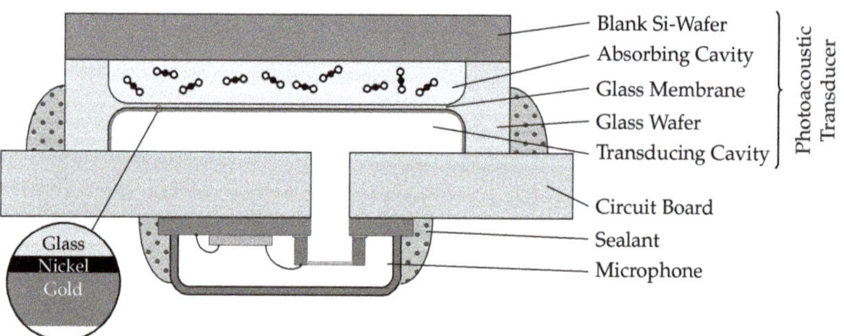

Figure 2. Proposed detector element comprising the photoacoustic transducer, a microphone, sealant compounds and a carrier PCB (not to scale).

The singularized transducers were then treated by means of physical evaporation and deposition to form a nickel-gold coating as a nontransparent, reflecting layer for the infrared light. In this way, it was ensured that no IR light could hit and potentially disturb the microphone membrane and, additionally, that reflection caused a second pass of the light through the absorption cavity, increasing the absorption inside.

2.3. Experimental Setup

Figure 2 also depicts a sketch of the whole photoacoustic detector unit used in this research. Below the photoacoustic transducer, which was described in Section 2.2, a printed circuit board (PCB) carrying a MEMS-microphone is shown. The used analog microphone (IM73A135, Infineon Technologies AG, Munich, Germany) was soldered on the backside of a PCB for electrical connection. The photoacoustic transducer was mounted on the top side of the PCB, whilst ensuring that its transducing cavity was connected to the microphone membrane via a small channel in the PCB.

As an infrared light source of the emitter unit, a small MEMS hotplate embedded in a $4 \times 4 \times 2.25$ mm^3 package was used. The top wall of the package consisted of an infrared filter, which was tuned to match the dominant CO_2 absorption band at its center wavelength of 4.26 μm. The IR emitter was soldered onto a second PCB which was mounted on the sensor evaluation system so that its surface was directly facing the photoacoustic transducer on the detector unit.

Figure 3 shows a photograph of both sides of the detector PCB carrying a MEMS microphone, photoacoustic transducer, connectors, and the transparent epoxy glue, which was used to attach the photoacoustic transducer and to acoustically seal the gaps between components and PCB.

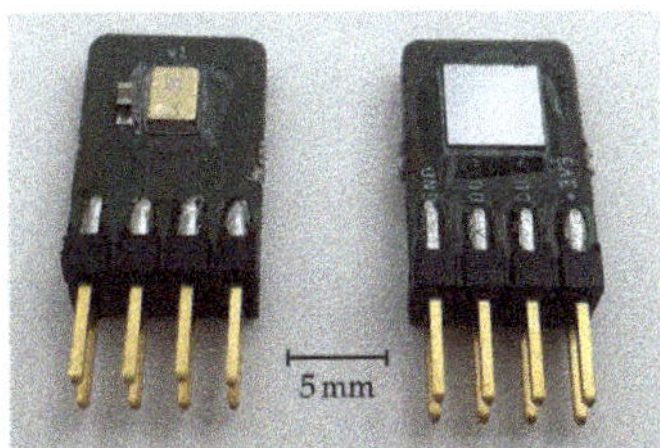

Figure 3. Detector module PCB with photoacoustic transducer (**right**, front side) and the MEMS microphone (**left**, back side).

The distance between the upper surface of the emitter package and the detector could be set between 3.0 mm to 15.5 mm in intervals of 2.5 mm. To maximize the optical transmission from emitter to detector, aluminum tubes with matching lengths and an inner diameter of 8 mm were used as reflectors in between the sensor elements. The sensing area of the measurement system is surrounded by a gasket, which, together with a 3D-printed box with inner dimensions of $22 \times 52 \times 35$ mm^3, forms an enclosed box connected to a gas measurement system where different CO_2 concentrations can be set. A photograph of the whole sensor assembly is shown in Figure 4.

The emitter was driven with a square signal by an n-channel metal-oxide-semiconductor field-effect transistor (MOSFET) with a frequency of 27 Hz for a duration of 1 s in every measurement period which lasted 5 s. The microphone output was a differential signal which was first pre-amplified and filtered by a second-order bandpass filter. It was configured to have a bandpass between 16 Hz and 210 Hz and an amplification gain of G = 9 (ADA4084-2, Analog Devices Inc., Wilmington, MA, USA). In a second amplification stage, the signal was amplified with an audio amplifier (SSM2019, Analog Devices Inc., Wilmington, MA, USA) with a gain of G = 370. The PCB was attached to a Digilent Analog Discovery 2 USB oscilloscope (Digilent, Pullman, WA, USA), which was used as a data acquisition system for controlling the sensor.

The system acquired the microphone waveform differentially with a resolution of 14 bit at a sampling frequency of 10 kS/s. After the acquisition, the output waveform of the microphone and its amplification circuitry was processed in MATLAB (MATLAB version R2020b) by means of a discrete Fourier transformation in order to extract the specific spectral components from the discrete signal as described in [22]. For this calculation, a rectangular window with a period of 950 ms was applied. As the emitter excitation was carried out with a square signal, the microphone output resulted in a non-sinusoidal signal with multiple frequency components (raw signal in Appendix A, Figure A2). For this reason, the raw output value of the sensor system was defined as not only the amplitude of the spectral component at the excitation frequency of 27 Hz, but also the sum of its first and second harmonics. Adding the harmonics to the 27 Hz component was beneficial for the calculation of the sensor response and resulted in both, higher signal response and less signal noise. For further reduction of the sensor noise, a centered moving mean filter with a sliding window width of 12, equal to 1 min of sampling, was applied to the sensor output in postprocessing.

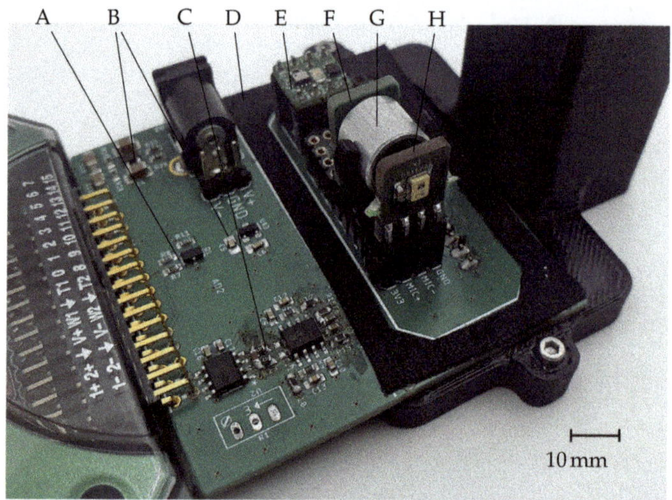

Figure 4. Sensor evaluation PCB with (A) MOSFET (B) power supply (C) amplifier stages (D) gasket (E) reference sensors (F) emitter module (G) reflector tube (H) detector module.

In addition to the actual sensor data acquisition, a secondary environmental data measurement system acquired information about relative humidity, temperature, pressure once in every measurement cycle. The CO_2 concentration was measured with an additional, external industrial-grade NDIR sensor (M1440, COMET SYSTEM, s.r.o., Rožnov pod Radhoštěm, Czech Republic) in the exhaust flow of the gas testing bench, which served as a reference signal.

3. Results and Discussion

Using the setup described in the previous section, we measured and analyzed the sensor sample at six different absorption lengths (3.0 mm to 15.5 mm in intervals of 2.5 mm) with decreasing CO_2 concentration steps ranging from about 9500 ppm down to 500 ppm. The measurement bench, which mixed CO_2 and dry, synthetic air (80% N_2 and 20 % O_2 ±2 %), allowed for a consistent gas flow of 850 cm^3 min^{-1} through the sensor enclosure. Each concentration step was maintained for 10 min, followed by a flush step of 0 ppm, allowing to evaluate the sensor's baseline after each concentration step. The temperature in the sensor enclosure during all measurements ranged from 31.8 °C to 34.3 °C with a mean temperature of 33.2 °C.

Figure 5 shows the absolute sensor signal with an absorption distance of 15.5 mm over a complete characterization procedure. Over the course of the measurement, the sensor signal repeatedly dropped in response to the application of the different CO_2 concentration steps. The figure also shows the output of the reference sensor, for which a lower noise level compared to our sensor could be observed. However, in this regard, it has to be noted, that despite our sensor being an early prototype aiming for low-cost production of an integrated MEMS solution, it yields almost similar results compared to the well-established NDIR reference sensor, which is a larger, fully developed product.

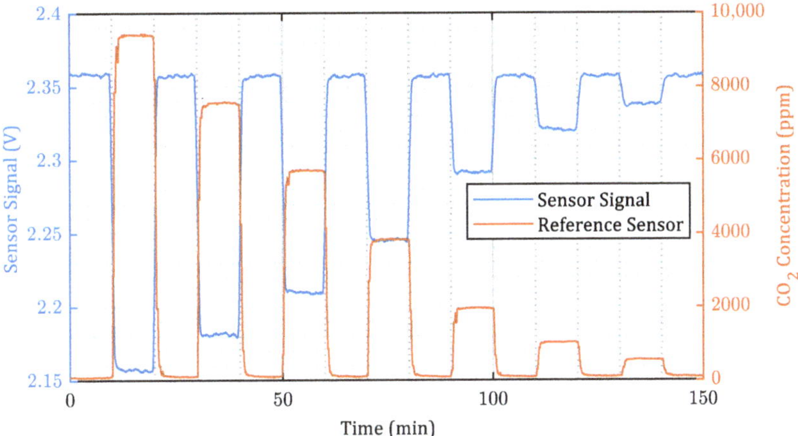

Figure 5. Measurement with an absorption distance of d = 15.5 mm showing 10 min steps with decreasing CO_2 concentration together with the output of a NDIR reference sensor (12-times moving mean filter applied).

The sensor prototype clearly responded with a lower signal to the increased number of CO_2 molecules inside the absorption path, as a portion of the infrared light intensity already was absorbed there rather than inside the pressure transducer. The signal level at the lowest CO_2 concentration applied (95 ppm) was found to be 2.37 V $\pm$ 0.70 mV. The signal response ranged from a decrease of about 197 mV (8.3%) at a concentration of 9400 ppm to a decrease of 18.8 mV (0.8%) at a concentration of 560 ppm. The sensitivity was calculated as signal response per decrease of 1000 ppm CO_2 and thus ranged from −21.5 mV/1000 ppm for 9500 ppm to −41.7 mV/1000 ppm for 500 ppm. The first 2 min before and after each concentration change were ignored in the data processing in order to ensure a stabilized CO_2 concentration.

We also characterized the same sensor prototype using six different absorption lengths. Figure 6 shows the relative signal response of all six sensor distances to seven different CO_2 concentrations. The signal of the sample with the shortest absorption length dropped between 2.9% at a concentration of 9500 ppm to 0.2% at a concentration of 570 ppm. As expected from the Beer-Lambert law, the measurements with the longest absorption distances resulted in the highest signal response and those with the shortest absorption path in the smallest signal response. The signals of the other sensor lengths are laid in between those two configurations. As already done for the data of Figure 5, all values have been calculated from signal regions with stable CO_2 concentrations and with a moving mean filter over 12 measurement periods applied. Although longer absorption paths provide higher sensitivity, they also result in larger setup sizes. For this reason, this study focused only on variants with an absorption path smaller or equal to 15.5 mm.

Figure 6 also indicates an increasing non-linearity in terms of the relative signal response with increasing sensor distance. We assume this is mostly based on the natural non-linearity of the Beer-Lambert law, if calculated with non-monochromatic light. With increasing absorption path length and CO_2 concentration, an increasing number of absorption lines in the CO_2 spectrum gets fully absorbed within the absorption path, leading to an amplification of the non-linear effect. A simulation with HITRAN for the given concentrations and sensor distances confirmed this theory [23,24]. However, the simulation did not take into account any reflections inside the aluminum tube, which multiply the effective path lengths. The results are shown in the Appendix A in Figure A3.

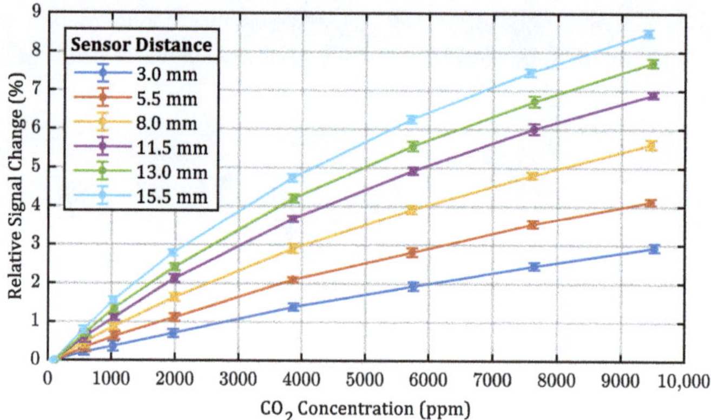

Figure 6. Signal response of six different absorption path distances ranging from 3.0 mm to 15.5 mm at different CO_2 concentrations, relative to the the lowest reachable concentration. Error bars represent standard deviation.

In order to get a meaningful sensor output, which can be compared to other CO_2 sensors, a basic quadratic regression curve was calculated from the measurements for the d = 15.5 mm sensor (Curve is depicted in the Appendix A, Figure A4). This calibration curve was then applied to the signal shown in Figure 5. The calibrated sensor response of this sample is depicted in Figure 7.

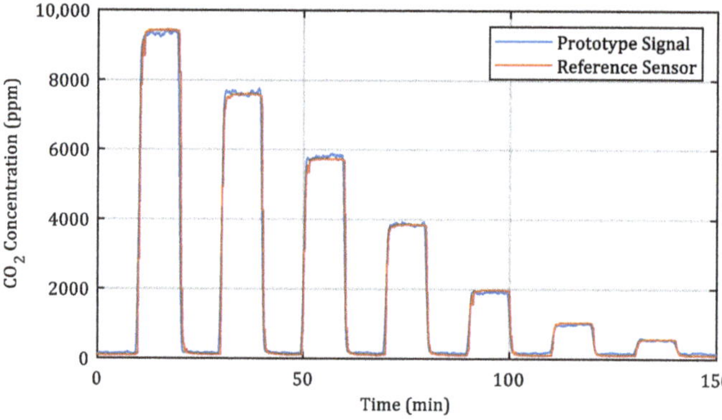

Figure 7. Calibrated sensor response absorption distance of d = 15.5 mm showing 10 min steps with decreasing CO_2 concentration together with the output of an NDIR reference sensor (12-times moving mean applied).

When applying the sensor's sensitivity to a commonly occurring CO_2 concentration of 1000 ppm (-39.8 mV/1000 ppm), the standard deviation of the 15.5 mm sample was found to be equivalent to 18 ppm.

4. Conclusions

We presented a novel approach for manufacturing wafer-level photoacoustic gas sensors and provided the first proof-of-concept measurement results. Basic characterization of the built sensor modules at various absorption distances showed promising sensitivity and noise levels even at short absorption path lengths.

The photoacoustic transducer can be manufactured using only materials and process steps, which are already standard in the semiconductor industry and are therefore reasonably priced in production. Moreover, as in this approach, the acoustic detector can be manufactured, tested, and handled separately from the photoacoustic cell. As a result, the exposition to harsh environments or the need for non-standard manufacturing processes is limited to a minimum. Earlier studies with pressure or sound transducers directly encapsulated in the gas-filled cell, either needed to use non-standard low-temperature encapsulation processes in gas bias atmospheres individually manufactured samples or harsh process environments [25–29].

The concept could be adapted for a wide range of gasses with absorption bands that overlap with the high-transmissivity region of silicon, which was used for the IR entrance window. The only changes needed for such an adaption would be the substitution of the gas inside the absorbing cavity with another target gas and the exchange of the IR source or filter to form a matching pair of emitter and detector. Possible target gas candidates could include Methane or fluorine-based refrigerants [28,30]. Further experiments on the photoacoustic transducers themselves could benefit from a variation of the inner pressure, as this would allow a higher relative sensor response [30].

To our knowledge, this is the first successful approach that combines elements of photoacoustic and NDIR sensing enabling low-cost and well-performing IR detectors for CO_2 sensors.

Author Contributions: Conceptualization, R.S., S.G. and M.E.; methodology, S.G. and M.E.; software, S.G. and S.E.; validation, S.G. and M.G.; formal analysis, S.G.; investigation, S.G., M.E. and R.S.; resources, S.G., C.v.K. and R.S.; data curation, S.G. and C.v.K.; writing—original draft preparation, S.G., R.S. and S.E.; writing—review and editing, S.G. and K.S.; visualization, S.G.; supervision, J.W.; project administration, R.S. and M.E.; funding acquisition, S.G. All authors have read and agreed to the published version of the manuscript.

Funding: The article processing charge was funded by the Baden-Württemberg Ministry of Science, Research and Art and the University of Freiburg in the funding program Open Access Publishing.

Institutional Review Board Statement: Not applicable.

Informed Consent Statement: Not applicable.

Data Availability Statement: Not applicable.

Conflicts of Interest: The authors declare no conflict of interest.

Appendix A

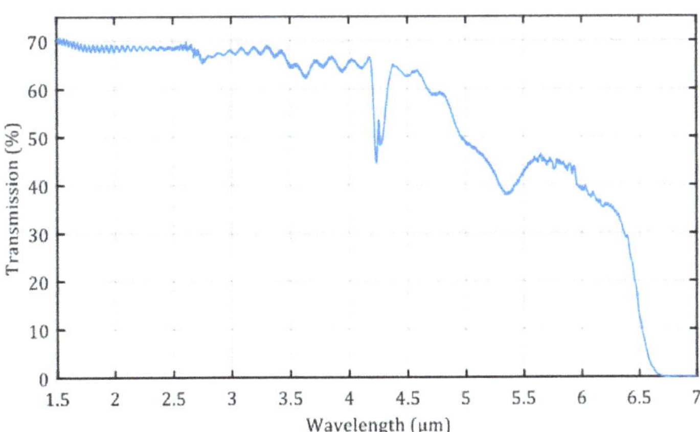

Figure A1. Background-compensated infrared transmission spectrum of the photoacoustic transducer. A characteristic dip at 4.26 μm indicates the presence of a high concentration of CO_2 inside the cavity of the device.

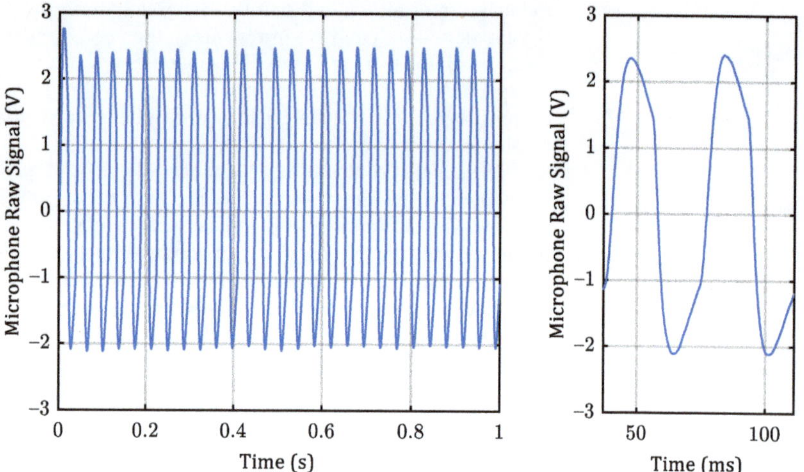

Figure A2. Raw microphone output signal over 1 s of sampling (**left**) and two periods (74 ms) of the signal (**right**).

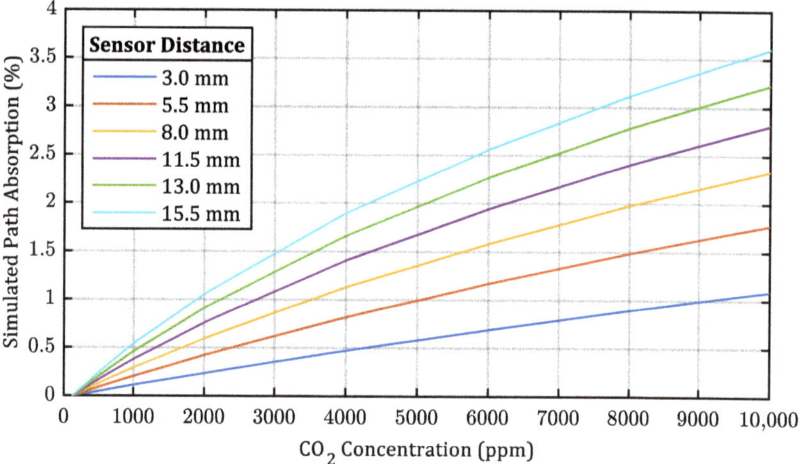

Figure A3. Additional absorption, simulated with HITRAN, for applied CO_2 concentrations in the sensor path, normalized to the absorption of each absorption path at the measurements' lowest CO_2 concentration. Only the direct distance was calculated. Reflections inside the reflector tube were neglected.

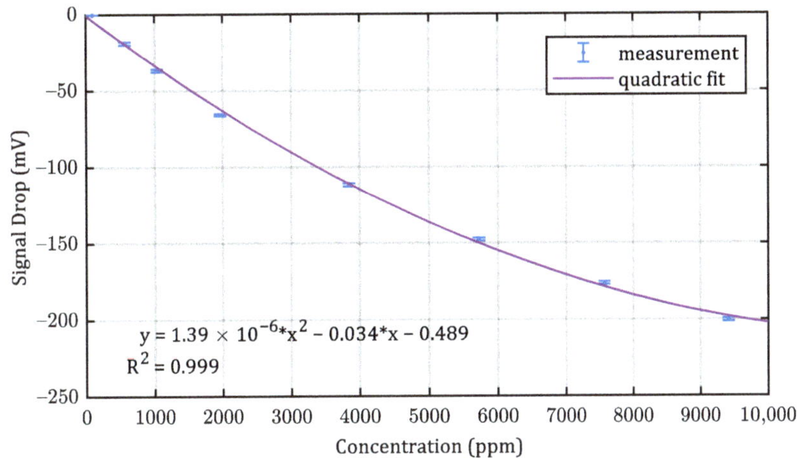

Figure A4. Absolute signal response and quadratic regression curve of d = 15.5 mm sample to applied CO_2 concentrations. Error bars represent standard deviation.

References

1. Kirby, B.P.; Lewis, S.P.; Lewis, P.D. Do You Think Air? Public interest in Air Pollution. *Ergon. Hum. Factors* **2021**, 8.
2. Gas Sensor Market Size, Share & Trends Analysis Report By Product (Oxygen/Lambda Sensor, Carbon Dioxide Sensor), By Type (Wired, Wireless), By Technology, By End Use, By Region, And Segment Forecasts, 2021–2028. Available online: https://www.grandviewresearch.com/industry-analysis/gas-sensors-market (accessed on 28 December 2021).
3. Paleologos, K.E.; Selim, M.Y.; Mohamed, A.M.O. Indoor air quality. In *Pollution Assessment for Sustainable Practices in Applied Sciences and Engineering*; Butterworth-Heinemann: Oxford, UK, 2021; pp. 405–489. [CrossRef]
4. Fang, L.; Clausen, G.; Fanger, P.O. Impact of Temperature and Humidity on the Perception of Indoor Air Quality. *Indoor Air* **1998**, *8*, 80–90. [CrossRef]
5. Wolkoff, P. Indoor air humidity, air quality, and health—An overview. *Int. J. Hyg. Environ. Health* **2018**, *221*, 376–390. [CrossRef]
6. Zhou, K.; Ye, Y.H.; Liu, Q.; Liu, A.J.; Peng, S.L. Evaluation of ambient air quality in Guangzhou, China. *J. Environ. Sci.* **2007**, *19*, 432–437. [CrossRef]
7. Satish, U.; Mendell, M.J.; Shekhar, K.; Hotchi, T.; Sullivan, D.; Streufert, S.; Fisk, W.J. Is CO_2 an indoor pollutant? Direct effects of low-to-moderate CO_2 concentrations on human decision-making performance. *Environ. Health Perspect.* **2012**, *120*, 1671–1677. [CrossRef]
8. Chen, R.Y.; Ho, K.F.; Chang, T.Y.; Hong, G.B.; Liu, C.W.; Chuang, K.J. In-vehicle carbon dioxide and adverse effects: An air filtration-based intervention study. *Sci. Total Environ.* **2020**, *723*, 138047. [CrossRef] [PubMed]
9. von Pettenkofer, M. *Über den Luftwechsel in Wohngebäuden*; Cotta'sche Buchhandlung: Munich, Germany, 1859.
10. Korotcenkov, G.; Cho, B.K. Instability of metal oxide-based conductometric gas sensors and approaches to stability improvement (short survey). *Sens. Actuators B Chem.* **2011**, *156*, 527–538. [CrossRef]
11. Haensch, A.; Barsan, N.; Weimar, U. Chemoresistive CO_2-Sensoren basierend auf Seltenerdoxycarbonat-beladenem Zinndioxid. In *Automobil-Sensorik*; Tille, T., Ed.; Springer: Berlin/Heidelberg, Germany, 2016; pp. 65–78. [CrossRef]
12. Ersoez, B.; Bauersfeld, M.L.; Wöllenstein, J. Ionogel—Based Composite Material for CO_2 Sensing Deposited on a Chemiresistive Transducer. *Proceedings* **2017**, *1*, 314. [CrossRef]
13. Ersöz, B.; Schmitt, K.; Wöllenstein, J. Electrolyte-gated transistor for CO_2 gas detection at room temperature. *Sens. Actuators B Chem.* **2020**, *317*, 128201. [CrossRef]
14. Gomes, M.; Duarte, A.C.; Oliveira, J.P. Detection of CO_2 using a qaurtz crystal microbalance. *Sens. Actuators B Chem.* **1995**, *26*, 191–194. [CrossRef]
15. Muraoka, S.; Kiyohara, Y.; Oue, H.; Higashimoto, S. A CO_2 Sensor Using a Quartz Crystal Microbalance Coated with a Sensitive Membrane. *Electron. Commun. Jpn.* **2014**, *97*, 60–66. [CrossRef]
16. Hodgkinson, J.; Smith, R.; Ho, W.O.; Saffell, J.R.; Tatam, R.P. Non-dispersive infra-red (NDIR) measurement of carbon dioxide at 4.2 µm in a compact and optically efficient sensor. *Sens. Actuators B Chem.* **2013**, *186*, 580–588. [CrossRef]
17. Bell, A.G. LXVIII. Upon the production of sound by radiant energy. *Lond. Edinb. Dublin Philos. Mag. J. Sci.* **1881**, *11*, 510–528. [CrossRef]
18. Röntgen, W.C. On tones produced by the intermittent irradiation of a gas. *Lond. Edinb. Dublin Philos. Mag. J. Sci.* **1881**, *11*, 308–311. [CrossRef]

19. Huber, J.; Wöllenstein, J. Kompaktes photoakustisches Gasmesssystem mit Potential zur weiteren Miniaturisierung. *tm-Tech. Mess.* **2013**, *80*, 448–453. [CrossRef]
20. Kühnemann, F. Photoacoustic Detection of CO_2: 8. In *Carbon Dioxide Sensing*; Gerlach, G., Ed.; Wiley-VCH Verlag GmbH & Co. KGaA: Weinheim, Germany, 2019; pp. 191–213. [CrossRef]
21. Golay, M.J.E. Theoretical consideration in heat and infra-red detection, with particular reference to the pneumatic detector. *Rev. Sci. Instrum.* **1947**, *18*, 347–356. [CrossRef] [PubMed]
22. Goertzel, G. An Algorithm for the Evaluation of Finite Trigonometric Series. *Am. Math. Mon.* **1958**, *65*, 34–35. [CrossRef]
23. Kochanov, R.V.; Gordon, I.E.; Rothman, L.S.; Wcisło, P.; Hill, C.; Wilzewski, J.S. HITRAN Application Programming Interface (HAPI): A comprehensive approach to working with spectroscopic data. *J. Quant. Spectrosc. Radiat. Transf.* **2016**, *177*, 15–30. [CrossRef]
24. Gordon, I.E.; Rothman, L.S.; Hargreaves, R.J.; Hashemi, R.; Karlovets, E.V.; Skinner, F.M.; Conway, E.K.; Hill, C.; Kochanov, R.V.; Tan, Y.; et al. The HITRAN2020 molecular spectroscopic database. *J. Quant. Spectrosc. Radiat. Transf.* **2022**, *277*, 107949. [CrossRef]
25. Eberl, M.; Jost, F.; Kolb, S.; Schaller, R.; Dettmann, D.; Gassner, S.; Skorupa, F. Miniaturized photoacoustic CO_2 gas sensors—A new approach for the automotive sector. In *AmE 2019—Automotive Meets Electronics, Proceedings of the 10th GMM-Symposium, Dortmund, Germany, 12–13 March 2019*; VDE Verlag: Berlin, Germany, 2019; pp. 1–5.
26. Huber, J.; Weber, C.; Eberhardt, A.; Wöllenstein, J. Photoacoustic CO_2-Sensor for Automotive Applications. *Procedia Eng.* **2016**, *168*, 3–6. [CrossRef]
27. Huber, J.; Enriquez, J.A.; Escobar, A.; Kolb, S.; Dehé, A.; Jost, F.; Wöllenstein, J. Photoakustischer Low-Cost CO_2-Sensor für Automobilanwendungen. In *Automobil-Sensorik*; Tille, T., Ed.; Springer: Berlin/Heidelberg, Germany, 2016; pp. 79–96. [CrossRef]
28. Weber, C.; El-Safoury, M.; Eberhardt, A.; Schmitt, K.; Wöllenstein, J. *4.2.2 Miniaturisiere Photoakustische Detektoren für den Nachweis Fluorhaltiger Kältemittel*; Tagungsband; AMA Service GmbH: Wunstorf, Germany, 2019; pp. 322–327. [CrossRef]
29. Schjølberg-Henriksen, K.; Wang, D.T.; Rogne, H.; Ferber, A.; Vogl, A.; Moe, S.; Bernstein, R.; Lapadatu, D.; Sandven, K.; Brida, S. High-resolution pressure sensor for photo acoustic gas detection. *Sens. Actuators A Phys.* **2006**, *132*, 207–213. [CrossRef]
30. Wittstock, V.; Scholz, L.; Bierer, B.; Perez, A.O.; Wöllenstein, J.; Palzer, S. Design of a LED-based sensor for monitoring the lower explosion limit of methane. *Sens. Actuators B Chem.* **2017**, *247*, 930–939. [CrossRef]

 *sensors*

Article

A Compact Fiber-Coupled NIR/MIR Laser Absorption Instrument for the Simultaneous Measurement of Gas-Phase Temperature and CO, CO_2, and H_2O Concentration

Lin Shi [1], Torsten Endres [1,*], Jay B. Jeffries [1,2], Thomas Dreier [1] and Christof Schulz [1]

[1] IVG, Institute for Combustion and Gas Dynamics—Reactive Fluids, University of Duisburg-Essen, 47057 Duisburg, Germany; lin.shi@uni-due.de (L.S.); jay@jeffries.org (J.B.J.); thomas.dreier@uni-due.de (T.D.); christof.schulz@uni-due.de (C.S.)

[2] High Temperature Gasdynamics Laboratory, Stanford University, Stanford, CA 94305, USA

* Correspondence: torsten.endres@uni-due.de

Citation: Shi, L.; Endres, T.; Jeffries, J.B.; Dreier, T.; Schulz, C. A Compact Fiber-Coupled NIR/MIR Laser Absorption Instrument for the Simultaneous Measurement of Gas-Phase Temperature and CO, CO_2, and H_2O Concentration. *Sensors* **2022**, *22*, 1286. https://doi.org/10.3390/s22031286

Academic Editors: Krzysztof M. Abramski and Piotr Jaworski

Received: 27 December 2021
Accepted: 4 February 2022
Published: 8 February 2022

Publisher's Note: MDPI stays neutral with regard to jurisdictional claims in published maps and institutional affiliations.

Abstract: A fiber-coupled, compact, remotely operated laser absorption instrument is developed for CO, CO_2, and H_2O measurements in reactive flows at the elevated temperatures and pressures expected in gas turbine combustor test rigs with target pressures from 1–25 bar and temperatures of up to 2000 K. The optical engineering for solutions of the significant challenges from the ambient acoustic noise (~120 dB) and ambient test rig temperatures (60 °C) are discussed in detail. The sensor delivers wavelength-multiplexed light in a single optical fiber from a set of solid-state lasers ranging from diodes in the near-infrared (~1300 nm) to quantum cascade lasers in the mid-infrared (~4900 nm). Wavelength-multiplexing systems using a single optical fiber have not previously spanned such a wide range of laser wavelengths. Gas temperature is inferred from the ratio of two water vapor transitions. Here, the design of the sensor, the optical engineering required for simultaneous fiber delivery of a wide range of laser wavelengths on a single optical line-of-sight, the engineering required for sensor survival in the harsh ambient environment, and laboratory testing of sensor performance in the exhaust gas of a flat flame burner are presented.

Keywords: combined NIR/MIR laser absorption; laser multiplexing in a mid-IR single-mode fiber; simultaneous multispecies (CO, CO_2, H_2O) in situ measurements

1. Introduction

Laser absorption measurements provide reliable characterization of high-temperature reactive flows such as practical combustion processes [1]. In addition to measurements of the concentration of specific target species, numerous schemes for the determination of temperature, pressure, and gas-phase velocity have been reported, and successful results are well-documented in a variety of review articles [1–5]. The work cited in these reviews shows that the development of solid-state semiconductor diode and quantum-cascade lasers has enabled a wide variety of small portable instruments applicable to numerous practical combustion applications, including a wide range of research and industrial facilities. Use of such instruments in existing industrial facilities and large-scale industrial research test rigs can present significant engineering challenges for the survivability of the sensor in the quest for low-noise laser absorption measurements [1,4].

The purpose of the gas sensor designed here is to enable in situ measurements of the post-combustion gases (CO, CO_2, H_2O) as well as temperature in a full-size high-pressure gas turbine burner test rig. The test facility offers a quite harsh environment both inside and outside the combustor. Inside, the combustion chamber is located inside a high-pressure vessel (up to 25 bar) with inlet hot compressed air (up to 770 K) similar to the intake of an aircraft gas turbine combustor can. In the combustion chamber, the temperature is even higher (2000 K) due to the combustion heat release. These conditions challenge sensitive species-selective laser absorption measurements, as the high-temperature distributes the population for any gas species over a very wide range of quantum states and the high-pressure broadens each absorption transition, resulting in a complex spectrum with few (if any) isolated lines. Outside the test facility, the walls of the test rig are quite hot (>150 °C) due to the flames near the combustor walls, which challenges the design of optical access windows and components near the combustor housing. The hot walls of the test rig produce an ambient temperature near 60 °C. Even more challenging is the extreme acoustic noise (up to 120 dB) surrounding the test rig. This acoustic noise exceeds specification of common electronic equipment, produces strong vibrations, and requires remote operation of the equipment. Thus, the thermal and optical engineering solutions for sensitive, species-selective laser absorption measurements that overcome the challenges of the harsh ambient and test environments are an important part of the research reported here.

The design of the sensor reported here builds on a vast library of previous literature. In the past 30 years, there has been an enormous amount of research developing absorption spectroscopy for practical gas-sensing applications [1–5], and the subsequent research literature is exploited to develop the sensor reported here. A complete review of the relevant research is beyond the scope of this introduction, and only a few selected research citations important for the new sensor design are highlighted here.

1.1. Selected Background Literature for Absorption Sensing of CO, CO_2, and H_2O

The absorption of water vapor in the $2\nu_1$ and $2\nu_3$ vibrational overtone bands and $\nu_1 + \nu_3$ combination band overlaps in wavelength with the emission range of mature diode lasers developed in the near-infrared (NIR) for telecommunication. Water vapor, a primary product of hydrocarbon combustion, was a natural target of the early diode laser sensor development for combustion, and there is a rich literature concerning it. More than 20 years ago, Allen's review [6] highlighted numerous combustion applications of water vapor sensing. In the work reported below, two frequently used and well-studied NIR transitions near 7185.59 cm^{-1} (1391.7 nm) and 6806.03 cm^{-1} (1469.3 nm) are exploited. More recently, Goldenstein et al. demonstrated high-temperature sensitivity using this line pair under high-pressure and high-temperature conditions realized in shock-heated H_2O/N_2 mixtures [7], and that work provides the fundamental spectroscopy needed to design the sensor reported here. In addition, the ratio of absorption of water vapor using this pair of laser wavelengths was used to monitor temperature in a rotating detonation engine, illustrating its suitability for harsh environments [8]. The lower-state energy E'' of this pair of transitions differs by 2246 cm^{-1}, and thus the ratio of absorption is sensitive to temperature in the 700 to 2400 K range needed for the gas turbine combustion sensing.

CO is an intermediate species, as hydrocarbons are oxidized on the way to becoming the primary product CO_2. Thus, the CO/CO_2 ratio is an important indicator for the complete combustion of hydrocarbon fuel. This ratio can increase dramatically if temperature and combustion time is not sufficient for full oxidation in specific operating scenarios of practical devices leading to unwanted toxic emissions of CO. Therefore, measurements of the CO/CO_2 ratio are invaluable to the designer of new combustor concepts. The second overtone vibrational band of CO (near 1600 nm) also overlaps with the telecommunications wavelength lasers, the first overtone vibrational band (near 2300 nm) overlaps with extended NIR diode lasers, and research on sensors using transitions from both of these bands appear in the literature (e.g., Wagner et al. [9] used the R20 line of CO near 2313 nm to measure spatially resolved CO profiles in atmospheric laminar counter-flow diffusion flame). Unfortunately, these transitions are relatively weak, and the concentration of the intermediate combustion product CO is often much lower than the primary product species CO_2. Thus, the much stronger transitions in the fundamental vibrational band near 4.6 µm are used here for CO detection. Wavelength-tunable solid-state lasers in the mid-infrared (MIR) with a tuning range that overlaps the CO fundamental band transitions have become available with the development of quantum cascade lasers [10]. CO detection is also complicated by the strong overlap of the fundamental band transitions of CO with those of CO_2. In the sensor described below, wavelength ranges with different amounts of CO_2 interference are used to infer the CO absorption.

These laser wavelengths have been used previously to detect CO, e.g., by Spearrin et al. [11], who characterized a QC laser for CO detection in the harsh environment of a pulsed detonation combustor at 4.85 µm (the P(20) transition at 2059.91 cm^{-1}) using wavelength-modulation spectroscopy (WMS) with the normalized second-harmonic signal detection (WMS-2f) technique, rendering the measured signal independent of the incoming laser intensity. The encountered gas pressures varying between around 6 and more than 20 bar. With the same technique, the same research group developed a NIR-absorption instrument for the detection of carbon dioxide (CO_2) with an isolated transition (R(26)) in the $\nu_1 + \nu_3$ band at 3733.48 cm^{-1} (2.68 µm), with the laser first characterized in shock-tube experiments with pressures between 3 and 12 bar and a large temperature range between 1000 and 2600 K [12]. The instrument was then utilized for CO_2 detection in a pulse detonation engine (PDE) using ZrO_2 optical fibers for beam delivery.

Nwaboh et al. [13] used a quantum-cascade laser operated in the intrapulse mode to probe the P(1) line of CO at 2139.4 cm^{-1} (4.67 µm) in gravimetrically prepared gas mixtures to perform an uncertainty analysis for direct absorption spectroscopy-based mole fraction measurements of CO. A reproducibility of 1% and an uncertainty of 4% were demonstrated.

Vanderover et al. [14] used a single MIR QCL covering four selected rovibrational transitions of CO (R(9), R(10), R(17), and R(18) at 2179.77, 2183.22, 2179.24, and 2182.36 cm^{-1}, respectively) in the fundamental vibrational band for temperature and concentration measurements from peak absorbance ratios and peak absorbance, respectively. The fractional temperature sensitivity and specific detectivity were demonstrated for gas mixtures prepared in a gas cell (298 K) and a shock tube (1000–3600 K).

Previously, a preliminary version of the sensor used a single quantum-cascade laser (QCL) near 2190.02 cm^{-1} (4566 nm) to detect CO in a full-scale gas turbine burner test rig [15,16]. The CO concentrations varied considerably after ignition, and the laser absorption measurements were in qualitative agreement with the temporal behavior observed by a sampling probe mounted in the free-space exhaust flow just downstream of the optical access port.

1.2. Simultaneous Detection of Multiple Species at Atmospheric Pressure

A wavelength-multiplexing scheme that combines multiple lasers onto a single path through the test gases and then separates the lasers by wavelength was exploited very early in the development of laser absorption sensing of combustion gases in the harsh environment of a pulse detonation engine [17]. Diffraction gratings were used to combine multiple lasers onto a common measurement line-of-sight and to demultiplex the wavelengths for individual detection. This approach enables a continuous and simultaneous acquisition of multiple wavelengths. However, this early sensor relied on negligible interference absorption in the ambient air in the optical paths between sensor and engine and between engine and detectors. For the sensor designed here, the light from four lasers (wavelengths spanning from 7185.59 cm^{-1} (1.3917 µm) in the NIR to 2059.91 cm^{-1} (4.8546 µm) in the MIR) is multiplexed onto a common beam path and then coupled into an optical fiber (single mode in the MIR) in a nitrogen-purged enclosure, which allows the light to be delivered to the combustion rig without background absorption of the hot, humid ambient air surrounding the test facility. To our knowledge, wavelength-multiplexing systems using fiber delivery have not previously spanned such a wide range (NIR–MIR) of laser wavelengths (although relevant work is described below using a hollow-core fiber bundle as an alternative solution).

Other lasers are also available in the MIR; for example, using DFB and interband cascade (IC) lasers (not operated simultaneously), Wei et al. [18] were able to measure temperature, CO concentration (via two CO transitions at 2008.53 and 2006.78 cm^{-1}, respectively), and CO_2 concentration (via probing a CO_2 transition at 2384.19 cm^{-1}) in an atmospheric-pressure pilot-stabilized C_2H_4/air jet flame. Profiles of these scalars were obtained by translating the sender/receiver units as a whole—and thus the analysis beam—horizontally and vertically through the fixed burner flame.

Nau et al. [19] also used simultaneously a DFB (2.3 μm) and an IC laser (3.1 μm) in a fiber-based absorption spectrometer to measure temperature and concentrations of CO, CH_4, C_2H_2, and H_2O in an entrained flow gasifier. A wavelength division multiplexing approach using a single ZrF_4 fiber was used to measure both wavelength regions simultaneously. Weng et al. [20] recently demonstrated simultaneous detection of multiple species (i.e., hydrogen cyanide (HCN), acetylene, and water) by scanning the water vapor spectra using a single NIR laser (1.5 μm). Highly detailed water spectra modelling based on the measured absorption spectra of H_2O at different temperatures allowed quantification of the small overtone and combinations bands of HCN and acetylene in the same spectral region.

1.3. Simultaneous Detection of Multiple Species at Pressures Higher Than Atmospheric

Open literature presentations of solid-state-laser-based simultaneous measurement of multiple species and temperature in the harsh environment of turbulent combustion at high temperature and pressure are scarce. Spearrin et al. [21] recorded two of the major hydrocarbon combustion species, CO and CO_2 in a scram-jet combustor using both direct absorption and wavelength modulation spectroscopy with a temporal resolution 200 μs (for DA) or 5 ms (for WMS). Through the selection of suitable temperature-sensitive transitions in the MIR at 2059.91 cm^{-1} (4.854 μm) and 2394.6 cm^{-1} (4.176 μm) for CO and CO_2, respectively, they extended the temperature-sensitive detection range to 800–2400 K in this high-velocity hydrocarbon combustion flow. The laser beams for both CO and CO_2 detection were coupled into a bifurcated hollow-core optical fiber for transfer to the engine site, where the collimated and collinear beams traversed the optically accessible test section before being split and focused on two separate detectors. Mounting the lasers on vibration-insulated breadboards and the detectors to water-cooled plates and purging all beam paths with N_2 (including the hollow-core fibers) minimized disturbances from the harsh environment of the flow facility.

In the work of Peng et al. [22], a diode laser absorption instrument was developed combining four diode lasers at NIR and MIR wavelengths for the measurement of water vapor temperature (3920.1 and 4029.5 cm^{-1}), CO (P(21) at 2055.4 cm^{-1}), and CO_2 (R(92) at 2394.4 cm^{-1}) in a single-ended beam configuration. The beams were directed through the combustion flow channel, while the back-scattered light from the opposite wall was captured by a receiving hollow-core fiber. To enable the beams of all four diode lasers to pass collinearly through the reactive flow, they were fiber-coupled in free space onto a 4-to-1 multimode hollow-core silver-coated glass fiber bundle.

Cassady et al. [23] developed a very compact and shielded MIR-absorption instrument for the simultaneous measurement of water vapor temperature (via a low- and high-E'' transition of 3920.06 cm^{-1} (2.551 m) and 4029.59 cm^{-1} (2.482 μm), respectively) and H_2O, CO (2059.92 cm^{-1} (4.854 μm)), and CO_2 (2390.52 cm^{-1} (4.175 μm)) concentrations with a high sampling rate of 44,000 samples pre second from the reactive flow in the annular gap of a rotating detonation engine (RDE, operating pressure 2–8 bar) using four co-aligned MIR laser beams and WMS. They were able to perform time-resolved measurements while running the engine at lean to stoichiometric equivalence ratios.

1.4. Aim of This Work

The goal of the present work is to develop a laser-absorption sensor for simultaneous CO, CO_2, and H_2O measurements in the exhaust of a gas turbine combustor test rig. The instrument is an extension of a single-wavelength transmitter/receiver sensor used for preliminary measurement of CO concentrations in the exhaust flow of a single full-scale gas turbine burner located at the Siemens Energy test center in Ludwigsfelde (Germany) [15,16]. Challenges include the high pressure and temperature inside the combustion zone and the high temperature and acoustic noise in the ambient humid gases near the test rig. The high temperature and pressure in the combustor exhaust produce a complex absorption spectrum with a paucity of isolated lines, where most transitions are strongly overlapped by pressure broadening. The overlap of the fundamental absorption bands of CO and CO_2 exacerbates this problem. The acoustic noise (up to 120 dB) exceeds the operating limits of typical electronic devices, and the sensor must be packaged to protect electronics and optics. The high ambient temperature (up to 60 °C) also exceeds typical electronics and optical equipment specifications, and the sensor package is thus assembled on water-cooled breadboards. The packaged sensor must be small enough for easy transport with installation by two operators with only minor adjustment for optical path alignment at the test site.

While our previous work [15,16] focused on the challenging measurement environment associated with gas turbine combustor testing and solved issues such as optical access and thermal and acoustic isolation of the measurement equipment, the focus in this work is on enhancing the measurement capabilities of the sensor. In the present work, a combined NIR/MIR laser-based absorption spectrometer is described for the simultaneous measurement of line-of-sight-averaged gas-phase CO, CO_2, and H_2O in combustion exhaust, and temperature is inferred from the ratio of two H_2O absorption transitions. The sender unit has a small footprint due to separately placing electronics and optical parts on two stacked temperature-controlled boards and mounting inside a N_2-purged box to avoid background water vapor absorption. The instrument uses two MIR lasers near 2059.91 cm^{-1} (4.8546 µm, centered around the P(20) CO transition) and 2190.02 cm^{-1} (4.5562 µm, centered around the R(12) CO transition), previously explored by Spearrin et al. [11]. Simultaneously, two NIR diode lasers probe two water vapor transitions (7185.59 and 6806.03 cm^{-1}) with differing temperature sensitivities previously used by Goldenstein et al. [7,8]. The beam delivery can be either free-space or via optical fibers compatible with the various wavelengths emitted by the installed laser sources. The receiver optics and detectors are also mounted in a metal box purged with an expanding flow of dry nitrogen for keeping the interior cool and free of water vapor.

1.5. Organization of This Paper

This manuscript is organized as follows: After a short introduction of the spectroscopic basics of laser absorption spectroscopy, a discussion of some specific details such as line selection, and the used fitting approach, a detailed outline is given of the experimental setup, including sending, auxiliary, receiving, free-beam guiding unit, and burner. Finally, the initial demonstration of the setup is described by presenting measurements on a standard McKenna burner to validate temperature and species-concentration values (CO, CO_2, and H_2O) for literature-known operating conditions.

2. Theoretical Background

2.1. Absorption Measurements in High-Temperature and -Pressure Environments

High temperature spreads the population of the target species into a large ensemble of quantum states, and this diluted population means less absorption for any specific optical transition. At the same time, high pressure broadens optical transitions and blends the individual transitions so that an absorption measurement with a narrowband laser often includes contributions from multiple transitions. Thus, the high-temperature, high-pressure measurement conditions require understanding of the temperature- and pressure-dependent absorption spectrum of the target species and the potential interference from other components of the gas along the optical path. For the sensor design and data analysis, we rely on modeling of the absorption spectrum using the HITRAN/HITEMP database [24,25] augmented with prior laboratory work on the specific laser wavelengths used for detection (see more details and citations in the Introduction section and below).

2.2. Basics of Laser Absorption

The theory of direct absorption spectroscopy is well-understood and is only briefly reviewed to clarify the notation and units [26,27]. For a more detailed introduction, we refer to textbooks on the subject [28,29]. Measurements are based on the Beer–Lambert law, which describes the relation between incident laser light intensity I_0 and transmitted laser light intensity I_t at wavenumber $\tilde{\nu}$ as it passes through a gas medium on a path length L as:

$$\left(\frac{I_t}{I_0}\right) = e^{-\alpha_{\tilde{\nu}}} = e^{-k_{\tilde{\nu}}L}, \tag{1}$$

where the spectral absorbance $\alpha_{\tilde{\nu}}$ is the product of L and the spectral absorption coefficient $k_{\tilde{\nu}}$ (cm^{-1}), which is defined according to:

$$k_{\tilde{\nu}} = p\, x_i\, S(T,\tilde{\nu}_0)\, \phi_{\tilde{\nu}} \tag{2}$$

where p (bar) is the pressure, x_i the mole fraction of the absorbing species i, $S(T,\tilde{\nu}_0)$ (cm^{-2} bar^{-1}) the line strength dependent on temperature T and line-center wavenumber $\tilde{\nu}_0$ (cm^{-1}), and $\phi_{\tilde{\nu}}$ (cm) the line-shape function, which is normalized with $\int_{-\infty}^{+\infty} \phi_{\tilde{\nu}}d\tilde{\nu} \equiv 1$.

For a single transition, the absorbance can be integrated as:

$$A_i = \int_{-\infty}^{+\infty} \alpha_{\tilde{\nu}}\, d\tilde{\nu} = p\, x_i\, S_i(T)\, L. \tag{3}$$

The line strength is given by:

$$S_i(T) = S_i(T_0)\frac{Q(T_0)}{Q(T)}\left(\frac{T_0}{T}\right)\exp\left(-\frac{hcE_i''}{k}\left(\frac{1}{T}-\frac{1}{T_0}\right)\right)\frac{1-\exp\left(\frac{-hc\tilde{\nu}_{0,i}}{kT}\right)}{1-\exp\left(\frac{-hc\tilde{\nu}_{0,i}}{kT_0}\right)}, \tag{4}$$

where T is an arbitrary temperature, T_0 is a reference temperature, h is the Planck constant (J s), c is the speed of light (cm/s), k is the Boltzmann constant (J/K), E_i'' is the lower-state energy (cm^{-1}), and $Q(T)$ is the partition function, which is also temperature-dependent and can be approximated with the following polynomial:

$$Q(T) = a + bT + cT^2 + dT^3. \tag{5}$$

Coefficients (a, b, c, d) for CO, CO_2, and H_2O from Ref. [30] are used.

The less congested absorption spectrum of water vapor (as compared to CO_2) is used to infer temperature from the ratio of two absorption transitions with different ground state internal energy. Along a common optical path, the water mole fraction and pressure are the same for both transitions; thus, the ratio of the two integrated absorbance values can be simplified to the ratio of the respective line strengths:

$$R = \frac{A_1}{A_2} = \frac{S_1(T)}{S_2(T)} = \frac{S_1(T_0)}{S_2(T_0)}\exp\left(-\frac{hc}{k}\left(E_1''-E_2''\right)\left(\frac{1}{T}-\frac{1}{T_0}\right)\right)\frac{1-\exp\left(\frac{-hc\tilde{\nu}_{0,1}}{kT}\right)}{1-\exp\left(\frac{-hc\tilde{\nu}_{0,1}}{kT_0}\right)}\frac{1-\exp\left(\frac{-hc\tilde{\nu}_{0,2}}{kT_0}\right)}{1-\exp\left(\frac{-hc\tilde{\nu}_{0,2}}{kT}\right)}. \tag{6}$$

If the two selected transitions have similar wavelengths, the ratio of two integrated absorbances can be approximated while maintaining high accuracy as follows:

$$R = \frac{S_1(T_0)}{S_2(T_0)}\exp\left(-\frac{hc}{k}\left(E_1''-E_2''\right)\left(\frac{1}{T}-\frac{1}{T_0}\right)\right). \tag{7}$$

To infer the highest temperature measurement accuracy, the value should be as large as possible over the expected temperature range.

$$\left|\frac{\frac{dR}{R}}{\frac{dT}{T}}\right| = \left(\frac{hc}{k}\right)\frac{\left|E_1''-E_2''\right|}{T} \tag{8}$$

From Equation (7), the temperature can be obtained by:

$$T = \frac{\frac{hc}{k}\left(E_2'' - E_1''\right)}{\frac{hc}{kT_0}\left(E_2'' - E_1''\right) + \ln\frac{S_2(T_0)}{S_1(T_0)} + \ln R}. \tag{9}$$

In Equation (2), the line-shape function $\phi_{\tilde{\nu}}$ (cm) is a convolution of Doppler and collisional broadening [31]:

$$\phi(\tilde{\nu}) = \phi_D(\tilde{\nu}_0)\frac{a}{\pi}\int_{-\infty}^{+\infty}\frac{e^{-y^2}}{a^2 + (w-y)^2}dy = \phi_D(\tilde{\nu}_0)V(a,w), \tag{10}$$

where $V(a,w)$ is the Voigt function that can be numerically approximated [32]. The Voigt parameter as a measure for the relative significance of Doppler and collisional broadening is defined as w indicates the non-dimensional line position $\phi_D(\tilde{\nu}_0)$ is the Doppler line center magnitude at $\tilde{\nu}_0$ and y is an integration variable:

$$a = \frac{\sqrt{\ln 2}\Delta\tilde{\nu}_c}{\Delta\tilde{\nu}_D}. \tag{11}$$

$$w = \frac{2\sqrt{\ln 2}(\tilde{\nu} - \tilde{\nu}_0)}{\Delta\tilde{\nu}_D}, \tag{12}$$

$$\phi_D(\tilde{\nu}_0) = \frac{2}{\Delta\tilde{\nu}_D}\sqrt{\frac{\ln 2}{\pi}}, \tag{13}$$

$$y = \frac{2u\sqrt{\ln 2}}{\Delta\tilde{\nu}_D} \tag{14}$$

The collision-broadened line width $\Delta\tilde{\nu}_c$ depends on the pressure and the product of the sum of the mole fraction for each collision partner species B and its collisional broadening coefficient $2\gamma_B$:

$$\Delta\tilde{\nu}_c = P\sum_B x_B 2\gamma_B, \tag{15}$$

which varies with temperature

$$2\gamma_B(T) = 2\gamma_B(T_0)\left(\frac{T_0}{T}\right)^N, \tag{16}$$

where T_0 is the reference temperature and N is the temperature coefficient. The temperature-dependent Doppler broadening is defined by:

$$\Delta\tilde{\nu}_D = \tilde{\nu}_0\left(7.1623 \times 10^{-7}\right)\left(\frac{T}{M}\right)^{0.5}, \tag{17}$$

where M is the molecular mass in g/mol.

When temperature has been determined by the two-color ratio method and pressure and optical path length are known, the species mole fractions can be determined from the absorption of a single transition of known spectroscopic parameters. In cases, where no isolated transition can be measured, the concentration of the target species is inferred from model calculations matching the measured absorption spectrum.

3. Wavelength Selection and Data Analysis

An important part of wavelength-multiplexed sensor design is the selection of laser wavelengths to target sections of the absorption spectrum of the target gas. In the ideal case, each laser is chosen to scan in wavelength across an individual transition of the target species that is free of interference from other chemical components of the gas along the laser line-of-sight. As noted before, in high-pressure, high-temperature hydrocarbon combustion gases, such isolated absorption transitions are not found for CO, CO_2, and H_2O detection. For water vapor, the pressure-broadened target transitions are blended with other water vapor transitions. For CO and CO_2, the situation is more complicated, as the fundamental absorption bands of these two species strongly overlap, and the CO concentration is typically much lower than CO_2 in combustion product gases. For all three species, determination of the zero-absorption transmitted laser intensity is not straightforward. Center wavenumbers and wavelengths, line strengths, and lower-state energies of the four target transitions utilized are listed in Table 1. The two NIR water vapor transitions have been previously used by several authors, e.g., [1,7,8,33], and the two MIR transitions have been used for CO and CO_2 by the Hanson group [27].

Table 1. Spectroscopic parameters of chosen main transitions.

Line	Species	$\tilde{\nu}_0$/cm^{-1}	λ/nm	S (296 K)/(cm^{-2}/bar)	E''/cm^{-1}	Transition
1	H_2O	7185.59	1391.67	1.93×10^{-2}	1045.59	Q(6)
2	H_2O	6806.03	1469.29	6.32×10^{-7}	3291.2	P(17)
3	CO	2059.91	4854.58	8.65×10^{-1}	806.38	P(20)
4	CO	2190.02	4556.17	7.04	299.77	R(12)

Using the HITEMP database [25], Figure 1 shows spectra simulations in the region of the selected wavelengths at 1 bar with a temperature of 2000 K and an optical pathlength of 60 mm. The assumed mole fractions of CO, CO_2, and H_2O are 0.1, 8, and 18%, respectively, typical for the burned gas effluent from hydrocarbon combustion. Figure 1a,c show the region around the two target water lines used for thermometry. The water vapor lines near 1391.67 nm (7185.59 cm^{-1}) and 1469.29 nm (6806.03 cm^{-1}) are free of significant interfering absorption from other components of the gas flow such as CO and CO_2. However, several water transitions overlap within the depicted scan range. Most importantly, note the absorbance never goes to zero even at pressures as low as 1 bar, making the determination of the zero-absorption baseline an important task. The absorbance shown in the scan range of Figure 1b is primarily CO absorption with only modest contributions from CO_2, while the scan range in Figure 1d is the reverse. Thus, simultaneous fitting of the two MIR scan ranges can return the CO and CO_2 concentrations, especially since the gas temperature is known from the ratio of the two water vapor measurements.

The zero-absorption baseline is determined for the sensor by first measuring the laser transmission over the scanned-wavelength region, with nitrogen purging of the measurement line-of-sight before and after the measurement. However, detection efficiency and laser intensity transmitted for the benign no-combustion measurement can vary from the combustion gas measurement by differences in optical alignment and potential beam steering in the hot combustion gases (especially in the target gas turbine combustor test rig). Therefore, we assume a linear loss term η (baseline scaling factor) for each laser intensity, which is determined by iterative spectral fitting over the scan range using the algorithm described in Figure 2.

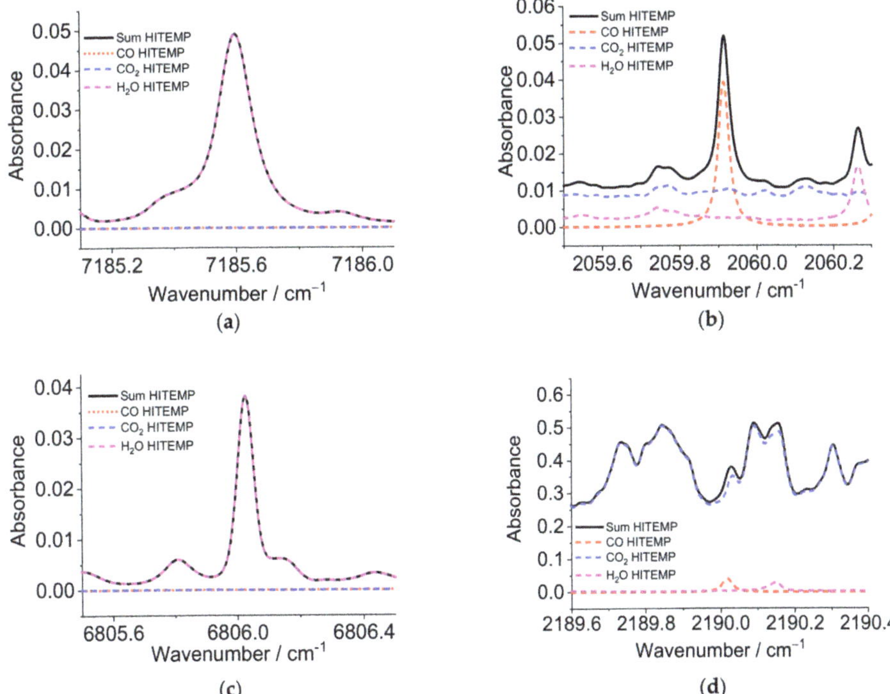

Figure 1. Simulated absorbance spectra for each of the four wavelength ranges used at 1.01 bar, 2000 K, 0.1% CO, 8% CO_2, 18% H_2O mole fraction with an optical path length of 60 mm for the spectral regions near 7185.59, 6806.03, 2059.91, and 2190.02 cm^{-1}, respectively. The NIR spectra (**a**,**c**) consist of several water absorption lines, and each spectrum is dominated by a single strong transition. The MIR spectrum (**b**) is dominated by an isolated CO transition that builds on a CO_2 and water background. MIR spectrum (**d**) consists mainly of several CO_2 absorption lines, with some overlap of isolated CO and water transitions. The large number of CO_2 absorption lines in this region and their superposition results in a significant baseline for the CO measurement.

The purpose of the algorithm in Figure 2 is the calculation of the transmitted laser intensity. The algorithm determines the integrated absorbance A for each line, the line-shape function $\phi(\tilde{\nu})$, and the baseline scaling factor η. Initial guesses are needed. For the integrated absorbance A (Equation 3) for each individual transition, line-center wavenumber $\tilde{\nu}_0$, and the collisional broadening coefficient $\Delta\tilde{\nu}_c$ values from the HITEMP database [25] are used. The relationship between laser scanning time and frequency is characterized by using an etalon placed in a third beam path (cf. Figure 4), so that the line-shape function, absorbance, and the measured incident and transmitted laser intensity can be converted from the time domain to the frequency domain. The simulated transmitted intensity versus frequency $^{S}I_t(\tilde{\nu})$ is obtained with the Beer-Lambert law and baseline measured without combustion $^{M}I_0(\tilde{\nu})$; to account for non-absorption losses with combustion, this baseline intensity is scaled by a fit parameter, η. After comparison of the simulated and the measured transmitted intensities versus frequency $^{S}I_t(\tilde{\nu})$ and $^{M}I_t(\tilde{\nu})$, the gas properties are determined from best-fit parameters.

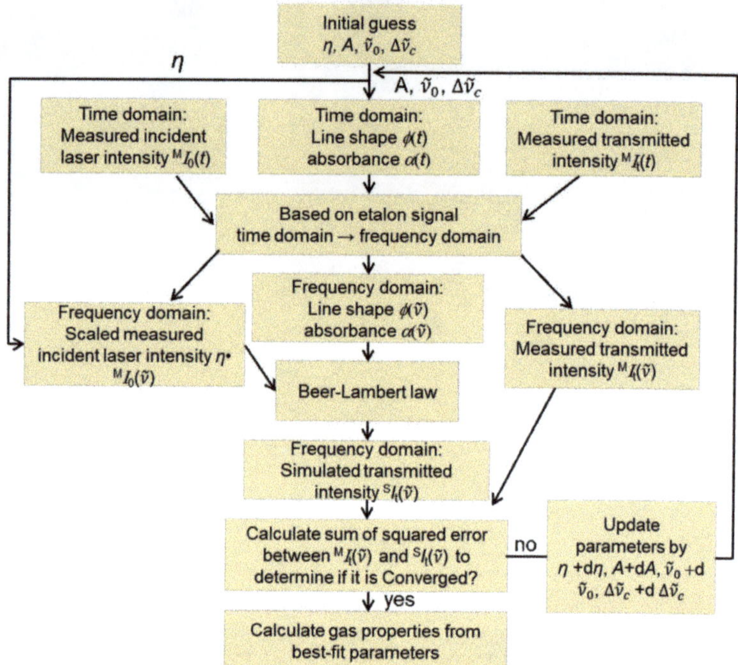

Figure 2. Algorithm for iterative spectra fitting by comparing simulation and measurement of the transmitted intensity versus time (i.e., wavelength) of a wavelength-scanned laser.

4. Experiment

4.1. Layout of the NIR/MIR Four-Color Laser Absorption Instrument

Figure 3 shows the schematic of the instrument consisting of the sending, auxiliary, receiving, and free-beam guiding units. The sending, receiving, and free-beam guiding units are enclosed in acrylic glass or metal boxes and purged with dry nitrogen to protect all optics from dust deposition and avoid spurious absorption from water vapor and CO_2 outside the probe region. Indium trifluoride (InF_3) single-mode MIR-transmitting optical fibers (Thorlabs, P3-32F-FC-1) are used to deliver the laser light from the sending unit to the free-beam guiding unit composed of optical windows and purged lens tube components with a varying diameter (Thorlabs lens tubes, 1″ and 1/2″ diameters). The attenuation of the InF_3 fiber in the MIR range is <0.35 dB/m and <0.52 dB/m in the NIR range. However, it should be noted that fiber for NIR radiation does not behave as a single-mode fiber, which causes limitations in the ability to focus the beam and the quality of the beam profile, but these effects do not cause any crucial limitation for the application here.

The laser beams leaving the sending unit (see below) are collimated by an off-axis paraboloid (Thorlabs, RC02APC-P01, Bergkirchen, Germany) mounted on an *xy* translation stage fixed at the free-beam-guiding lens tube. Inside the lens tube, the light beams are steered with a gold-coated mirror and split into two parts by an optically polished CaF_2 wedge (3 degrees, Korth Kristalle GmbH, Altenholz, Germany). The reflected part is transmitted through a wedged window (1°) towards the auxiliary unit. The transmitted laser beams are guided via 1/2″-diameter lens tubes through the flame and then towards the receiver unit. Each lens tube is closed with wedged (0.5°) CaF_2 windows (Thorlabs, WW50530) and apertures (Thorlabs, SM05D5D) and flushed with dry nitrogen to avoid absorption by room air water vapor. The flat flame is operated using a McKenna burner (Holthuis & Associates, Sebastopol, CA, USA) mounted on a lab jack (Thorlabs, L490/M) for height adjustment.

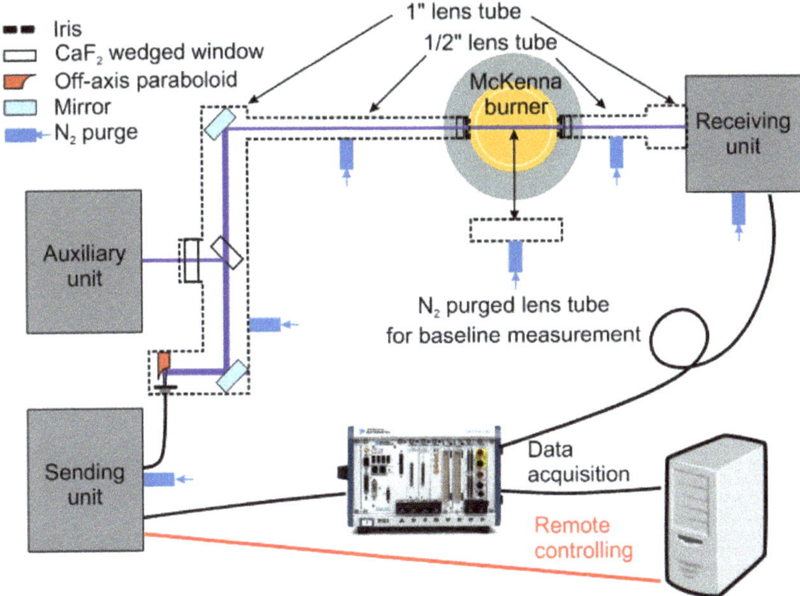

Figure 3. Optical layout of MIR/NIR laser absorption instrument.

All detector outputs in the sending and receiving units are acquired by two DAQ cards (PXIe-6124, National Instruments, Austin, TX, USA) in the PXI system, which also worked as a function generator for the current modulation of the lasers. The system also controls and remotely operates stepper motors for mechanical translation stages.

4.2. Sending Unit

Figure 4 shows a more detailed sketch of the optical layout inside the sending unit. The cw-QCL (Alpes Laser, 40 mW) with emission wavelength around 4.85 µm is housed together with a collimator in a sealed laser mount. The laser is fixed on a home-made water-cooled heat sink. A second cw-QCL (Hamamatsu L12004-2190H-C, 20 mW, Herrsching, Germany) with an emission wavelength centered around 4.56 µm is packaged in a sealed laser mount on a water-cooled heat sink (Hamamatsu A11709-02). An aspheric ZnSe lens (Hamamatsu, A11331-02) inside a precision zoom unit (Thorlabs, SM1ZM) can also be translated perpendicular to the laser beam path (*z* direction) with an *xy* translation mount (Thorlabs, ST1XY-D). The QCLs are thermoelectrically cooled with Peltier elements. To attain stable performance, the heat sinks are connected in series and water cooled with a chiller (Thermo Fisher Scientific, 200 W, Waltham, MA, USA). Two free-space adjustable narrowband optical isolators (Thorlabs, I4500W4 and I4730W5) are used to shield the lasers from back reflections, thus reducing intensity noise and mode hopping. The free beams from both lasers are collinearly combined on a narrowband filter (C1, Spectrogon, NB5040-155 nm, Täby, Sweden) that is used as a long-pass beam combiner transmitting the 4.85 µm beam and reflecting the 4.56 µm beam. In comparison with a 50:50 beam splitter, this filter achieves better overall optical efficiency of 75% for each beam. However, the filter may cause ghosting and beam distortion due to etaloning and limited surface optical quality, respectively.

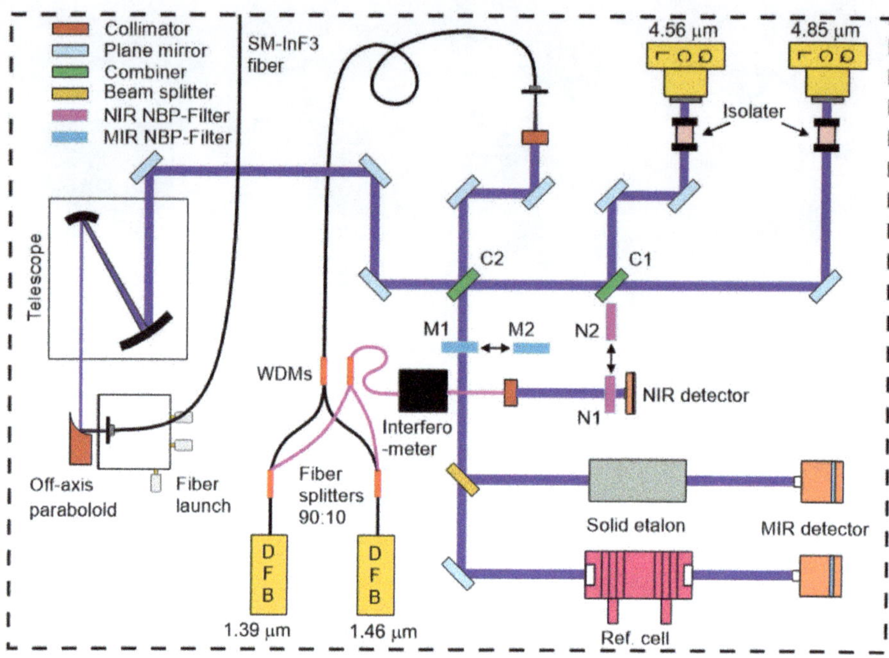

Figure 4. Optical layout of the sending unit.

For the water transitions, two polarization-maintaining (PM) fiber-coupled DFB (distributed-feedback) lasers with emission wavelengths of 1.392 and 1.469 μm (NEL, 20 mW) are clamped into butterfly laser mounts (Arroyo, 203) and thermoelectrically cooled with Peltier elements. For a compact arrangement, they are stacked onto a platform mounted above other optical elements. To prevent damage of long sections of optical fibers, they are coiled into storage reels (Thorlabs, FSR1) separately fixed on the cover. With two 1×2 PM fiberoptic couplers (Thorlabs, PN1310R2A1 and PN1480R2A1), the two beams at 1.39 and 1.47 μm are separately split into two PM fractions of 90% and 10% each. To monitor the mode quality and to obtain frequency markers to convert scan time to relative wavelength, the 10% beams are combined using a single-mode (SM) fiber wavelength-division multiplexer (Thorlabs, WD202C-APC, WP9850A) and together propagate through a SM fiber-coupled interferometer (Micron Optics, FFP-I, 0.8 GHz). Its output is collimated and free-space recorded by an amplified InGaAs photodetector (Thorlabs, 0.8 mm^2 active area, max. 11 MHz bandwidth). In front of the detector, two narrow bandpass filters (Thorlabs, FB1400-12 as N1, FB1480-12 as N2 in Figure 4) are placed in a dual-position filter holder slider with resonant piezoelectric motors (Thorlabs, ELL6K) such that either 1.39 or 1.47 μm can be selected for etalon analysis to determine the wavelength tuning characteristics. The motor of the filter slider is remotely controlled during the measurement.

The 90% beams are combined using another PM wavelength-division multiplexer (Thorlabs, WP9850A). The coupled beams with the two NIR wavelengths are collimated with an aspherical lens (Thorlabs, C280TMD-C) in a fiber-launch system (Thorlabs, KT110/M). They are subsequently combined with the two collinear propagating MIR beams by a second bandpass filter (Spectrogon, BP-4700-600 nm) operated as combiner C2, with most of the MIR beams being transmitted and the NIR beams reflected off the front surface. For wavelength and tuning characterization, a fraction of the incoming free-space MIR beams is reflected towards a solid germanium etalon and a reference absorption cell, respectively. The detectors record each laser beam through respective narrow bandpass filters (Laser Components, SNB-4860-001793 as M1 and Spectrogon, NB4560-135 nm as M2 in Figure 4) installed in another dual-position filter slider equipped with a resonant piezoelectric motor (Thorlabs, ELL6K). The filtered beams are split into two parts: One passes through the low-pressure reference cell to determine the spectral position of the narrow CO absorption line during a wavelength scan whose peak is identified with the absolute wavelength listed in the HITEMP database. The other part is guided to the 7.62-cm-long solid Ge etalon (free spectral range (FSR) 0.016 cm^{-1}). The known fringe spacing enables conversion of tuning time into relative laser wavelength. Information from both channels can be combined to convert scan time into absolute wavelength.

All collinearly combined NIR and MIR beams are then steered to a parabolic mirror (Thorlabs) and are focused into a SM MIR InF$_3$ patch fiber (Thorlabs, P3-32F-FC-1). Due to its small numerical aperture of 0.26, a six-fold reflective beam telescope (Thorlabs, BE06R/M) reduced the beam diameter to approximately 100 µm. This telescope significantly reduced the coupling losses in the fiber. By precisely adjusting the positions of the collimators in front of the PM fiber outputs of the NIR lasers (and the QCL at 4.56 µm), the difference in beam divergence of the four collimated beams can be slightly minimized to achieve better focus quality and reduce coupling losses.

All four lasers are driven by individual controllers (Arroyo, 6300 series) including temperature stabilization and injection current control. The optical setup is installed on a water-cooled aluminum optical board (450 × 600 mm^2, Thorlabs), which rested on four aluminum profiles with passive vibration isolators (Figure 5). Below this plate, two fans provided air circulation within the unit. The remaining space accommodated controllers and power supplies for lasers and detectors.

Figure 5. Side view of the sending unit mounted on an optical breadboard with the laser controllers below.

4.3. Receiving Unit

The receiving unit is illustrated in Figure 6. An aperture (Thorlabs, SM05D5D) and a CaF$_2$ wedged window (Thorlabs, WW50530) are fixed inside of a 12.7 mm diameter adjustable lens tube to allow the laser beam to couple directly into the N$_2$-purged environment after passing the burner. Thus, the absorption path is confined within the flame gases and minimizes interference of ambient water vapor in the beam path outside of the flame. Optics and detectors of the receiving unit are contained inside an aluminum box

(10-mm wall thickness) with a 25.4 mm diameter optical access port originally designed for shielding the probe from excessive acoustic noise and thermal load when placed close to the gas turbine test site. Lens tubes and the whole receiver box are purged with dry nitrogen supplied through a Vortex tube (Vortec, 208-25H, Cincinnati, OH, USA) that simultaneously keeps the internal atmosphere cooled to approx. 293 K even at elevated environmental temperature.

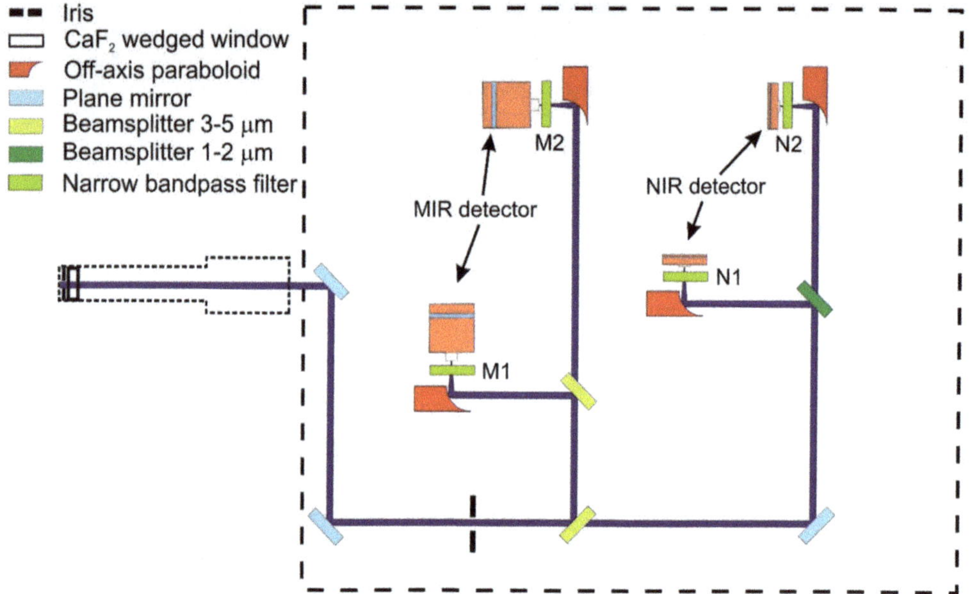

Figure 6. Schematics of the receiving unit.

The four incoming collinearly propagating transmitted laser beams are split by a 45:55 pellicle beam splitter (Thorlabs, BP145B4), originally designed for the 3–5 µm wavelength range, which also has a transmittance for the NIR-lasers of 85% and 63% near 1392 and 1469 nm, respectively. In the MIR detection beam path, a second 3–5 µm 45:55 pellicle beam splitter delivers a reflected and transmitted beam towards two TE-cooled IR photovoltaic detectors (Vigo Systems, 2 mm diameter active area, 1 MHz bandwidth), where they are focused with parabolic mirrors (Thorlabs, MPD019-M01). Two narrow bandpass filters (Spectrogon, NB4560-135 nm as M1 and Laser Components, SNB-4860-001793 as M2 in Figure 6) transmit the respective beams and block unwanted background emission from hitting the detectors. The fraction of the NIR beams transmitted through the first pellicle beam splitter to the NIR detection side is again divided into two separate beams with a third 45:55 pellicle beam splitter for 1–2 µm (Thorlabs, BP145B3) and directed to two InGaAs free-space amplified photodetectors (Thorlabs, PDA20CS2, 3.14 mm^2 active area, max. 11 MHz bandwidth). Narrow-bandpass spectral filters (Thorlabs, FB1480-12 as N1 and FB1400-12 as N2 in Figure 6) are mounted in front of the respective detectors. All the optics are fixed on a threaded aluminum breadboard (300 × 300 mm^2) inside the closed box.

4.4. Auxiliary Measurements

A split-off fraction of the incoming beams enters the auxiliary unit (details Figure 7) through a separate side window in the lens tube of the sending unit, where they are again split by a 45:55 pellicle beam splitter (coated for 3–5 µm; Thorlabs, BP145B4). Depending on the wavelength, they are then focused with parabolic mirrors (Thorlabs, MPD00M9-M01, f = 15 mm) on an InGaAs free-space amplified photodetector (Thorlabs, PDA20CS2,

0.8 mm^2 active area, max. 11 MHz bandwidth) or a TE-cooled IR photovoltaic detector (Vigo Systems, 1 mm active diameter, 1 MHz bandwidth), respectively. A flip mirror in front of the beam splitter can in addition switch the optical path towards a MIR wavemeter (Bristol 621B) for monitoring the absolute wavelength.

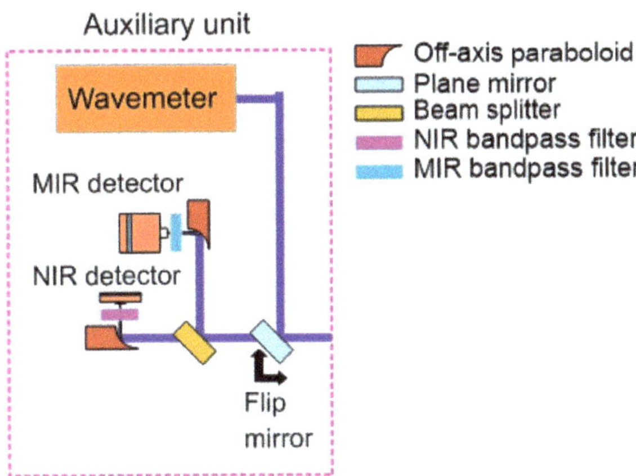

Figure 7. Schematics of the auxiliary unit.

4.5. Burner Facility

Flat methane/air premixed flames are stabilized on a water-cooled McKenna burner at atmospheric pressure. It is equipped with a central sintered bronze matrix of 60 mm diameter for delivering the fuel/air mixture and an outer 6 mm-wide ring-shaped matrix for providing a coannular flow of dry nitrogen. All gases are metered by calibrated mass flow controllers (Bronkhorst, Kamen, Germany), while the cooling water flow for the burner matrix is circulated at 16 °C by a thermostat (Huber, Minichiller 600 OLÉ, Offenburg, Germany). For vertical translation with respect to the spatially fixed optical path for the absorption measurements, the burner is mounted on a lab jack. The height above burner is measured with a caliper, whose two arms are fixed to the lab jack and the breadboard, respectively, with a zero value when the traversing analysis beam is just halfway clipped by the burner surface. Premixing of fuel and oxidizer well upstream of the burner inlet improved the flatness of the flame, as is visible from flame chemiluminescence.

5. Results

To validate the performance of the wavelength-multiplexed NIR/MIR absorption sensor, measurements are conducted in the burned gases above the flat flame at conditions identical to measurements presented by Weigand et al. [34], where gas-phase temperatures were obtained by coherent anti-Stokes Raman scattering (CARS) of molecular nitrogen, and CO, CO_2, and H_2O concentrations are calculated assuming local thermodynamic equilibrium; three stoichiometric and slightly fuel-rich flames labeled No. 17, 18, 19, and 20 ae studied and provide validation data for measurements with the sensor developed here. Table 2 lists flow rates (in standard liters per minute, slm) of CH_4, air, and co-flow N_2 and the corresponding equivalence ratio ϕ for the premixed CH_4/air flames. Also listed are the adiabatic flame temperatures (T_{ad}) and the measured N_2 CARS temperatures, with the latter varying in the range 1883–2100 K. The three flames have equilibrium concentrations of CO ranging from 0.31 to 4.3%, CO_2 varying between 6.5 and 9.1%, and water vapor at approximately 19% (Table 2).

Table 2. Operating conditions of the four premixed CH_4/air flames from Weigand et al. [34] examined in this work. Adiabatic flame temperatures and species mole fractions from equilibrium calculations (EQ) as well as measured CARS temperatures as listed from [34]. CARS temperatures were measured at 15 mm HAB. Measured laser absorption temperature and species mole fractions from this work are listed below the reference values.

Flame No.	CH_4/slm	Air/slm	Co-flow N_2/slm	ϕ	T_{ad}/K	T/K CARS Laser Abs	$x(H_2O)$ EQ Laser Abs	$x(CO_2)$ EQ Laser Abs	$x(CO)$ EQ Laser Abs
17	2.55	24.14	10.55	1.0	2226	2009 2226	0.1877 0.1906	0.0917 0.0868	0.0031 0.0031
18	2.55	22.00	9.7	1.1	2211	1934 2211	0.1897 0.1842	0.0780 0.0709	0.0237 0.0169
19	2.55	20.20	8.99	1.2	2137	1883 2137	0.1863 0.1912	0.0653 0.0590	0.0426 0.0343
22	3.42	32.4	14.15	1.0	2226	2100 2226	0.1860 0.1750	0.0891 0.0925	0.0055 0.0059

The measurement line-of-sight is parallel to the burner surface and crosses the center of the burner at a height of 15 mm (matching the height above the burner for the CARS measurements). At this height, temperature gradients in the exhaust above the flame front are negligible [35]. The four lasers are tuned at 10 Hz with individual inverse sawtooth current ramps to cover each of the respective absorption line shapes from which integrated line areas are calculated. The signal voltages from each detector are acquired at 10 MHz and averaged for 30 s.

Temperatures are measured with two-line thermometry using water vapor transitions with center wavenumber positions 7185.59 and 6806.03 cm^{-1} from Table 1. Figure 8 shows the transmitted signal (upper row) and the corresponding absorbance (lower row) over the relative wavenumber for these two lines for flame no. 17 in Table 2. As can be seen from Figure 8a,c, the "corrected" baseline that is scaled by measured reference baseline without absorption matches very well in the far wings of the lines due to the large tuning range of the NIR lasers (>1 cm^{-1}) and the fact that the transitions are isolated from interference of neighboring lines from other species. The absorption line at 7185.59 cm^{-1} has significant absorption at room temperature from ambient water vapor. Although the sender unit is also covered by an acrylic glass box on the optical table, the observation of a small absorbance indicates imperfect purging of the sensor unit. In contrast, the high energy of the lower state of the transition near 6806.03 cm^{-1} reduces the sensitivity to interference from ambient water vapor, and its baseline scan is free of the corresponding interference absorbance (Figure 8c). As can be seen from Figure 8b,d, the fitted absorbance line shape agrees with the measured one fairly well, with a maximum residual of 0.5% of the peak absorbance.

For CO detection in the MIR, interference by CO_2 and water vapor present a challenge. Since the water absorption lines are not significantly disturbed by other interfering species and temperature is measured using two-line thermometry, any water vapor interference can be calculated from the corresponding absorbance of one of the water transitions.

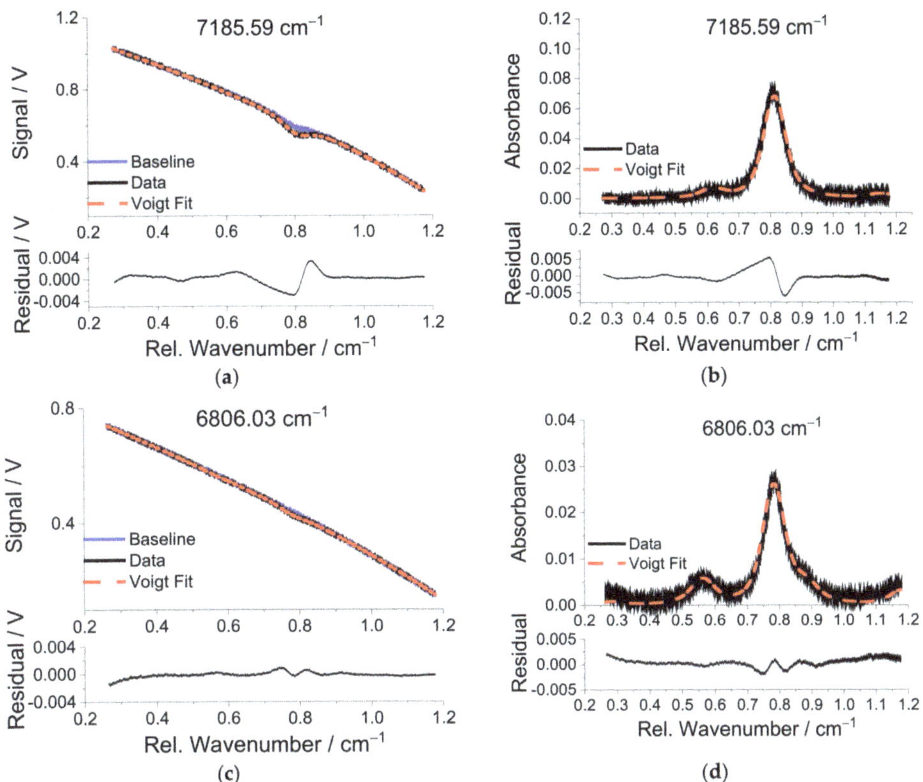

Figure 8. Water vapor transmitted signal (black lines in (**a**,**c**)) and calculated absorbance (black lines in (**b**,**d**)) over relative wavenumber at the H_2O center wavelengths 7185.59 and 6806.03 cm^{-1}, respectively, for the premixed flame no. 17 (Table 2). The separately measured baseline (blue lines in (**a**,**c**)) and fitted Voigt functions (dashed red lines) are also shown. The residuals shown in the lower section of each figure are between original data and fitted line shapes.

For the MIR scan of the CO lines, the interference of CO_2 must be considered. Using the temperature determined from the water vapor, the CO and CO_2 concentrations are simultaneously fit to the two MIR scanned-wavelength absorbances. The CO line near 2059.91 cm^{-1} experiences only slight interference from CO_2, while the spectral region around the CO line near 2190.02 cm^{-1} at elevated temperatures is dominated by CO_2 absorbance. Figure 9b and d show that the Voigt fits match the measurements well, with maximum residuals below 2% of the absorbance. Although there is a significant residual in the baseline in Figure 9d, the best-fit spectral profile matches well with the measured data suggesting a fairly complete CO_2 spectroscopic model extracted from the HITEMP data base in this spectral range. However, the residual between measurement and modeled spectrum still is not "flat", signifying discrepancies in CO_2 spectral line positions and relative intensities in the data base.

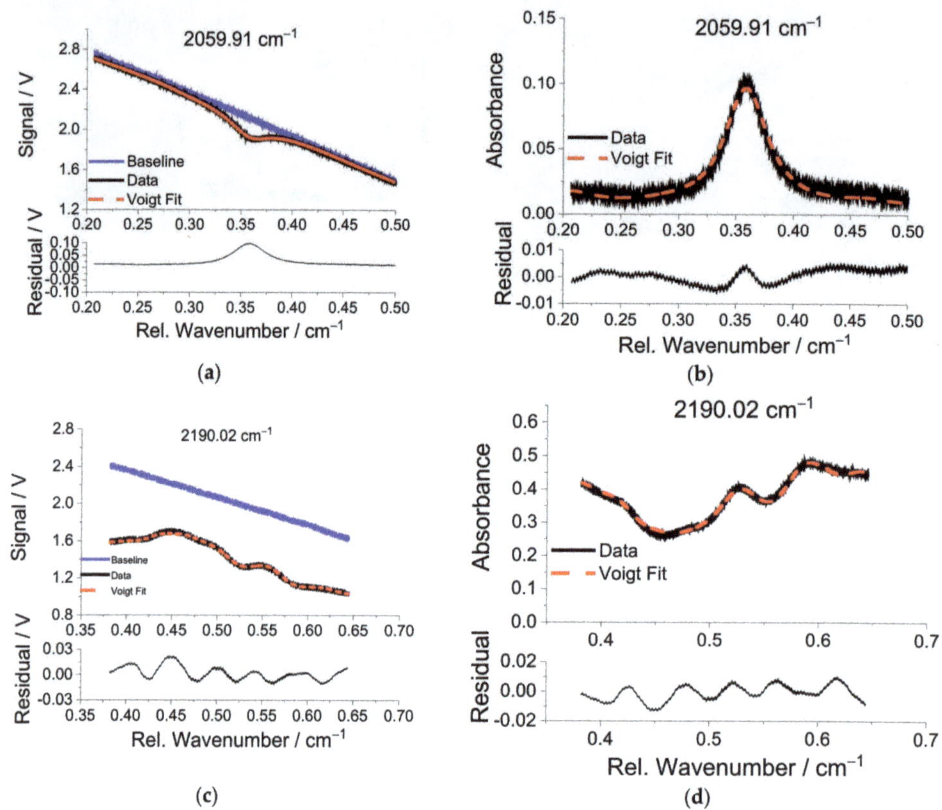

Figure 9. Transmitted signal and calculated absorbance originating from CO, CO_2, and H_2O over the relative wavenumber at the CO center wavelengths 2059.91 cm^{-1} (**a**,**b**) and 2190.02 cm^{-1} (**c**,**d**) for the premixed flame no. 17.

Table 2 and Figure 10 compare the measurements of the laser absorption sensor with the CARS temperatures and equilibrium concentrations for CO, CO_2, and H_2O [34]. Laser absorption recovers line-of-sight averaged temperature and concentration values, while the CARS technique provides spatially resolved measurements (within a roughly cylindrical probe volume of a few millimeter in length and a diameter of approx. 500 μm). Laser absorption temperature measurements are higher than the reported CARS measurements for lower flow rates and lower than CARS temperature for the high-flow-rate flame (No. 22). These statements lose their significance when the measurement uncertainties (plotted in Figure 10) are considered; note the error bars of CARS and the absorption measurements (for details, see figure caption) strongly overlap. Nerveless, it can be concluded that the temperature can be successfully measured using the sensor within the measurement uncertainties.

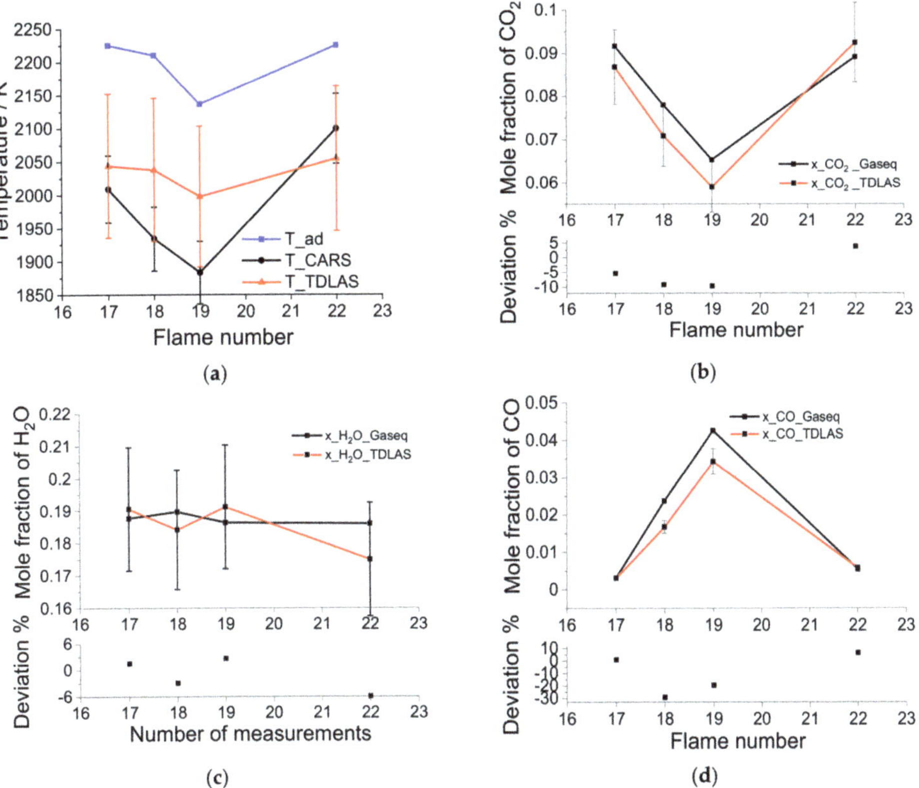

Figure 10. Measured temperatures (**a**) and concentration of CO_2 (**b**), H_2O (**c**) and CO (**d**) in comparison to the temperatures measured by CARS and the simulated equilibrium concentrations from Ref. [34]. The measurement uncertainties for the CARS measurements were assumed to be 2.5% according to Ref. [34]. The measurement uncertainty for temperature measurements using two-line absorption thermometry with 1.4-μm lasers is discussed in the relevant literature as being between 2.8% [7] and 6.7% [36]. In Ref. [29], the temperature deviation of identical burners is estimated at 2.5% as an additional source of error. Therefore, and as no detailed accuracy analysis was performed as part of this study, we estimated the error for our temperature measurements to be ~5%. Mole fraction measurements using laser absorption methods under best conditions (only CO_2 in N_2 at room temperature) have been reported in the literature with accuracies up to 4% [37]. In Ref. [36], an accuracy of 10% was achieved for the determination of water vapor mole fraction in the exhaust gas of a flat Mckenna-type burner. In Ref. [7], an accuracy of 4.7% for the determination of shock-tube-heated water vapor was reported. Since the conditions and the used setup in Ref. [36] best match ours, we estimated the error for our mole fraction measurements to be ~10%.

Laser absorption measurements of the concentrations of CO, CO_2, and H_2O shown in Figure 10b–d agree well with calculated equilibrium values from Gaseq [35]. For only the two rich operating points (flame #18 and #19), strong deviations between the equilibrium calculations and the CO concentration results are observed, which are larger than can be explained by the measurement uncertainty. In this case, there are apparently unconsidered systematic errors. Since the dynamic range of the CO signal is very large, small nonlinearities in the detector could cause these systematic errors at higher concentrations. Another explanation could be inaccuracies in the spectroscopic line parameters for the selected CO transitions or deficiencies in the determination of the CO concentration via the equilibrium calculation using GasEq. Since these deviations occur at concentrations of CO that are irrelevant for the operation of gas turbines (only a few ppm CO may be generated) and the chosen laser was designed for this low-CO range, this deviation does not represent a limitation for the applicability of the sensor.

6. Summary

The design and demonstration of a compact, modular fiber-coupled laser absorption instrument for simultaneous remote measurement of temperature, concentrations of water vapor, carbon dioxide, and carbon monoxide is reported. The instrument is designed for remote operation in test facilities with extremely harsh ambient conditions. The assembly of the sender and receiver units on water-cooled breadboards allows them to be enclosed with a purged atmosphere to protect against the high-temperature humid gases and the extreme acoustic noise of the gas turbine combustion test facility. The gas-phase temperature is determined by two-line thermometry using two well-known NIR transitions near 7185.59 cm^{-1} (1.3917 μm) and 6806.03 cm^{-1} (1469.3 nm) without spectral interference from other primary product species; the H_2O concentration is then inferred from the total absorbance of either transition. The concentrations of CO and CO_2 are detected using two MIR laser wavelengths near 2059.91 cm^{-1} (4.8546 μm) and 2190.02 cm^{-1} (4.5562 μm). These two spectral regions have quite different amounts of CO_2 interference, enabling simultaneous modeling of the absorbance measured by both MIR lasers using the temperature determined by water vapor absorbance to infer the CO and CO_2 concentrations.

The two NIR and two MIR lasers are multiplexed into an optical fiber (single-mode in the MIR) for delivery from the sending unit to the measurement line-of-sight. The four laser beams are combined into an indium trifluoride (InF_3) single-mode optical fiber, delivering all lasers colinearly overlapped along the absorption path length. We believe the efficient, low-noise laser-multiplexing into the InF_3 fiber is unique to this sensor.

An algorithm is developed for finding the baseline for wavelength-scanned direct absorption spectroscopy; even when within the scan range of each laser, no true zero-absorption baseline can be recovered.

The sensor operation is validated by measurements in the burned gases in the exhaust of a premixed flat flame previously studied by CARS [34]. The best-fit laser absorption measurements of gas temperature and CO, CO_2, and H_2O concentration agree well with the CARS measurements and equilibrium calculated concentrations. The validation experiment illustrates that the combined NIR/MIR laser absorption sensor is suitable and ready for field applications.

Author Contributions: Conceptualization, T.E., T.D.; methodology, L.S., T.E.; software, L.S.; validation, T.E., J.B.J.; formal analysis, L.S.; investigation, L.S.; resources, C.S.; data curation, L.S.; writing—original draft preparation, L.S.; writing—review and editing, T.E., T.D., J.B.J., C.S.; visualization, L.S., T.E.; supervision, T.E., J.B.J., T.D., C.S.; project administration, T.E.; funding acquisition, C.S. All authors have read and agreed to the published version of the manuscript.

Funding: This work was funded by the Federal Ministry for Economic Affairs and Energy of Germany, project 03ET7011L, and Siemens Energy AG. JBJ's visit to Duisburg was sponsored by the German Research Foundation (DFG) in the context of a Mercator Fellowship within project 229633504.

Data Availability Statement: Not applicable.

Acknowledgments: We acknowledge support by the Open Access Publication Fund of the University of Duisburg-Essen.

Conflicts of Interest: The authors declare no conflict of interest.

References

1. Goldenstein, C.S.; Spearrin, R.M.; Jeffries, J.B.; Hanson, R.K. Infrared laser-absorption sensing for combustion gases. *Prog. Energy Combust. Sci.* **2016**, *60*, 132–176. [CrossRef]
2. Hanson, R.K. Applications of quantitative laser sensors to kinetics, propulsion and practical energy systems. *Proc. Combust. Inst.* **2011**, *33*, 1–40. [CrossRef]
3. Lackner, M. Tunable diode laser absorption spectroscopy (TDLAS) in the process industries—A review. *Rev. Chem. Eng.* **2007**, *23*, 65–147. [CrossRef]
4. Schulz, C.; Dreizler, A.; Ebert, V.; Wolfrum, J. Combustion Diagnostics. In *Handbook of Experimental Fluid Dynamics*; Tropea, C., Foss, J., Yarin, A., Eds.; Springer: Berlin/Heidelberg, Germany, 2007; pp. 1241–1316.
5. Bolshov, M.; Kuritsyn, Y.; Romanovskii, Y. Tunable diode laser spectroscopy as a technique for combustion diagnostics. *Spectrochim. Acta Part B* **2015**, *106*, 45–66. [CrossRef]
6. Allen, M.G. Diode laser absorption sensors for gas-dynamic and combustion flows. *Meas. Sci. Technol.* **1998**, *9*, 545–562. [CrossRef]
7. Goldenstein, C.S.; Spearrin, R.M.; Schultz, I.A.; Jeffries, J.B.; Hanson, R.K. Wavelength-modulation spectroscopy near 1.4 µm for measurements of H_2O and temperature in high-pressure and -temperature gases. *Meas. Sci. Technol.* **2014**, *25*, 055101. [CrossRef]
8. Goldenstein, C.S.; Almodóvar, C.A.; Jeffries, J.B.; Hanson, R.K.; Brophy, C.M. High-bandwidth scanned-wavelength-modulation spectroscopy sensors for temperature and H_2O in a rotating detonation engine. *Meas. Sci. Technol.* **2014**, *25*, 105104. [CrossRef]
9. Wagner, S.; Klein, M.; Kathrotia, T.; Riedel, U.; Kissel, T.; Dreizler, A.; Ebert, V. Absolute, spatially resolved, in situ CO profiles in atmospheric laminar counter-flow diffusion flames using 2.3 µm TDLAS. *Appl. Phys. B* **2012**, *109*, 533–540. [CrossRef]
10. Faist, J.; Capasso, F.; Sivco, D.L.; Sirtori, C.; Hutchinson, A.L.; Cho, A.Y. Quantum cascade laser. *Science* **1994**, *264*, 553–556. [CrossRef]
11. Spearrin, R.M.; Goldenstein, C.S.; Jeffries, J.B.; Hanson, R.K. Quantum cascade laser absorption sensor for carbon monoxide in high-pressure gases using wavelength modulation spectroscopy. *Appl. Opt.* **2014**, *53*, 1938–1946. [CrossRef]
12. Spearrin, R.M.; Goldenstein, C.S.; Jeffries, J.B.; Hanson, R.K. Fiber-coupled 2.7 µm laser absorption sensor for CO_2 in harsh combustion environments. *Meas. Sci. Technol.* **2013**, *24*, 055107. [CrossRef]
13. Nwaboh, J.A.; Werhahn, O.; Schiel, D. Measurement of CO amount fractions using a pulsed quantum-cascade laser operated in the intrapulse mode. *Appl. Phys. B* **2011**, *103*, 947–957. [CrossRef]
14. Vanderover, J.; Oehlschlaeger, M.A. A mid-infrared scanned-wavelength laser absorption sensor for carbon monoxide and temperature measurements from 900 to 4000 K. *Appl. Phys. B* **2010**, *99*, 353–362. [CrossRef]
15. Shi, L.; Görs, S.; Endres, T.; Akyildiz, E.; Dreier, T.; Jeffries, J.; Witzel, B.; Klapdor, E.V.; Schulz, C. Laser-absorption CO detection at 4.56 µm in the exhaust plume of a full-scale gas turbine burner. In Proceedings of the 8th European Combustion Meeting, Dubrovnik, Croatia, 18–21 April 2017. [CrossRef]
16. Schulz, C.; Dreier, T.; Endres, T.; Görs, S.; Shi, L. *Final Report on the BMWi Project, Development of Combustion Technologies for Climate-Friendly Power Generation, Project 3C: Evolution of Fiber Optic Measurement Methods for Application at the Clean Energy Center*; Institute for Combustion and Gas Dynamics (IVG), University of Duisburg-Essen: Duisburg, Germany, 2019. [CrossRef]
17. Sanders, S.T.; Baldwin, J.A.; Jenkins, T.P.; Baer, D.S.; Hanson, R.K. Diode-laser sensor for monitoring multiple combustion parameters in pulse detonation engines. *Proc. Combust. Inst.* **2000**, *28*, 587–594. [CrossRef]
18. Wei, C.; Pineda, D.I.; Paxton, L.; Egolfopoulos, F.N.; Spearrin, R.M. Mid-infrared laser absorption tomography for quantitative 2D thermochemistry measurements in premixed jet flames. *Appl. Phys. B* **2018**, *124*, 123. [CrossRef]
19. Nau, P.; Kutne, P.; Eckel, G.; Meier, W.; Hotz, C.; Fleck, S. Infrared absorption spectrometer for the determination of temperature and species profiles in an entrained flow gasifier. *Appl. Opt.* **2017**, *56*, 2982–2990. [CrossRef]
20. Weng, W.B.; Alden, M.; Li, Z.S. Simultaneous Quantitative Detection of HCN and C_2H_2 in Combustion Environment Using TDLAS. *Processes* **2021**, *9*, 2033. [CrossRef]
21. Spearrin, R.M.; Goldenstein, C.S.; Schultz, I.A.; Jeffries, J.B.; Hanson, R.K. Simultaneous sensing of temperature, CO, and CO_2 in a scramjet combustor using quantum cascade laser absorption spec. *Appl. Phys. B* **2014**, *117*, 689–698. [CrossRef]
22. Peng, W.Y.; Goldenstein, C.S.; Spearrin, M.; Jeffries, J.B.; Hanson, R.K. Single-ended mid-infrared laser-absorption sensor for simultaneous in situ measurements of H_2O, CO_2, CO, and temperature in combustion flows. *Appl. Opt.* **2016**, *55*, 9347–9359. [CrossRef]
23. Cassady, S.J.; Peng, W.Y.; Strand, C.L.; Dausen, D.F.; Codoni, J.R.; Brophy, C.M.; Hanson, R.K. Time-resolved, single-ended laser absorption thermometry and H_2O, CO_2, and CO speciation in a H_2/C_2H_4-fueled rotating detonation engine. *Proc. Combust. Inst.* **2021**, *38*, 1719–1727. [CrossRef]
24. Rothman, L.S.; Gordon, I.; Babikov, Y.; Barbe, A.; Benner, D.; Bernath, P.; Birk, M.; Bizzocchi, L.; Boudon, V.; Brown, L.R.; et al. The HITRAN 2012 molecular spectroscopic database. *J. Quant. Spectrosc. Radiat. Transf.* **2013**, *130*, 4–50. [CrossRef]
25. Rothman, L.S.; Gordon, I.; Barber, R.; Dothe, H.; Gamache, R.R.; Goldman, A.; Perevalov, V.I.; Tashkun, S.A.; Tennyson, J. HITEMP, The high-temperature molecular spectroscopic database. *J. Quant. Spectrosc. Radiat. Transf.* **2010**, *111*, 2139–2150. [CrossRef]
26. Zhou, X.; Liu, X.; Jeffries, J.B.; Hanson, R.K. Development of a sensor for temperature and water concentration in combustion gases using a single tunable diode laser. *Meas. Sci. Technol.* **2003**, *14*, 1459–1468. [CrossRef]
27. Ren, W.; Farooq, A.; Davidson, D.F.; Hanson, R.K. CO concentration and temperature sensor for combustion gases using quantum-cascade laser absorption near 4.7 µm. *Appl. Phys. B* **2012**, *107*, 849–860. [CrossRef]
28. Stewart, G. *Laser and Fiber Optic Gas Absorption Spectroscopy*; Cambridge University Press: Cambridge, UK, 2021.

29. Hanson, R.K.; Goldenstein, C.S.; Spearrin, R.M. *Spectroscopy and Optical Diagnostics for Gases*, 1st ed.; Springer: Cham, Switzerland, 2016.
30. Gamache, R.R.; Kennedy, S.; Hawkins, R.; Rothman, L.S. Total internal partition sums for molecules in the terrestrial atmosphere. *J. Mol. Struct.* **2000**, *517*, 407–425. [CrossRef]
31. Hanson, R. *Introduction to Spectroscopic Diagnostics for Gases*; Stanford University: Stanford, CA, USA, 2001.
32. Mclean, A.B.; Mitchell, C.E.J.; Swanston, D.M. Implementation of an Efficient Analytical Approximation to the Voigt Function for Photoemission Lineshape Analysis. *J. Electron. Spectrosc. Relat. Phenom.* **1994**, *69*, 125–132. [CrossRef]
33. Yang, H. Tunable Diode-Laser Absorption-Based Sensors for the Detection of Water Vapor Concentration, Film Thickness and Temperature. PhD Thesis, University Duisburg-Essen, Duisburg, Germany, 2012.
34. Weigand, P.; Lückerath, R.; Meier, W. Documentation of Flat Premixed Laminar CH_4/air Standard Flames: Temperatures and Species Concentrations. Available online: http://www.dlr.de/vt/datenarchiv (accessed on 26 December 2021).
35. Morley, C. *Gaseq: A Chemical Equilibrium Program for Windows*, Version 0.79; 2005. Available online: http://www.gaseq.co.uk/ (accessed on 26 December 2021).
36. Yu, X.L.; Li, F.; Chen, L.H.; Chang, X.Y. A Tunable Diode-laser Absorption Spectroscopy (TDLAS) Thermometry for Combustion Diagnostics. In Proceedings of the 15th AIAA International Space Planes and Hypersonic Systems and Technologies Conference, Dayton, OH, USA, 28 April–1 May 2008; American Institute of Aeronautics and Astronautics: Reston, VA, USA, 2008.
37. Werhahn, O.; Koelliker Delgado, J.; Schiel, D. Calibration-free Determination of Amount of Substance Fractions—Potentials for Quantum Cascade Lasers in Gas Analysis. *Tech. Mess.* **2005**, *72*, 396–405. [CrossRef]

MDPI

St. Alban-Anlage 66

4052 Basel

Switzerland

Tel. +41 61 683 77 34

Fax +41 61 302 89 18

www.mdpi.com

Sensors Editorial Office

E-mail: sensors@mdpi.com

www.mdpi.com/journal/sensors

www.ingramcontent.com/pod-product-compliance
Lightning Source LLC
LaVergne TN
LVHW071306170726
843515LV00006BA/1505